AF681558

DUST EXPLOSIONS AND FIRES

POWDER TECHNOLOGY SERIES

Edited by
J. C. WILLIAMS
Chairman of Postgraduate
School of Studies in Powder Technology
University of Bradford

Dust Explosions and Fires

K. N. PALMER

Head of Industrial and Toxic Hazards Section
Fire Research Station

LONDON
CHAPMAN AND HALL

First published 1973
by Chapman and Hall Ltd
11 *New Fetter Lane, London EC4P 4EE*
Printed in Great Britain by
Willmer Brothers Limited, Birkenhead

SBN 412 09430 4

© 1973 K. N. Palmer

All rights reserved. No part of this book may be reprinted, or reproduced or utilized in any form or by any electronic, mechanical or other means, now known or hereafter invented, including photocopying and recording, or in any information storage and retrieval system, without permission in writing from the Publisher.

Library of Congress Catalog Card Number 73-7341

Distributed in the U.S.A.
by Halsted Press, a Division
of John Wiley & Sons, Inc
New York

Contents

Foreword

Dust clouds will often burn with explosive violence while the parent bulk material presents no remarkable flammable hazard. Thus homely materials such as sugar, flour, and other foodstuffs present a real danger when they have to be handled on an industrial scale, indeed their very familiarity tends to mask their danger. Industrial dusts cover foodstuffs, fodder, pharmaceuticals, fuels, cosmetics, metal powders among other things, and a book dealing with their hazards is long overdue.

Mr Palmer has spent many years carrying out research himself or supervising research in this field and his work has brought him into close contact with industrial processes. Mr Palmer has an international reputation in this subject and he is uniquely qualified to write such a book.

The present work, the first of its kind, gives an exhaustive treatment of the subject in that it surveys the research that has been carried out on explosion hazards, and it also deals with industrial processes together with the preventive measures that can make for the safe handling of dusts.

This book is essential reading for anyone concerned with the industrial processing of dusts as it is a comprehensive work on the subject. The book brings together for the first time the research results which are scattered throughout the world's literature and the research student will find in it a most valuable bibliography.

D. I. LAWSON
formerly Director of the Fire Research Station

Editor's Foreword

This book is part of a series of monographs in Powder Technology, the aim of which is to present up to date reviews of specialized aspects of the subject by research workers who are themselves actively engaged in the subject about which they are writing. Previous books have dealt with 'Particle Size Measurement', 'The Packing of Solid Particles' and 'Flowing Gas-Solids Suspensions'.

Engineers concerned with the design and operation of plant for handling particulate solids show a continued and growing concern about the problems of dust fires and explosions, and it is hoped that this book will meet their needs by setting out the nature and causes of the problems and giving recommendations for good practice to minimize the danger involved in such process. It is also intended that the book will show up those areas in which our knowledge is still deficient and in which further research is needed both in universities and in the research and development departments of industrial companies. The author of this book is Mr K. N. Palmer who is well-known for his contributions to knowledge on these subjects both as a research worker and as an adviser to industrial companies.

J. C. Williams

Preface

Dust explosions and fires are known hazards in many industries but the literature on the subject is fragmentary and there is a need of a unified account. This book attempts to give a comprehensive picture of the present state of knowledge concerning dust explosions and fires, involving both practical and theoretical aspects. The subject has in the past received less attention than it merits and it is to be hoped that when the serious gaps in knowledge are made apparent, stimulus will be given to future study both in the laboratory and on industrial plant.

The aim has been to present clearly the problem of dust explosion and fires in factories and to describe how it arises, how the hazard may be assessed, the means of obtaining protection, as well as present knowledge on the technical background.

The book contains an introductory account of the general problem in factories, the test methods used in various countries to assess the explosion and fire hazards of dusts, together with interpretation of the results, and a critical discussion of research investigations into the combustion of dusts both from laboratory and full industrial scale. Means of obtaining explosion and fire protection are described, with a separate account of dust fires, and the application of protection to numerous common industrial processes is detailed. Particular attention is paid to dusts which have traditionally been regarded as an explosion hazard, either because of their widespread use or their severe explosibility. The basic steps in the safe design of a dust handling industrial plant are listed, the measured explosion properties of several hundred dusts are tabulated, and various legal and other authoritative requirements are indicated.

The subject matter of the book is applicable to a wide range of industries in which there are factories handling powders and dusts. Specialized aspects associated with coal mining have not been dealt with in detail, but the more general aspects common to other industries have been included.

The book is of direct concern to the designers, managers and operators of plant in factories, to safety and fire fighting authorities, and to those concerned in research and teaching on the combustion of solids.

In those parts of the book concerned with the evaluation of hazard and the protection of plant it is only practicable to deal with general cases. Where a specific application is to be considered particular attention should be paid to local conditions and plant layout, as hazards can arise from the juxtaposition of different plant and facilities. Pointers to possible problems of this type are given in the book, but in case of doubt on a specific project expert advice should be sought. The book does not in any way supplant expert professional advice but is a source of information intended to help readers in their handling of the subject, and enable them to usefully participate in discussions with others.

The author is deeply grateful to colleagues at the Fire Research Station and to other experts in various countries for their help and encouragement during the preparation of the text. Special mention of Mr J. C. Williams is necessary, because of his great help in commenting on the text and in general organization. Grateful acknowledgment is extended to the Organizations and Companies who have provided photographs to enliven the text. Thanks are also due to the copyright holders of many of the figures, who have kindly agreed to their use in this book.

Borehamwood
January 1973

K. N. Palmer

1 *Introduction*

Combustion of dusts

Many solid materials are combustible, that is they can be made to burn if a source of ignition is applied. The ease with which ignition occurs, or the size of the source required to initiate burning, depends on the nature and dimensions of the solids. Generally solids are ignited and burn more readily as their size decreases, and when reduced to powders or dusts the conditions are most favourable. Not only are dusts and powders relatively readily ignited, but they also burn more rapidly. Two of the reasons are that air, or oxygen, gains easier access to the whole mass if it is in powder form, and heat cannot be drained from the burning surfaces into the interior of the solid material when this is in the form of small particles.

A favourable arrangement for rapid burning of dust is when the particles are separated from each other by relatively large distances, so that the air has ready access to each particle, as when the dusts are dispersed as a cloud or suspension in the air. The optimum condition is when the particles are separated to have sufficient air for complete combustion, but they are close enough for the heat release to support the burning of adjacent particles. If carried out under controlled conditions, the burning can be made to produce heat rapidly and efficiently with relatively little loss of combustible. An example is a well designed pulverized fuel furnace where the burning rate of the fuel is high and complete combustion is the aim. Under uncontrolled conditions, the rapid rate of burning can be accompanied by pressure effects and flame spread causing catastrophic damage to property and injury to personnel. Undoubtedly fires in dust layers, rather than suspensions in air, are less spectacular, but nevertheless can lead to large financial loss.

Definitions

In the context of dust explosions and fires there is no agreed definition of the term 'dust'. In practice the terms 'dust' and 'powder' have been used indiscriminately, some industries favouring one term and some the other. For present purposes the two terms will be regarded as interchangeable.

For convenience there need to be limitations on the diameters of particles entitled to be called dusts or powders. Following the British Standard [1], particles of diameter less than 1 micron can be termed 'smoke', so that dust particles have diameters greater than 1 micron (10^{-3}mm). It is also convenient, for present purposes, to accept the definition of 'powder' [1] as containing particles with a maximum dimension of less than 1000 microns (16 B.S. mesh sieve). The term 'dust' was restricted to particles passing a 200 B.S. mesh sieve (76 microns), and the term 'grit' was defined for coarser powders [1]. This restriction on the term 'dust', and the use of 'grit', has not been followed here in relation to dust explosions and fires.

Where information relative to the subject has involved particles coarser than 1000 microns it has been accepted, although not strictly applying to dusts or powders.

The term 'explosible' has been used to describe a dust which is able to cause a dust explosion. The term 'explosibility' indicates the extent to which a dust is explosible. 'Explosible' has been preferred to 'explosive' to avoid confusion with high explosives, whose explosion properties are different from those of dusts dispersed as a cloud and ignited. A dust explosion propagates by the combustion of particles with surrounding gas, whereas explosives do not require a surrounding atmosphere. The properties of explosives are outside the present context.

Industrial hazard

In industry many of the dusts handled are combustible. These may be produced either as end-products, or as by-products. Examples of the former include flour, sugar, starch, plastics and metals, coal, dyestuffs and pharmaceuticals. Dusts produced as by-products include wood sawdust, textile fluff and dust from many processes where solid surfaces are polished, cleaned, stripped or cut. Wherever combustible dusts are handled in industry there is a risk that some form of ignition source may be present and this could in turn lead to a fire or explosion.

Generally an explosion occurs only where the dust is dispersed in the air, or can be dispersed by some means whilst the source of ignition is present, whereas dust fires develop in heaps, layers, storage bins, and other deposits. Transition from fire to explosion and vice versa can readily occur. If burning dust is suddenly disturbed and a suspension is formed in the air, a 'flash' is easily produced which in turn can disturb more dust and cause a more violent effect. In contrast, if a suspension already formed is ignited, and causes an explosion, burning dust particles are likely to fall out of suspension on to nearby surfaces and spread fire. In either event the damage can be considerable. The basic philosophy of protection is that safety precautions shall be designed to be adequate, but as economical as possible, that any fire or explosion which should occur will be safely dealt with without causing loss or injury, and that catastrophic situations are prevented. It will be clear that basically the problem is an industrial one, and unlikely to be a serious hazard in domestic premises, and that dust explosions are particularly hazardous because of the speed with which they develop and the impracticability of improvising protection during the incident.

Design considerations

The possibility of dust explosions and fires should be considered as a routine when combustible dusts are involved, and should not wait until there are incipient signs of trouble or a major incident in the plant. Generally it is more economical to provide against the hazard at the design stage so that precautions can be built into the plant. If the protection has to be applied after the plant has been constructed, the costs may be substantially higher. In addition, restrictions may have to be placed on the type of plant unit that can be installed in particular places, especially work rooms, and may involve splitting of the process so that certain operations are carried out separately elsewhere. This again leads to less economic working, and to problems of transportation which would not arise if the operation had been planned logically from the start.

Sometimes difficulties are encountered, as when the materials involved in a process are changed, and it is proposed to use dusts of greater explosibility than in the past, in the same plant. Under these conditions modifications of the plant would have to take place, and the types of protection that could be employed may be restricted by the situation of the plant. Relatively expensive methods of protection may be required which might also have limitations resulting in the plant being run at below optimum conditions. Ultimately, a compromise has to be reached between solutions involving substantial initial capital expenditure, but with low running costs, and alternatives where the initial outlay may be lower but the running costs much higher. No general guidance can be given, but the principal alternatives in methods of protection are considered here so that the design engineer can choose that which, in his own circumstances, has the greatest attraction. What cannot be done, where combustible dusts are concerned, is to assume either that a source of ignition will never be present, or that in any case an explosion or fire will never develop, even though one may not have occurred in the past. A sufficiently large number of dust explosions of severities ranging from minor to catastrophic have occurred in a wide range of industries to illustrate that the hazard should be considered in plant design.

Sufficient information is available to help the engineer, although not of course to present standardized answers for all circumstances. The question of personal judgment therefore inevitably becomes involved, and this can only be on a rational basis if the applicability and limitations of the information are known. Factors such as these must be borne in mind by the engineer, and others concerned, because they will want safety margins which are adequate and economic. It is only by considering all the evidence that an informed judgment can be made.

Legal and other requirements

Although it is logical to consider protection against dust explosions and fires in terms of plant design there are other requirements which must also be met. In many countries there are national laws which apply and there may also be local legal requirements. If the plant is covered by insurance, the premium will be related to the risk, and notification should be given either of the planning of new plant or of any modification of existing plant which could affect the risk rating. Manufacturing companies may also have internal requirements relating to safe working which must be complied with.

The information given here is intended only to serve as a broad guide, and expert opinions are essential for ensuring that legal and other requirements have been fully met.

Information on technical, legal and insurance matters is available from the Fire Protection Associations in many countries (Appendix 1). Governmental and technical agencies and institutes are other sources of information, as well as insurance organizations. Expert advice should always be sought in specific instances of doubt.

In the United Kingdom the principal Act relating to requirements connected with dust explosion hazards in industries other than coal mining is the Factories Act 1961. Further regulations made, or deemed to have been made, under this Act are also relevant.

Section 31 of the Factories Act relates to provisions for plant employing flammable dusts. Other sections of direct relevance are Sections 40 to 47 inclusive (Provision and Maintenance of Means of Escape), Section 48 (Safety in Case of Fire), Section 49 (Escape Instruction), Section 50 (Enabling Special Regulations for Fire Prevention to be made), Section 54 (Dangerous Conditions and Practices), Sections 80 and 81 (Notification of Accidents and Dangerous Occurrences), Section 148 (Persons empowered to inspect Premises).

The sections mentioned above are intended as a guide only, and where the Act is to be interpreted it should be consulted in its full original.

Other legislation which may be regarded as supplementing the Factories Act for certain dangerous processes is as follows:

Celluloid manufacture etc. regulations (SRO 1921 No. 1825).

Magnesium (Grinding of castings and other articles) Special Regulations (SRO 1946 No. 2017).

Chemical Works Regulations (SRO 1922 No. 731).

In addition to the Factories Act 1961 there may be other legislation covering particular plant or processes. In specific instances expert advice should be sought.

General approach

In the succeeding chapters information is presented in the following sequence. Firstly an outline is given of the characteristics of dust explosions and fires and the important factors described. The tests used in various countries in establishing the explosibility of dusts are described and are followed by discussion of the interpretation of their results. The more technical aspects of dust explosion flame propagation, both experimental evidence and proposed mechanisms, are then presented in order to give the reader a fuller understanding of the background and also to indicate the areas of uncertainty. The more frequent sources of ignition and the accepted methods of dealing with dust explosions are covered, as these topics are important to plant designers and management. A chapter on dust fires follows, because these present specific problems. Consideration is then given to applying the information already presented to actual industrial operations, and the individual properties and hazards of specific groups of dusts are dealt with. Finally, a suggested outline procedure in plant design as regards dust explosions and fire problems is laid down. The appendices contained summarised data for reference.

Reference

[1] British Standard 2955, (1958).

2 *Outline of Practical Problem*

Knowledge of the industrial problems associated with dusts, as regards explosion and fire, is less widespread than that for flammable liquids and gases. The latter are generally recognized as being fire risks, and are popularly associated with the possibility of explosion. The comparable hazard present with dusts, particularly in industry where they are handled on a large scale, is less well known. There are certain industries where exceptions can be made, for example those traditionally associated with dusts such as flour milling and coal mining. But throughout the broad span of manufacturing industry, the awareness of dust explosion risks is less than that for liquids and gases.

In order to set the problem of dust explosions and fires in perspective a survey will be given of the principal important aspects. This will assist in establishing the nature of the problem, and will show the need for the various tests and protection measures discussed subsequently.

The characteristics of dust explosions and fires will be described, and brief consideration will then be given to explosible dusts and the industries involved. The effect of dust explosions on industrial plant and buildings is of basic importance and finally an outline is given of the main methods of protection against fire and explosion.

Historical development

Although dust explosions have been recorded for over a century, appreciation of their true character was relatively slow in developing. In particular great difficulty was experienced in accepting that an explosion could be caused by the dust alone, and that the presence of a flammable gas to support the explosion was not necessary. For instance, an account of a flour dust explosion in Turin in 1785 was published shortly afterwards [1] in which it was postulated that ignition first occurred in flammable gas given off by the disturbed dust, although it was recognized that the dust itself contributed to the subsequent explosion. Over 100 years elapsed before it was generally accepted that dust explosions in coal mines could be initiated and sustained by coal dust alone, without needing support from flammable mine gas. When the position was finally clarified, early in the present century, explosions in numerous other industrial dusts had occurred, and the way was open for a more rational assessment of the explosion hazard and the need for adequate precautions.

A review of the position in 1922 gave details of numerous explosions in various industrial dusts, and estimates of life and economic losses. [2] A record of important dust explosions in the United States and Canada since 1860 [3], excluding those in coal mining, gave details of the causes of explosions, the extent of damage and casualties, and numerous photographs of damage to plant and buildings. Comparable statistics have not been published for the number of explosions and loss of life and property in the United Kingdom. The number of incidents notifiable by law, in the U.K., involving dusts, in recent years has averaged about two or three per month. A technical review, covering data up to 1959, gave general references [4] particularly to dust explosions in factories. A review of mining explosions in India [5] covered both dust and gas, and explosions involving both. The causes and effects of the explosions were similar to those reported earlier in the West.

Characteristics of dust explosions

EXPANSION EFFECTS

A dust explosion may be envisaged as combustion of a dust cloud which results either in a rapid build-up of pressure or in uncontrolled expansion effects. The gas in which the dust is suspended takes part in the combustion, and hence in considering the properties of dust explosions, both the nature of the dust (i.e. the fuel) and the gas (i.e. the oxidant) are important. It is the expansion effect, or the pressure rise, if expansion is restricted, which presents one of the main hazards in dust explosions. In some industrial plant, such as pulverized fuel furnaces, dust clouds are continuously burned under controlled conditions so that in normal working the pressure is not permitted to rise dangerously. This process is not regarded as a dust explosion, although there may be similarities in the mechanism of the combustion.

The expansion effects in dust explosions arise principally because of the heat developed in the combustion and, in some cases, to gases being evolved from the dust because of the high temperatures to which it has been exposed. If the gases are not consumed in the explosion they would remain afterwards and would contribute to the final gas volume, and hence would augment the expansion effects. The heat generated in a dust explosion is eventually lost to the surroundings and so the expansion and pressure effects are transient. However, it is a characteristic of explosions that flame speeds are high, whereas the time required for gas temperatures after the explosion to revert to the level of the surroundings could be many seconds. In industrial plant, the rate of heating is likely to greatly exceed the rate of cooling whilst the explosion is developing, and so expansion and pressure effects are able to become manifest. Indeed, in some cases, under controlled laboratory conditions, the rate at which the explosion propagates is so rapid that maximum pressures are obtained which are close to the theoretical values calculated on the basis of no heat loss whilst the explosion is developing. Heat loss to the surroundings is clearly not likely to be the basis for adequate protection of industrial plants against dust explosions.

DEFLAGRATION AND DETONATION

In dust explosions the rate of propagation of flame through the dust cloud is fast compared with rates of flame spread in fires. In explosions, in small enclosures, flame speeds range from hundred of metres per second to the slowest which are of the order one metre per second. The most rapid burning dust fires are below this speed range. The type of explosion usually occurring under industrial conditions is termed a 'deflagration' and an important property is that the flame speed is less than the velocity of sound in the gaseous products of combustion. There is another mode of combustion, termed 'detonation,' in which the flame speed is equal to the velocity of sound in the gaseous products and is accompanied by a shock wave. This velocity is of the order one thousand metres per second. Whether or not a true stable detonation can occur in a dust explosion in industrial plant has not yet been established. In theory, because of the energy released during the explosion, detonations should be possible and, indeed, explosions having velocities similar to those of detonations have been reported. The majority of evidence relates to explosions in coal-mine galleries, at least 100 m in length, but the explosions were initiated by powerful sources which in themselves were capable of generating strong shockwaves. Whether or not detonations can develop from relatively small ignition sources in industrial plant, has not yet been proved, but in any case a detonation would initially start as a deflagration. It is general practice in considering protection against dust explosion hazards, in industrial plant, to assume that deflagrations rather than detonations occur. This procedure has proved to be satisfactory in practice, which is fortunate because deflagrations are simpler to deal with than are detonations.

There are several reasons for deflagrations being an easier problem than detonations. In a deflagration, as the flame moves away from the ignition source, its speed is typically of order 1 or 10 m/s, whereas the compression waves, or shockwaves, arising from the expansion effects will travel through the unburnt dust cloud at sonic velocity (i.e. about 330 m/s). In effect, therefore, a signal is sent ahead of the flame giving warning that expansion, or a pressure rise, is occurring. The shockwaves rapidly traverse the interior of the plant so that apart from relatively small gradients due to friction, the pressure rise throughout the plant is uniform. Explosion vents or pressure detectors can then operate during the early stages of the explosion and steps taken to prevent the maximum theoretical explosion pressure from developing. In a detonation, the flame accompanies the shockwave and hence no advance warning signal is received by an explosion vent or pressure detector. No time is therefore available for the operation of safety measures before the arrival of the flame; the detonation would continue unchecked.

A transient high pressure is also associated with the shockwave in a detonation; after the shockwave has passed through a closed vessel the pressure would ultimately fall to the maximum obtained in a deflagration. For closed vessels, with no venting or loss of gas, maximum explosion pressures from deflagrations can be as high as 150 lb/in^2 (about 1000 kN/m^2); the detonation pressures would be several times higher. Detonations are more hazardous than deflagrations and, with gases and vapours, considerably more stringent precautions have to be taken against detonation. With dust explosions, and with the types of industrial plant in common use to-day, the evidence suggests that precautions against deflagration in dusts are adequate and they need not be extended to cover detonation. However, industrial plant is continually increasing in size as new designs are developed, and increase in dimensions particularly in length of ducting is one factor likely to increase the chances of detonation. Detonation should also be borne in mind if dusts can mix with more powerful oxidants than air, or with fast burning gases such as hydrogen.

FLAME SPEEDS

The flame speed in a dust explosion is not constant but depends on a number of factors. Two of the principal factors are the chemical compositions of the dust and the oxidant, but with the present state of knowledge the flame speed cannot be related simply to composition, including moisture content, or to heat of combustion. Furthermore, the flame speed depends upon the particle size of the dust, the smaller the particles the higher the flame speed, providing that with fine particles the dust is evenly dispersed in suspension. The flame speed also depends on the turbulence of the gas in which the dust is dispersed; broadly, increased turbulence leads to higher flame speed. The relationships between flame speeds and properties of the dusts including particle size, and the turbulence of the gas have not yet been derived by theory.

To illustrate the weakness of the present theoretical position, the case can be considered of the maximum flame speed of a given dust, of given particle size, dispersed in air in the almost complete absence of turbulence. The most favourable condition for flame propagation, i.e. the maximum flame speed, might be expected when the ratio of the fuel to oxidant was stoichiometric. A stoichiometric mixture is a balanced mixture, that is it contains just sufficient fuel to consume all the oxidant. In fact experience shows that the maximum flame speed would occur when a substantial excess of fuel was present, and that as the fuel concentration was further increased, the flame speed remained near the maximum value and only decreased significantly when the fuel concentration was several times that at stoichiometric. This behaviour is in contrast to that of combustible gases where maximum flame speeds do occur with mixtures near to stoichiometric. Obviously with dust dispersions all the fuel cannot be burned and a surplus must remain after the explosion. Why this behaviour should occur with dust has not been satisfactorily explained, and it is an illustration of the backward state of the theory of dust flame propagation. This backwardness has important practical implications because much of the evidence on which recommendations for the safe handling of dusts in industrial plant is based, has to be empirical. Where conditions arise in practice which have not been covered by previous experimental investigations, an element of uncertainty is introduced. This uncertainty cannot then be gauged or minimized by calculations based on theoretical principles.

IGNITION

For an explosion to occur, the dust cloud must be ignited. Apart from a few special instances where the act of dispersion of dust may ignite it, a separate source of ignition must be present and the concentration of dust in the cloud must be favourable. As the main stimulus for the measurement of dust explosion properties has been the need to provide information on the safe use and handling of dust in industry, the laboratory tests of ignitability of dust have been designed to incorporate sources of ignition likely to occur in practice. Two frequent sources would be a hot surface or a spark. Consequently, the minimum ignition temperature and the minimum ignition energy are the ignition properties most commonly measured in the routine testing of dusts.

It is common experience that ignition can occur in a combustible material in contact with, or close to, a sufficiently hot surface. It might be expected that the ignition temperature would be a fundamental property of the dust, and in fact minimum ignition temperatures of dust clouds have been published (see Appendix 2). But the minimum ignition temperature is not a fundamental characteristic of a dust cloud and it depends upon the size and shape of the apparatus used to measure it as well as the rate of rise in temperature of the dust. Minimum ignition temperatures have therefore to be determined in a standardized form of apparatus for meaningful comparisons between dusts. Because the value depends on the design of apparatus used it would not necessarily apply to equipment of different size and shape. The minimum ignition temperature also depends upon the particle size and moisture content of the dust, and so a specific value for different samples of a material is often not possible. This point is considered in more detail later. Another factor of considerable practical importance, which is often overlooked, is that dust can be ignited by surfaces at temperatures lower than the minimum ignition temperature for dust clouds, if the dust is in a layer deposited on the hot surface. Under these conditions the time of contact of the dust with the surface is greater, permitting a slower rate of rise in temperature to lead to ignition than would be possible in a dust cloud which is usually in motion past the hot surface. Time is also provided for the drying of a moist dust. When the layer has become ignited, it can lead to ignition of a cloud either by generating a temperature above that of the underlying hot surface, or by producing a flame, or by becoming dislodged as a burning cloud from which flame propagates into further dispersed dust. The ignition temperatures of dust layers will be considered in greater detail later, but they again depend upon the rate of temperature rise and the thickness of the layer. Hence, again the experimental conditions have a marked effect on the value of minimum ignition temperatures, which are not unique properties of individual materials.

Another frequent source of ignition of dust clouds in industry is a spark. There are a number of ways by which these can be produced, e.g. by electricity, friction, impact, or by hot cutting or welding. A characteristic of all these forms of spark is that a small particle or volume of gas at a high temperature, is produced for a short interval of time. Although the temperatures of sparks are well above the minimum ignition temperatures listed for dust clouds, the size of the spark is much smaller than that of the hot surface used for ignition temperature measurement. Because of the smallness of the source, and its short duration, much higher temperatures are necessary to transfer the energy necessary to ignite dust clouds. It is much easier for experimental purposes to measure the energy delivered in an electric spark than that delivered by friction or other thermal processes. Consequently the routine method of measurement of minimum ignition energies of dust clouds is to pass an electric spark of known energy through the cloud and to observe whether ignition occurs. As with other explosion properties the minimum ignition energy of a dust cloud depends on the dust concentration, particle size, moisture content, etc., and in addition there is evidence, which will be considered later, that the design of the electrical circuit producing a spark can also affect the minimum ignition energy. Routine determinations of minimum ignition energies are made by a standardized procedure which allows for comparison between different dusts without yielding fundamental information on the ignition process.

EXPLOSIBILITY LIMITS

If flame is to propagate through a dust cloud, the concentration of dust must lie within the lower and upper explosibility limits. The lower explosibility limit may be defined as the minimum concentration of dust in a cloud necessary for sustained flame propagation. It is a fairly well defined quantity, can be determined reliably in small scale tests, and is often listed as the minimum explosible concentration. Values are usually in terms of weight of dust per unit volume of air or gas (Appendix 2).

When the concentration of dust is raised above the lower explosibility limit and past the stoichiometric value, the flame speeds and vigour of the explosion increase as described above. As the dust concentration is further increased, however, the quenching effect of the surplus dust on the explosion becomes more marked and eventually a concentration is reached at which flame propagation no longer occurs. This concentration is the upper explosibility limit and may be defined as the dust concentration above which flame propagation does not occur. The upper explosibility limit is not easy to measure in practice, mainly because of the difficulty of ensuring that a dust is uniformly dispersed in the cloud and that regions of low concentrations do not form by chance movement of the dust particles. Because of difficulties in measurement relatively few values have been determined and these are much higher than would be expected by analogy from combustible gases. There are also difficulties in ensuring that in industrial plant dust clouds could be maintained at concentrations above the upper explosibility limit, and interest has been much less than with the lower explosibility limit, which in turn has only limited applications in the safe handling of dusts, for reasons given in greater detail later.

The lower and upper explosibility limits depend not only on the composition of the dust, but also on properties such as particle size, and moisture content as well as on the type of ignition source used in the test. The particle size and shape also affect the dispersability of the dust; if this is poor due to cohesive forces or to entanglement of fibrous particles, then the explosibility limits are also modified. Many of these factors do not arise in comparable experiments with flammable gases, for which the flammability limits are well defined, and with dusts the difficulties have contributed to the lack of knowledge of explosion processes.

EXPLOSION PRESSURES

When a dust explosion occurs in industrial plant, spectacular destruction may result if the explosion is initially completely confined within the plant which is ultimately too weak to withstand the full force of the explosion. Two of the factors influencing the violence of the explosion are the maximum explosion pressure and the maximum rate of pressure rise. These quantities govern the rate of pressure build-up. The maximum values obtained with dusts in small closed vessels, which are sufficiently strong not to burst or leak during the explosion, may be as high as 150 lb/in^2 (1000 kN/m^2) for maximum pressures and 15 000 lb/in^2s (100 MN/m^2s) for maximum rates of pressure rise. The values obtained under standard conditions depend on the explosion properties of the dust including particle size, moisture content, as well as on its basic composition. The values cannot be predicted from knowledge of the composition and thermal properties of the dust, but must be obtained by measurement. They may be compared with each other for different dusts and relative explosion severities can be predicted. Maximum explosion pressures are used as indications of the likely severity of an explosion. Where it is intended to contain the explosion within the plant, the strength of the plant must be so designed that it is well capable of withstanding the maximum pressure. The maximum rate of pressure rise, or alternatively the average rate of pressure rise, are often used in connection with the design of explosion relief venting. Where rates of pressure rise are high, products of combustion are being generated rapidly. If protection of plant is to be obtained by the relief of pressure through vents, it is necessary that combustion products should escape whilst they are being formed, otherwise unwanted build-up of pressure will occur. If the rate of formation of combustion products, i.e. the rate of pressure rise is high, the amount of venting will need to be greater than in explosions with lower rates of pressure rise. Dusts with very high rates of pressure rise need extremely careful handling for safety, and their explosion pressure characteristics may be the main advance warning to the user of their potential explosion hazards.

PRIMARY AND SECONDARY EXPLOSIONS

Dust explosions have the characteristic that the pressure waves transmitted through the atmosphere during the initial stages of the explosion can cause dust which is deposited to be thrown into suspension. This disturbance may either arise because of air movement over the dust deposit causing the dust to be blown into suspension, or to vibration of structures so that dust deposits are shaken off and fall into the air. With either mechanism, or both, further dust clouds are formed which can be ignited by the explosion flames already present. The initial stage of the process during which disturbance of the dust occurs, is customarily termed the primary explosion, and the subsequent stages when the newly disturbed dust becomes involved with the flame, is termed the secondary explosion.

In many dust explosions, in industry, the sequence has been as follows: a dust explosion has developed inside an item of plant (the primary explosion) and, because of inadequate explosion protection, flame has been discharged from the plant to the workroom. This discharge has occurred either because the plant ruptured or because relief venting has been badly installed. In either event the result was that flame has discharged from the plant and was accompanied by disturbance of the air in the workroom. Dust settled in the workroom then became dispersed and subsequently ignited by flame from the plant. This subsequent explosion is the secondary explosion and its destructive effects are frequently much greater than those of the primary explosion. A greater amount of dust becomes involved in the secondary explosion, and so the total amount of energy released is greater; also the building structure itself is often relatively weaker than the plant. The results of secondary explosions may well be collapse of buildings which of course leads to further damage of plant and to casualties (Plates 1 and 2).

The hazards arising from secondary explosions are now well established and their avoidance is one of the basic principles of dust explosion protection. The required precautions are obvious, to prevent emission of flames into the workroom from plant should an explosion occur inside the plant, and to avoid the presence of dust deposits in workrooms. This latter requirement means that clean working is necessary and that frequent and regular clearing of any dust spilled, or accidentally discharged is vital. Suitable means of ensuring good housekeeping are described later.

Although one mechanism by which primary explosions develop into more powerful secondary explosions is because of the discharge of flame from within plant, other mechanisms have occurred in industry. For instance the accidental spillage of a relatively small quantity of dust onto a source of ignition may give rise to a very mild primary explosion because of the complete lack of restraint and opportunity to build up explosion pressure. Nevertheless the disturbance to the atmosphere can be sufficient to cause other settled dust nearby to become disturbed. This dust then develops into the secondary explosion which can spread wherever an explosible dust suspension has formed.

Another example is the development of dust explosions in coal mines. Here the primary explosion is frequently a relatively mild ignition of fire damp, which is mainly methane gas, and this primary explosion in itself may have very limited power and capability for damage. However, the air movement caused by the burning of the gas can be sufficient to disperse coal dust which has settled in the neighbourhood. This dispersed dust can become ignited, forming the secondary explosion, which may develop into disastrous proportions unless the explosion protection is adequate. Dust explosions in coal mines differ from those usually encountered in industry because of the construction characteristics of the mines. These involve roadways or galleries of considerable length relative to their cross-sectional area, the walls of which constrain the effects of the primary explosion. The generation of dust clouds to form secondary explosions is thereby favoured because of the lack of opportunity of the blast waves from the primary explosion to escape other than along the roadways.

PRESENT TRENDS

Although the development of primary and secondary explosions has been recognized from past experience, recent trends in design of plant in modern industry are changing the pattern. The increased scale of much new industrial plant means that the volume of the plant giving rise to the primary explosion may be factors of 10, or more, greater than was customary only a few years ago. Hence the energy release in a primary explosion would be much greater and the protection requirements of the plant itself are becoming a serious problem. The difficulty is aggravated because increase in size of plant tends to cause structural weakness. Provision of mechanical strength is expensive and it may not be economic to provide more than the minimum for safety. On the credit side, there is a clear tendency for large modern plant not to be situated in enclosed spaces but to be built in the open. The risk of secondary explosions as originally defined, arising from the disturbance of dust external to the plant and causing damage to surrounding property, is reduced. However, the total amount of energy which may be released during an explosion in such large plant may be comparable to that obtained in the past as the sum of primary and secondary explosions. Information available at present on the safe methods of handling explosions on a large scale is extremely scanty, and a need for such information is likely to increase rapidly in future.

TOXIC GASES

The flames and pressure effects arising from dust explosions are a principal source of casualties, but there are also dangers arising from the combustion products of the explosions. These are deficient in oxygen and also contain toxic gases, particularly carbon monoxide. The concentration of toxic gases can be sufficiently high for a few breaths to be fatal, so that momentary unconsciousness due to the blast can result in death.

The toxic nature of the combustion products is particularly hazardous where the explosion has occurred in the working space with operatives present, as occurs in secondary dust explosions or in coal mine explosions. In coal mines, because of the necessary mechanical ventilation systems, toxic gases from a relatively minor explosion may circulate over a wide area and cause casualties in regions where the blast and heat effects from the explosion were negligible. The hazards from toxic gases are additional reasons for not permitting dust explosions to occur in working spaces, and emphasize the desirability of venting industrial plant to atmosphere outside the work room.

Toxic gases are also likely to be present inside plant, after a dust explosion. Plant must not be entered by operatives or fire fighters until either the atmosphere has, by test, been shown to be safe, or breathing apparatus is worn by all personnel entering.

Dust fires

The explosions which can be obtained with dispersed dusts can be spectacular and cause extensive damage, but only a minority of ignitions of dust result in explosions. The majority of ignitions occur in dust layers, deposits, or heaps, where the dust is not disturbed and a dust fire rather than an explosion is the result. There are several important aspects of dust fires which should be considered; these are, firstly that the dust which is ignited may in itself be a valuable material and that destruction of it by fire may cause substantial loss. Secondly, even if the dust is of relatively little value, and it may even be a waste material, it can cause fire to spread to other neighbouring materials, plant or buildings. There may also be a potential explosion risk, which could become reality if the burning dust were violently disturbed or otherwise dispersed into the air. Fires in burning dust should therefore be dealt with rapidly and calmly so as to minimize other damage.

A general account of dust fires will be given later, but some of the most outstanding points will be briefly discussed here. A fire in a dust layer or heap may burn with a flame or it may just glow and smoulder. If it is not burning with flame and the smouldering is buried within a heap or deep layer, the presence of the fire may be very difficult to detect. Little or no smoke may be produced under these conditions, and the absence of any acrid smell of burning, which is frequent with smouldering dusts having good access to air, may lead to a false sense of security. Smouldering can develop into flame under favourable conditions and hence lead to a rapid spread of fire across the surface of a dust layer or heap. Where the fire is smouldering within a heap, the rate of progress may be very slow so that the fire persists for weeks, months, or even years in large tips and waste heaps. The rate of burning accelerates considerably if the smouldering penetrates through to the outer surface of the heap. Flames are then able to establish themselves and to travel over the surface of the dust heap.

Detections of dust fires can therefore be very difficult, particularly if these are buried within heaps, and the first indication of fire may be a warming up of the surface of the heap giving the impression that self heating is occurring. If the fire is in a waste heap, of extensive areas and height, direct fire fighting may be ineffective and large scale methods may have to be used to control the outbreak. Such methods include digging out the burning part of the heap, excavating channels across the heap to sub-divide the volume and attempting to produce fire breaks. If the fire is on the surface of the heap, as would be the case if flames were present, control is easier provided that disturbance of the dust during fire fighting is avoided. For instance, low pressure water sprays are preferable to water jets for the majority of dusts, although some materials, particularly metals, should not be attacked with water at all (Chapter 9). When the surface fire is extinguished careful examination should be made of the residue to ensure that smouldering has not spread downwards into the heap from the surface. If unchecked this smouldering may propagate into the interior of the heap and build up an internal fire which could cause a further outbreak elsewhere on the surface. Alternatively, if the dust is a deposit in a length of ducting, the existence of such a deposit probably indicates that the plant has not been well designed, fire fighting may again be difficult because of limited access and laborious methods such as raking the dust out may have to be resorted to. The same general principle applies, however, namely that on no account should burning dust be violently disturbed or agitated as otherwise a dust explosion can result, and then in confined conditions this could accelerate and raise further dust leading to a serious explosion.

Dust fires are therefore to be treated with respect because, under certain circumstances, they give little signs of their presence, and are often difficult to detect, they have the capability of accelerating to larger fires quickly thus leading to the ignition of other materials, and considerable care is needed in fire fighting to avoid disturbance of the dust which could lead to explosion. Finally it is often difficult to ensure that all the burning dust is extinguished, particularly if the fire has been smouldering, and watch should be kept for may hours after extinction has apparently been obtained to ensure that smouldering dust does not lead to a further outbreak. In many cases dust fires occur in waste materials of little value, and the nuisance caused is far in excess of the direct monetary loss, although if precautions are not taken and fire subsequently breaks out a final serious loss of other materials, plant buildings etc., can result.

Explosible dusts and industries involved

Past experience in industry and in the laboratory testing of dusts has shown that a wide range of dusts can give rise to explosions. Not all materials that will burn in air can cause dust explosions, even if finely divided and dry; that is, not all combustible dusts are explosible. All explosible dusts must be combustible. The reason why some combustible dusts are not explosible has not been definitely established, but it is clear that it is not related directly to the heat of combustion or calorific value for the dust. Some dusts of relatively high heat of combustion, such as graphite and some anthracites, are not explosible whereas other materials of lower heats of combustion, such as wood sawdust, are readily explosible. Dusts which are not explosible in air at atmospheric temperatures may be able to cause explosions in a heated chamber, such as a furnace or oven. When combustible dusts are being handled it is necessary to either refer to previous records or to carry out laboratory tests to decide whether the dust is explosible.

Over the years a considerable amount of information on the explosibility of dusts has been accumulated both from practical experience and from laboratory tests, and this information is of great assistance in assessing dust explosion hazards. Clearly where new materials are involved, or mixtures of old materials, or old materials made by new processes where their characteristics may be different, recourse must be made to further laboratory testing.

The industries concerned in the manufacture or handling of explosible dusts are numerous, and further details will be considered later. However, some of the principal industries are as follows:

Agricultural
Chemical, including dyestuff's
Coal, mining and utilization
Foodstuffs, human and animal
Metals
Pharmaceuticals
Plastics
Woodworking

This list gives only a few examples and consists of those where the principal products present dust explosion risks. There is in addition a wide range of industries which produce explosible materials in their processes, although dusts and powders may not be principal products as they are, for instance, in flour and sugar. Examples of industries where explosible dusts are present, and have in fact caused explosions, are rubber dust in the footwear industry, esparto grass dust in the paper industry (Plate 3), and aluminium dust in the manufacture of refrigerators.

Although the principal product of the industry may not be a dust and presents no dust explosion hazard, consideration must also be given to intermediate materials used in the manufacture of the final product, and to the processes involved in the manufacture. These processes may give rise to by-products or waste materials which are explosible, and which in the factory may present as great a hazard as dusts in other industries where the principal product is explosible.

In attempting to assess the dust explosion risk, and the precautions which must be taken, due account must be given of the size and nature of the plant and operations in which the explosible dust is involved. For example, any process where explosible dusts are handled on the tonnage scale, and where dispersion of this dust in air may occur would need detailed consideration to ensure that adequate explosion precautions have been taken. On a small scale, where perhaps only a few kilograms per hour of dust are involved, but where this dust is in suspension in air and operatives are nearby (as in the polishing of metals accompanied by a dust extraction system) a dust explosion hazard is again present. Explosion protection is then advisable. Where the same quantity of dust is being manufactured under closely controlled conditions, and dispersion in air is most unlikely and ignition sources are not present as part of the process, there may be scope for the relaxation of explosion protection requirements. Such a situation might arise for instance where a pharmaceutical product is being synthesized and the process requires great care in the handling of materials, and where to prevent contamination the workroom is maintained scrupulously clean. Another example is where the dust has a high moisture content or is handled in a wet state.

In general, relaxation of explosion protection measures can only be justified where there is strict control over the plant processes, often because of reasons unconnected with explosion prevention, the scale of operations is small, and the dust is not subjected to manufacturing processes liable to cause it to ignite. Where explosible dusts are involved in industrial processes the correct attitude is to assume that explosion protection would be necessary, but the extent of the precautions must be related to the explosibility characteristics of the dust, the size of the plant and the types of operations carried out within it, the amount of dust and the period for which it is likely to be in suspension.

Although the first and most obvious step is to protect a plant itself against dust explosions, it is also most important to consider the situation of the plant in relation to other installations and to the presence of operatives. The reason is that in safeguarding the plant, it may be necessary to discharge flame into the neighbourhood and this must obviously be done safely. Thus adequate explosion protection involves not only the plant but also its environment.

Effect of dust explosions on plant and buildings

Numerous dust explosions have occurred in plant which was inadequately protected and which as a result has become a part or total loss. Much dust handling plant, unless specially strengthened, is very weak compared to the stresses exerted by a dust explosion. Some of the most vulnerable parts of a dust handling system, e.g. the collection unit, may be capable of withstanding pressures of only 7–15 kN/m^2

(1–2 lb/in^2) and if exposed to higher internal pressures may burst or disintegrate (Plate 4). In dust explosions, under the most unfavourable conditions, pressures up to 1000 kN/m^2 (140 lb/in^2) have been reported and so an uncontrolled explosion in plant is liable to cause severe damage. Some plant units are inherently stronger, such as mills and grinders, and these in themselves may be able to withstand explosions. Usually however, they are interconnected with ducting and collectors, of relatively light construction, which is structurally much weaker, and it is these latter items which suffer the most severe damage although the explosions have originated elsewhere in a stronger part of the system.

If plant is to withstand an explosion successfully all parts must be of adequate strength, not only those units in which an explosion might originate. Much dust handling plant is constructed to have mechanical strength sufficient only for normal working, and not for explosion conditions. This is of course justified where a non-explosible dust is handled, but where the material is explosible some form of explosion protection is necessary.

A particularly dangerous situation can arise if the dust handled by the plant is changed from a non-explosible to an explosible material, or to one of greater explosibility. Unprotected plant units may be unsuitable for the new situation; that is, sufficient protection could not be built in. Alternatively, the amount of explosion protection initially provided may not be sufficient for the new situation. Thirdly, additional legal and other requirements may have been incurred as a result of the change of process. If the dust being handled or processed is to be changed, re-design of the plant may be essential.

Although the main damage to plant caused by an explosion is due to pressure and expansion resulting from combustion, the cooling down period immediately after the peak of the explosion pressure has passed, can also be hazardous. If weak plant is adequately protected so that explosion pressures within it are kept to a safe low level, considerable quantities of hot gas may be contained within the plant immediately after the explosion. The hot gas will cool rapidly and if there are insufficient openings to atmosphere in the exterior of the plant the internal pressure may drop to substantially less than atmospheric. Collapse of the plant may then result. Thus although the positive pressure due to the explosion has been safely dealt with, the subsequent negative pressure could lead to extensive damage. The amount of venting required to prevent hazardous negative pressures is often only a small fraction of that required for dealing with the main positive explosion pressure, but it is important.

The provision of adequate explosion prevention and protection measures not only protects plant so that it may be put back into production with the minimum of delay after an incident, but also reduces the danger of missiles being generated during an explosion. The plant should be so constructed that missiles are not projected, and loose attachments to the plant are safely anchored. Because of the explosion, pressure waves may travel through the atmosphere and pick up or dislodge loose articles nearby. The immediate neighbourhood of plant handling explosible dusts should be inspected to minimize possible chance missiles, as these can cause serious injury to operatives and damage to neighbouring plant.

In the same way that much dust handling plant is incapable of withstanding the full effects of dust explosion, the strength of buildings in which the plant may be housed is also low, unless specially constructed. Factory buildings are therefore most unlikely to be able to withstand the full effects of dust explosions, and this underlines the need to prevent secondary dust explosions. Primary explosions can also occur in buildings e.g. dust dispersed and ignited in a store room, and explosion protection for the building may be necessary. In such a situation there would also be a serious fire hazard.

An unfortunate legacy from the past is that many buildings housing dust handling plant are multi-storey and were not originally constructed for the processes which they now house. Very serious dust explosions have occurred in multi-storey brick buildings, converted from other uses, and have led to heavy loss of life. Ordinary brick walls, not reinforced, are capable of withstanding internal explosion pressures of only 7–20 kN/m^2 (1–3 lb/in^2) without serious movement which can lead to collapse. In a multi-storey brick building an explosion on one floor caused the walls to move outwards, and as a result higher floors and the roof fell downwards leading to a total loss of building and contents (Plate 5). In this type of incident pieces of masonry may be projected for distances comparable with the height of the building, and debris can collapse on to any road or path nearby. A further hazard with multi-storey buildings is that communication between floors within the building is often unprotected so that explosion effects may traverse the building by these openings. This hazard is particularly frequent when an old building is taken over from another process, or even another industry, and dust handling plant is installed within it. Although for other reasons this procedure may be economically attractive, it leads to great difficulty in the safe design of plant.

In the opportunity arises, the best situation for dust handling plant is in the open air, with the minimum of enclosure, or in lightly constructed buildings with a minimum of weather protection. If permanent buildings do have to be used, these should be preferably steel framed, and not unsupported brick. Where old buildings have to be employed particular care should be paid to protect against explosions, whether secondary or primary. All intercommunication between the floor levels should be controlled, and should be external rather than internal. More detailed consideration of these points is given later.

Principles of explosion prevention and protective measures

Although it is good practice to eliminate as far as possible the formation of explosible dust clouds, and of sources of ignition, these steps give only partial protection. The reasons are plain. Industrial operations demand the movement of explosible dusts, and the formation of suspensions in air is inevitable. The most severe precautions against the introduction or generation of sources of ignition may be taken, but because of unforeseen mechanical or human failures, complete elimination of ignition sources cannot be relied upon, particularly where powered machinery is involved. Sooner or later the necessary conditions will occur for a primary dust explosion, namely generation of a suitable cloud of an explosible dust in the presence of a source of ignition. To avoid disaster reliance must be placed on the adequate functioning of explosion protection provided in advance. Explosion protection will be dealt with in greater detail later and may be summarized as follows.

(a) Prevention of ignition
(b) Suppression or containment of the explosion flame
(c) Allowing the explosion to take its full course, but to ensure that it does so safely.

The method of protection selected will depend upon a number of factors including the design of the industrial process, the running costs, the economics of alternative protection methods, the extent to which an explosion and its consequences can be foreseen, and the requirements of any authorities concerned.

As a basic technical principle the explosion hazard in a given plant with given materials increases as the amount of energy released by the explosion increases. The damage caused by the explosion of a large volume of combustible dust suspension may be more than proportionately greater than the damage from a smaller volume because larger structures are often relatively weaker. The escalation does not always occur because in a large volume there is always the possibility that parts of it may be at unfavourable dust concentrations for the explosion to propagate and this introduces an uncontrolled but positive safety factor. However, it is unwise to rely on this factor and in designing against an explosion hazard the safest method is to assume the worst conditions.

In trying to minimize the energy developed by an explosion, the most attractive method of protection, technically at least, is to prevent ignition of the dust cloud. As well as eliminating sources of ignition, an additional method which has been used with success is to reduce or eliminate the presence of the oxidant gas (usually the oxygen of the air). This may be done by introducing into the dust handling process an inert gas which either replaces the air originally present or at least dilutes it so that the level of oxygen is below that at which flames can be supported. This concentration level depends on the dust concerned, and also the inert gas used for diluting the air, and has to be determined in each case by an appropriate test.

In using inert gas to prevent the propagation of the explosion flames, protection is not necessarily given against fire in dust deposits within the protected volume. Thus, although dust explosion flames are prevented, unless the concentration of oxygen is reduced specifically for the purpose, protection may not be obtained against smouldering fires in dust deposits. These may well proceed slowly and relatively harmlessly whilst the process is in operation, but they present a major fire hazard when the process is stopped and air is allowed to enter the protected volume. Rapid build-up of fire can then occur and if dust becomes dislodged whilst burning an explosion may develop. However, provided this hazard is guarded against, and that the concentration of inert gas is controlled and adequately monitored, safe working is obtained. The advantages of using inerting are that explosions are avoided, by positive means, and that in ensuring that the inerting system is working properly, information on the general running of the plant is obtained which assists the rapid detection of malfunctioning of the process. The disadvantages of the inerting technique are that it may only be applicable to closed or semi-closed plant; otherwise the loss of inerting gas is expensive. Monitoring devices for the gas concentrations have to be installed and possible toxicity or suffocation risks to operatives have to be guarded against. The technique may therefore only be practicable and economic in a minority of instances, but it should be considered in the design stage of a new plant as a possible means of obtaining protection.

Once a dust explosion has begun to propagate, its energy may be minimized by either suppressing it rapidly or else by restricting its spread through the plant. Explosion flame speeds are such that any suppression mechanism must be effective within about 10 ms of the detection of the explosion. Commercial equipment is available which is able to provide this rapid response. If spread of the explosion is to be restricted the plant must be divided into small separate volumes, between which the explosion is unable to propagate, and that part of the plant within which the explosion occurs must be strong enough or sufficiently well protected to withstand explosion throughout its volume. The technique of suppression is used widely in industry and the suppression mechanism must always be available and ready to act. A process may continue for many years before a dust explosion develops, and after the flame has started to propagate, suppression will be needed within a small fraction of a second. The mechanism of suppression must therefore always be practically instantaneously available. Containment of explosions within industrial plant is frequently difficult because much modern plant is increasing in scale and sub-division into small working volumes would be impracticable. Whilst sub-division should always be encouraged, as a safety measure, economic considerations may preclude the use of containment alone as the major precautionary measure.

Because of difficulties in preventing ignition, or the unsuitability of the process to suppression or containment of explosion, recourse is often made to sub-division of the plant as far as is economic, coupled with explosion relief venting to prevent dangerous pressures. This method is probably the most commonly used. As industrial plant increases in scale, and is likely so to continue in the future, the smallest volumes into which it may be practicable to sub-divide the plant may still be large. This is a problem which is becoming more serious for the design engineer, but the amount of information on safety is not increasing correspondingly. The problem is important because in large sub-divisions the amount of energy released is high and can endanger operatives and neighbouring plant over a wide area. Secondly the plant itself may have high economic value, and the costs of repair and of the losses of production on a large scale will be expensive.

In the venting of primary explosions to atmosphere strict attention must be given to the safe dissipation of the explosion products. It is a characteristic of dust explosions that the volume of flame discharged from vents can be very large; much larger than would be expected by analogy from gas explosions. During explosions large quantities of unburnt or partly burnt dust can be discharged and, on meeting with the air outside the plant, are ignited by the explosion flame. With dusts considerable quantities can be disturbed from rest and ejected through vents during an explosion and thereby generate a large volume of flame outside the plant.

The vented flame and gases must be directed to a safe place away from operatives and neighbouring plant. This is commonly done by attaching a length of ducting to the vent, or by installing deflector baffles. If the plant is inside a building, the vented flame should be ducted to the exterior. In all cases the duct attached to the vent should be short, free from bends and other restrictions to flow, and be kept clean from dust during normal working. Further details are considered in Chapter 8. It is of importance that, wherever possible, vented plant should not discharge into the interior of a building because the conditions for a secondary explosion within the building would be favoured and the severity of the explosion as a whole would be greatly increased.

The design of the building in which explosible dusts are handled must also be considered. Where the plant is inside a building, and explosion suppression equipment or ducted vents have been installed to prevent burning dust from being discharged into the workroom, there is still the likelihood of occasional spillages and leaks of dust from the plant into the room. If dust is deliberately discharged from the plant, e.g. for bagging, the presence of dust within the building must be expected on occasions even with the most stringent precautions. Again, if a building is used for the storage of dusty material in sacks or drums, breakages and spillages can be expected. In all these cases there is a risk of dust being present within the room and the probability of a dust explosion must again be considered. Means of protecting buildings against such explosions are available; at this stage it is sufficient to emphasize that explosion and fire hazards may be present in buildings containing the plant used for the handling of the dust.

If adequate explosion protection has been installed, and the flame from plant or building has been safely directed to atmosphere, the hazardous pressure effect should have been adequately guarded against. However it should be borne in mind that because of the venting process, considerable quantities of burning material may be discharged over an external area. This discharged material may well continue to burn on reaching the ground and if other combustible materials are present may cause numerous widely scattered fires. The growth of the small fires into large ones may well be rapid unless prompt action is taken. The best precaution is obviously to ensure that combustible materials are not present in the vicinity of vents etc. which may discharge burning dust during an explosion. If this is not possible or if other plant or buildings may be affected by the discharge of burning dust the requirement for rapid access for fire fighting, together with the appropriate facilities, must be met.

If the explosion protection measures were adequate all operatives would have been completely protected from the direct effects of an explosion. Explosions often generate noise even if handled safely, and coupled with the suddenness of their onset they cause panic or distress to operatives previously unfamiliar with the process. The effects on operatives may be mitigated to some extent by instituting an emergency drill to be followed in the event of fire or explosion.

Finally, it cannot be too strongly emphasized that the most opportune time for the installation of safety precautions against dust explosions is at the design stage of the plant. At this stage the necessary modifications can be considered readily both as regards the effect upon the running of the plant and also their implications for neighbouring plant. At this stage the necessary legal and insurance considerations can also be made. Where it is a question of modifying existing plant or buildings for a new process, the requirements of explosion protection can be relatively more onerous. Particular care then has to be given to choice between the various possible methods of protection, and their relative capital and running costs.

References

[1] Morozzo, Count (1795), *The Repertory of Arts and Manufactures*, **2**, 416–32.

[2] Price, D. J. and Brown, H. H. (1922), *Dust Explosions*. National Fire Protection Association, Boston, USA.
[3] Report of Important Dust Explosions, (1957), National Fire Protection Association, Boston, USA.
[4] Brown, K. C. and James, G. J. (1962), Safety in Mines Research Report, 201.
[5] Deshmukh, R. T. and Kejriwal, B. K. (1966), *Indian J. Mines Metals and Fuels* **14**, 335–45.

3 *Tests for Dust Explosibility*

Reasons for tests

Where combustible dusts are being handled or produced in industry, it is necessary to know whether these dusts are explosible. At present there is no reliable method of predicting the explosibility from the composition or heat of combustion of dusts. Until further information becomes available to enable this to be done, direct tests of explosibility have to be made.

Laboratory tests have been set up in various countries to enable investigations of dust explosion properties to be made and these tests have become formalized. The test procedures accepted at present differ between countries because they have developed independently, to some extent, and international standardization has not yet been made.

The tests in the various countries have certain points in common. Those concerned particularly with explosibility properties provide for a small sample of dust to be dispersed in the presence of a source of ignition. In some countries tests of ignition in dust deposits are also undertaken. Although some large scale tests are made to study factors affecting propagation and control of explosion, evaluations of dust explosibility are generally made in the laboratory, which enable small samples of dust to be investigated. This is important industrially, because some dusts are expensive and are produced only in small quantities, in other cases at the time of testing, only the output from pilot plant or test rig may be available. By working on laboratory scale the time required for testing is minimized, and precautions are simplified if the dust is toxic or has other obnoxious properties.

The dust explosibility tests allow a rapid assessment to be made of the hazards to be expected with new materials, mixtures of materials previously tested, or existing materials manufactured by new processes. The results of the tests may either be implemented directly, or may be used in comparison with results already available for other materials whose explosion characteristics have been established in previous industrial experience. The tests may also be used to decide whether a new material shall be used in an industrial process, bearing in mind that if explosion precautions would be necessary, the use of that material could be uneconomic.

The explosibility tests are of laboratory scale, and the results obtained from them have therefore to be extrapolated to full scale for use in industry. The extrapolation introduces a measure of uncertainty, because the theoretical background for such scaling up is not yet available. Past experience in the interpretation of the tests results has given confidence that they are realistic, and that scaling is permissible; there have also been laboratory investigations leading to the same conclusion. The present practice is to regard dusts which respond vigorously to tests as presenting definite explosion hazard in industrial scale plant, and dusts which do not respond to the small scale tests as non-explosible on the large scale.

In intermediate cases difficulty may arise, such as where feeble flames are produced in the laboratory scale tests. These flames might either accelerate and become more vigorous on the industrial scale, or conversely might be unable to sustain propagation and extinguish themselves without doing serious damage. In laboratory scale apparatus, it is difficult to ensure that the flame produced during a test has propagated away from the influence of the source of ignition and would be able to continue propagation indefinitely throughout a dust cloud. The available evidence indicates that if flame propagation is obtained in laboratory tests, although it may only be weakly sustained, the dust should be regarded as explosible. Such dusts, which may be called marginally explosible, usually present a reduced explosion hazard compared with other dusts which respond vigorously to the laboratory scale tests. Some judgement is therefore required in the interpretation of the results of tests both as regards the size of the hazard to be met, and the precautions necessary to deal with it. These points are considered in more detail in Chapter 4.

Because the test methods used in various countries differ, it is convenient to consider practice in these countries separately, although some of the tests are common. In the United Kingdom, the United States, and the Federal German Republic there are comprehensive ranges of tests which have become standard in their respective countries. These tests are described in detail. Some other countries use tests from these ranges, and separate consideration has not been given here. Some miscellaneous tests for specific explosion properties from countries other than those specified above, are of technical interest and have been included. A summary of the information obtained from the tests is given in Appendix 2. This is intended as a guide for those concerned with the design and operation of industrial plant; for detailed consideration of a particular dust a reference back to the original publication is recommended.

Selection of dust sample

To ensure the usefulness of tests it is important that the sample of dust should be representative of the material concerned. The explosion properties of particular fractions can be measured under controlled conditions. The chemical composition of the dust is usually obtained from the supplier. The sample taken for testing may be only a very small part of the total dust present, and the difficulty in obtaining a representative sample under these conditions is well known. The difficulty is increased if the dust is known to be a mixture of materials. Techniques have been developed for the withdrawal of representative samples from larger amounts by mechanical means e.g. by using a spinning riffler. The method of quartering may be resorted to if other facilities are not available, but it is very operator dependent. Where a dust sample is to be taken from an industrial process in operation, a further complication can arise because a fine dust is likely to be more explosible than a coarse material. In general the dust from which the sample is taken should be representative of the process or condition in which it is present and also that most likely to be explosible. Examples would be fine materials rather than the standard product, freshly prepared materials if the explosibility is likely to be reduced by ageing (due to oxidation of particle surfaces), the product after milling rather than before, the product after drying rather than before. Where coarse materials are handled in bulk, considerable amounts of fines may be produced by attrition and these fines may be present in sufficient quantities to enable a sample of them rather than the coarse material to be taken. Usually a few kilograms (about 5 lbs) of dust is required for the testing, although the quantity may be less if the dust is finely divided and well mixed. Considerable care should be taken in the collection of the dust sample because the results of the tests depend on the composition of the dust, and the collection of the sample is usually not done by personnel engaged on the testing but by others, often less familiar with the tests.

If there is doubt concerning the place from which the dust is likely to be most hazardous, it is safer to take samples from several alternative positions, and then have each of the samples tested. A similar situation may arise if the composition of the dust is changed during processing through the plant, where again samples taken at various stages of manufacture should be tested. Because of the importance of ensuring that dusts submitted for testing are representative, expert advice should be taken in cases of doubt. Such advice can often be obtained from the laboratory where the testing is to take place.

Toxicity of dust

Before a dust sample is tested, its toxicity, and that of its combustion products, are considered. If the material does have toxic properties, appropriate steps are taken during the testing to ensure that the risk is minimized. Other possible hazardous properties are also considered at this stage. If the dust is known by its supplier to have hazardous properties, he should ensure that the package is adequately marked before despatch.

Where the hazard is very serious, as with radioactive or highly toxic material, consideration should be given to the construction of a special set of test apparatus which may be safely destroyed after use. This procedure is sometimes more economic than a thorough decontamination.

If test apparatus is to be set up permanently it is advisable to install mechanically ventilated cabinets or hoods for those items of equipment which are open to atmosphere and which have dust dispersed within them. Apart from such precautions being necessary with toxic dusts, the ventilation ensures that dust does not escape into the laboratory where it may contaminate other materials, or accumulate and generate a dust explosion hazard. A ventilation system also enables the combustion products to be dispersed rapidly and safely.

Test methods in various countries

UNITED KINGDOM

Dust explosibility testing is now carried out at the Fire Research Station, Boreham Wood, Herts. A detailed account of the test apparatus and methods has been published, together with a summary

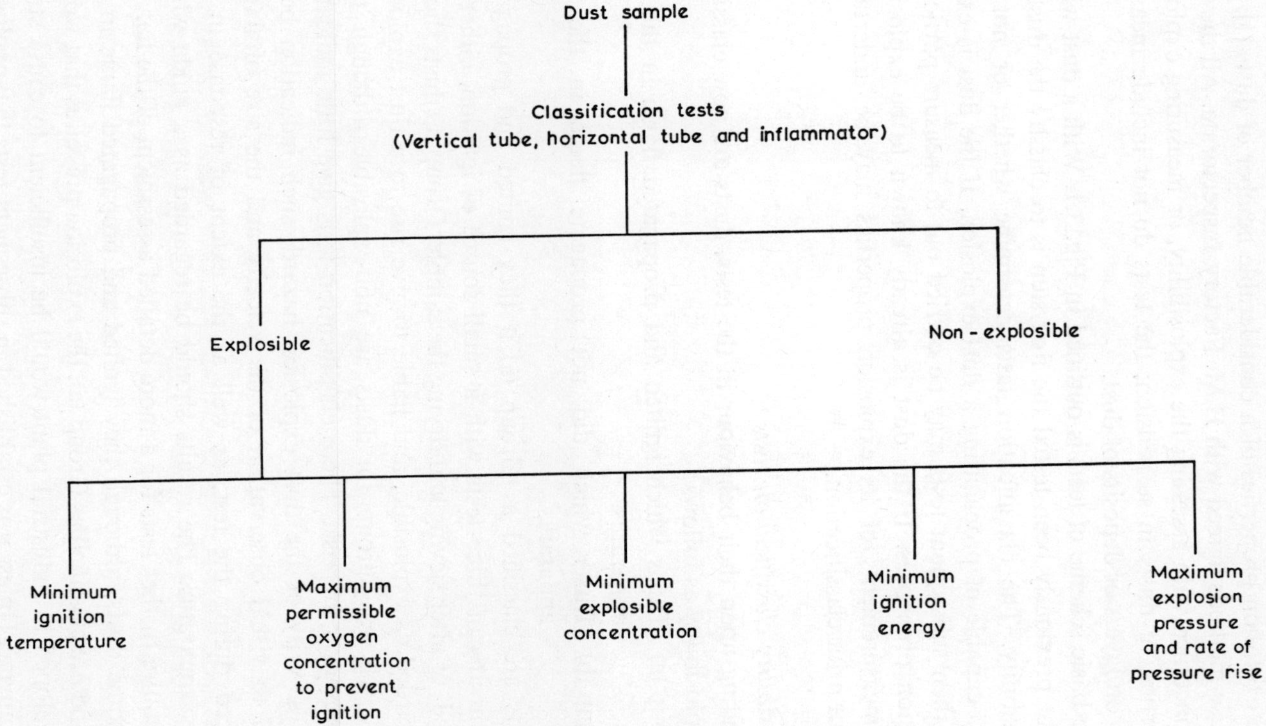

Fig. 3.1 Basic scheme of tests in United Kingdom.

of the explosion properties of a considerable number of dusts [1]. The tests use methods agreed with H.M. Factory Inspectorate. All the tests are concerned with assessing the explosibility, or measuring explosion properties, of dusts in suspension; the tests do not include measurements on layers or deposits of dust.

The basic scheme of tests is outlined in Fig. 3.1. With a dust which has not previously been tested the first step is to classify the dust for explosibility. The classification tests determine whether or not the dust is capable of propagating a dust explosion. If the dust is explosible, then subsequent tests may be carried out to measure particular explosion properties. If the dust is already known to be explosible, then measurements of its explosion properties may be undertaken without prior classification.

Explosibility classification tests

Depending upon their behaviour in the tests, dusts are now classified for explosibility as follows:

Group (a) Dusts which ignited and propagated flame in the test apparatus.

Group (b) Dusts which did not propagate flame in the test apparatus.

Dusts are classified as Group (a) if they ignited and propagated flame in any of three tests with a small source of ignition, either 'as received' or after sieving and drying the sample. Group (a) dusts should be regarded as explosible, and liable to give rise to a dust explosion hazard, whereas Group (b) dusts are non-explosible although they may present a fire risk. These classification tests give little indication of the severity of the dust explosion hazard, such indication being limited to visual observation of the velocity and the size of flames produced during the test, as well as the extent of response in the various apparatus. The results should be regarded as a guide which may require further tests for a more detailed assessment of the hazard. With some dusts, although they ignited and propagated flame in the tests, and are classified as Group (a), the explosion pressures that would be developed in industrial plant would be insufficient to cause structural damage. Unless it is established in subsequent tests that explosion pressures will be negligible, Group (a) dusts should be regarded as requiring explosion protection.

Originally, dust samples were tested with a wider range of apparatus and were classified as follows:

Class I : Dusts which ignite and propagate flame readily, the source of heat required for ignition being small.

Class II : Dusts which ignite readily with flame but require a larger source of ignition.

Class III : Dusts which do not ignite in the tests.

This classification has now been discarded, and the new and simpler criteria are now used in which dusts are listed as Group (a) or Group (b). As a general guide, dusts which previously were Class I may now be taken as Group (a), whereas those previously Class II and III are now regarded as Group (b).

The classification into Groups (a) and (b) only applies to dusts dispersed in air at, or near, ordinary atmospheric temperatures. Where a dust cloud is in a heated environment, such as a furnace or oven, other considerations may apply and an explosion hazard may arise with Group (b) dusts. In addition, if the dust is dispersed in a gas other than air, or in air at elevated pressure, special consideration should be given.

Before testing, no attempt is made to grind the dust. If it contains coarse material the sample is sieved through a 12 BS mesh. This procedure ensures that material is avoided which is appreciably coarser than the maximum diameter defined as dust (Chapter 1). Further sieving into various finer fractions is undertaken if the dust 'as received' does not propagate flame in the test apparatus, and the fractions are tested individually. The dust 'as received', and sieve fractions, can also be tested after drying in an oven at 105°C for 1 hour.

The results of tests apply to the particular sample of the dust, rather than to the material in generic terms, and are closely related to the conditions of the material from the particular plant from which it was taken. The importance of taking a representative sample of the dust in the first place is thus clear.

The basic characteristics of the test apparatus used for the classification of dusts according to explosibility are summarized in Table 3.1.

Several types of test apparatus are required because dusts have a wide range of dispersability and different means by which the dust cloud is formed are necessary, as well as scope for varying the quantity of dust and the pressure of the dispersing air.

The Vertical Tube apparatus (Hartmann apparatus developed at the US Bureau of Mines) is sketched in Fig. 3.2 and shown in Plate 6. It consists of a perspex combustion tube, 30.5 cm long, and 6.4 cm inside diameter, mounted over a brass dispersion cup, of 6.4 cm diameter and 1.6 cm deep in the centre. The volume of the apparatus is

Table 3.1 Classification test apparatus

Apparatus	*Direction of dispersion of dust*	*Igniting source*
Vertical tube	Vertically upwards	Electric spark or electrically heated wire coil
Horizontal tube	Horizontally	Electrically heated wire coil at 1 300°C
Inflammator	Vertically downwards	Electrically heated wire coil or electric spark.

1.2 litres. The dust under test is placed in the cup, and is dispersed by air passing upwards through a tube on the axis of the cup; to improve dispersion a mushroom-shaped deflector is mounted in the cup over the end of this tube. The dispersing air is released from a reservoir of 460 cm^3 capacity at gauge pressures up to 280 kN/m^2 (40 lb/in^2), either as a momentary or a continuous blast. The ignition source is a pair of brass electrodes mounted at various positions in the perspex tube, with a spark gap of approximately 0.6 cm. The spark is obtained from an induction coil. Observations are made as to whether flame propagation through the dust cloud occurs away from the source of ignition. The dust concentration may be changed by varying the quantity of sample and the pressure of the dispersion air. Tests are carried out for a range of dust concentrations and qualitative observations are made on whether the explosion of the dust is feeble or vigorous.

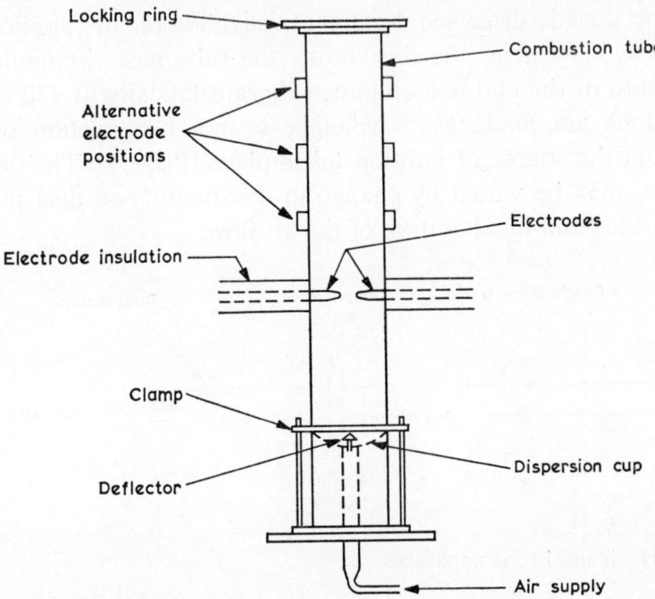

Fig. 3.2 Vertical tube apparatus.

As some dusts are not readily ignitable by an electric spark, the Vertical Tube apparatus may be modified to incorporate a wire coil igniter. In this case the perspex tube is replaced by a glass tube, of diameter 7.6 cm, and the dispersion cup is fitted with an adapter to take the wider tube. The ignition coil consists of $7\frac{1}{2}$ turns of 20 SWG alloy wire, formed in a helix of 0.8 cm internal diameter, and mounted on the axis of the tube about 10 cm above the dispersion cup. The temperature of the coil igniter is not measured. The test procedure is the same as for the spark ignition source.

The Vertical Tube apparatus, in either modification, is suitable for testing all types of dust, but particularly materials which are adherent, or of high density, or are available only in limited quantity.

The Horizontal Tube apparatus is constructed of glass, is 1.38 m long and with internal diameter 7.6 cm (Fig. 3.3). The tube is open at both ends and the dust is introduced through a hole in the upper surface of the tube 13 cm from one end, and deposited as a loose heap in the tube. The ignition coil is mounted 46 cm from the same end of the tube, and consists of a helix 18 mm long and 2.5 mm internal diameter. It is constructed from 32 SWG platinum wire, and is mounted on a ceramic former horizontally across the diameter of the tube. The dust is dispersed by a metered blast of air, applied for 2 seconds, which carries the dust down the tube past the igniter. The temperature of the coil is maintained thermostatically at 1300°C, and observations are made as to whether or not propagation of flame away from the source of ignition takes place (Plate 7). The dust concentration may be varied by change in the quantity of dust placed in the apparatus and by alteration of the air flow.

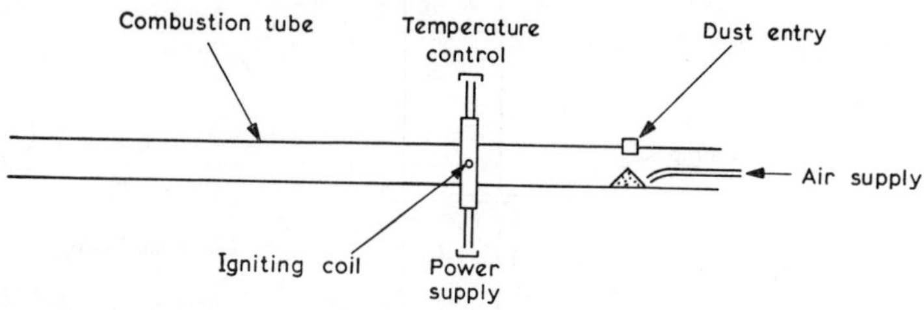

Fig. 3.3 Horizontal tube apparatus.

The apparatus is suitable for testing all types of carbonaceous materials, particularly if easily dispersed, but is not suitable for metal dusts which could damage the coil igniter. A disadvantage is that the dispersion of the dust and its residence time near the igniter are not independent; if vigorous dispersion is used the dust passes the igniter rapidly.

With the Inflamator apparatus the dust falls vertically downwards on to the igniting source, (Fig. 3.4). A vertical glass tube, of length 1.02 m and diameter 7.6 cm, is open at the top and closed at the bottom with a rubber bung. The tube is fitted with three side arms, at 25 cm intervals, for the position of the dust holder to be varied. Two further side arms, 13 cm from the bottom of the tube, enable a spark electrode or heated wire coil igniter to be inserted. The temperature of the coil igniter is not measured. The dust holder is 15 cm in length and 1.8 cm internal diameter, and is connected to an air reservoir of 460 cm^3 capacity. The reservoir can be pressurized up to 280 kN/m^2 (40 lb/in^2). When the dust is dispersed from the holder by the air it strikes a deflector plate mounted on the axis of the tube and then falls under gravity past the igniter. Observations are made as to whether flames propagate away from the source of ignition. The dust concentration may be varied by altering both the amount of dust and the pressure of

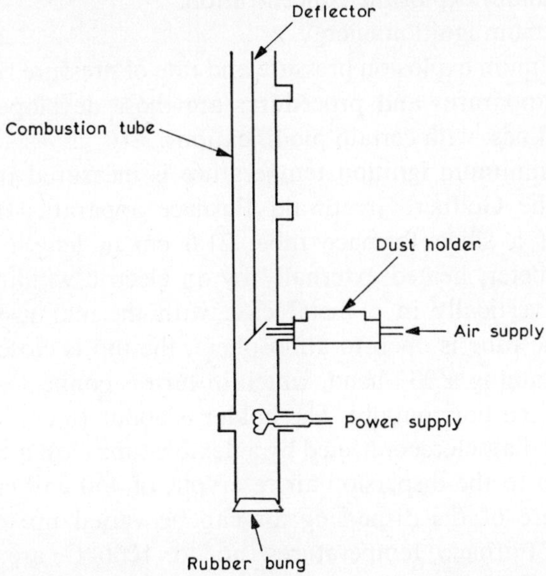

Fig. 3.4 Inflammator apparatus.

the dispersion air, and the dust may be introduced at different positions relative to the igniter. Both carbonaceous and metal dusts may be tested in the apparatus and it is particularly useful for materials which are coarse or fluffy (Plate 8). An advantage of the apparatus is that the dust falls past the ignition source relatively slowly, even when a considerable air pressure is used for dispersing the dust.

Dusts which ignite and propagate flame in the Vertical Tube apparatus, the Horizontal Tube, or the Inflammator, either 'as received' or after sieving and drying are classified as Group (a) dusts.

A classified list of Group (a) and Group (b) dusts has been published [2].

Measurement of explosion parameters of dusts

If a dust is shown to be explosible, further information on the extent of the explosion hazard may be required when considering suitable precautions for the safe handling of the dust. The following properties of the dust, or explosion parameters, can be determined.

(1) Minimum ignition temperature.

(2) Maximum permissible oxygen concentration of the atmosphere to prevent ignition in a dust cloud.

(3) Minimum explosible concentration.
(4) Minimum ignition energy.
(5) Maximum explosion pressure and rate of pressure rise.

The test apparatus and procedures are those developed at the US Bureau of Mines, with certain modifications.

1. The minimum ignition temperature is measured in a modified version of the Godbert-Greenwald Furnace apparatus [1] (Fig. 3.5), consisting of a silica Furnace tube, 21.6 cm in length and 3.7 cm internal diameter, heated externally by an electric winding. The tube is mounted vertically in a metal case with thermal insulation. The bottom of the tube is open to atmosphere, the top is closed by a glass adaptor containing a 90° bend, which in turn is connected to the dust holder, mounted horizontally. The holder is about 10 cm in length and 1 cm internal diameter, connected by a flexible tube closed with a spring-loaded clamp to the dispersion air reservoir, of 460 cm^3 capacity. The gauge pressure of the dispersing air can be varied up to 70 kN/m^2 (10 lb/in^2). Furnace temperatures up to 1000°C are maintained thermostatically by a temperature controller governed by a thermocouple attached to the Furnace tube. In the tests the dust is dispersed through the Furnace and observations are made as to whether or not flame appears at the bottom open mouth of the Furnace (Plate 9). If ignition occurs, the Furnace temperature is lowered until two tempera-

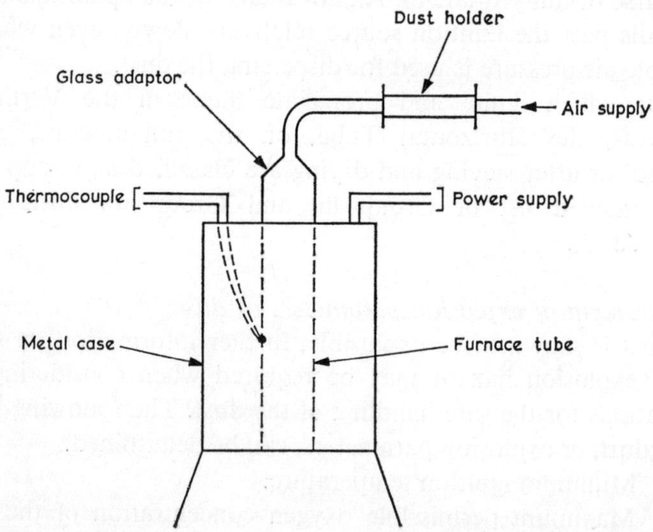

Fig. 3.5 Furnace apparatus.

tures differing by 10°C are obtained so that ignition takes place at the higher temperature only. The quantity of dust in the holder and the air dispersion pressure are varied during the determination. The lowest Furnace temperature at which ignition occurs is taken as the minimum ignition temperature. Some combustible dusts which are not explosible, i.e. Group (b) dusts, can ignite with flame when dispersed through the Furnace apparatus and their minimum ignition temperature may be measured.

2. The maximum permissible oxygen concentration of the atmosphere to prevent ignition in a dust cloud needs to be known when obtaining explosion protection by operating a plant with an inerted atmosphere. An inert gas, such as nitrogen, is introduced into the plant so that the oxygen concentration is reduced below that of air. The extent of oxygen reduction, or the maximum permissible oxygen concentration for safe working, must be measured by test. In applying the results a safety margin is necessary, and also monitoring devices on the plant to ensure that the inerted atmosphere is maintained. The apparatus used at present is basically the Furnace apparatus (Fig. 3.5) with some modification. The adaptor at the top entrance to the Furnace is replaced by another which is fitted with a side arm through which the air/inert gas mixture enters. The mixture then flows through the Furnace and emerges from the base into a glass tube extension which prevents backward flow of air into the apparatus. The air/inert gas mixture is also used to disperse the dust into the Furnace. The inert gas is usually nitrogen but could be varied depending upon the dust under test. The temperature of the Furnace is maintained at 850°C. By varying the composition of the air/inert gas mixture, the maximum permissible oxygen concentration is found.

3. The minimum explosible concentration is measured in the Vertical Tube apparatus using a continuous electric spark as the source of ignition. A weighed quantity of the dust is placed in the dispersion cup, and a paper diaphragm with a small hole in the centre is attached across the top of the combustion tube by a locking ring. The dust is dispersed in the usual way and the criterion for indicating an explosion is propagation of flame away from the igniting source; propagation is usually accompanied by bursting of the diaphragm. When explosion occurs the amount of dust is reduced and testing continued until no explosion is observed in at least ten tests at the same dust concentration. The mass of the minimum quantity of dust

with which explosion occurs, divided by the total volume of the tube, is taken as the minimum explosible concentration in air.

4. The minimum ignition energy is measured using an electric spark as the igniting source. Tests are carried out in a modified version of the Vertical Tube apparatus. Electrodes, with fine points, are set with spark gaps of 3 to 6 mm and connected to the secondary winding of a transformer, an input of 240 V giving an output of 10 kV. Energies up to several joules are obtained by charging condensers of various capacities to different voltages and discharging them through the primary of the transformer. Sparks of known energies are passed between the electrodes connected to the secondary of the transformer while a dust cloud is formed around them. Tests are repeated until the minimum energy of the spark is obtained at which flame propagates through the dust cloud, but no ignitions are obtained in at least 10 tests with slightly lower spark energy. From the capacity of the condensers (C), and the voltage to which they are charged (V), the

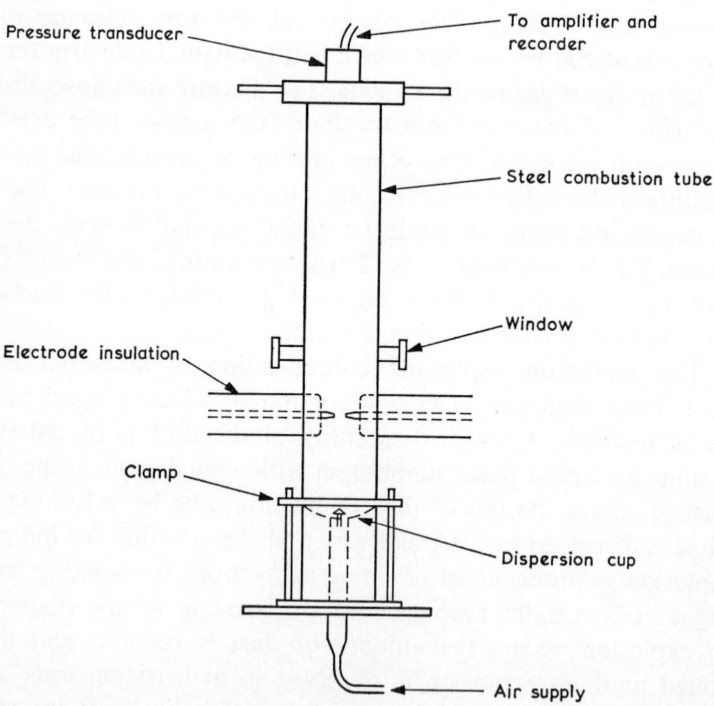

Fig. 3.6 Explosion pressure test apparatus.

energy is calculated as $\frac{1}{2}$ CV^2 joules, assuming no energy losses in the transformer, and are expressed in millijoules.

5. The maximum explosion pressure and maximum rate of pressure rise are measured in a stronger version of the Vertical Tube apparatus. The equipment is sketched in Fig. 3.6 and consists of a vertical explosion tube of steel, to the same dimensions as the apparatus in Fig. 3.2. There are fittings on the side of the tube for the ignition source, which is either an electric spark or a wire coil, and also two small observation windows (Plate 10). The top of the tube is closed, with a fitting for a pressure transducer. A weighed quantity of the dust under test is placed in the dispersion cup at the base of the explosion tube, and its concentration is calculated assuming uniform dispersion. The dust is dispersed by the release of air from a 49 cm^3 reservoir at a gauge pressure of 840 kN/m^2 (120 lb/in^2) by means of a solenoid valve. A non-return valve, fitted between the solenoid valve and the explosion vessel, prevents the combustion gases from escaping during the explosion. The output from the pressure transducer is amplified and recorded, the variation of pressure with time is usually as shown in Fig. 3.7. The small initial pressure rise p_1 at the beginning of the test, often appearing as a plateau, is due to the entry of the dispersing air into the explosion vessel. The rise in pressure due to the

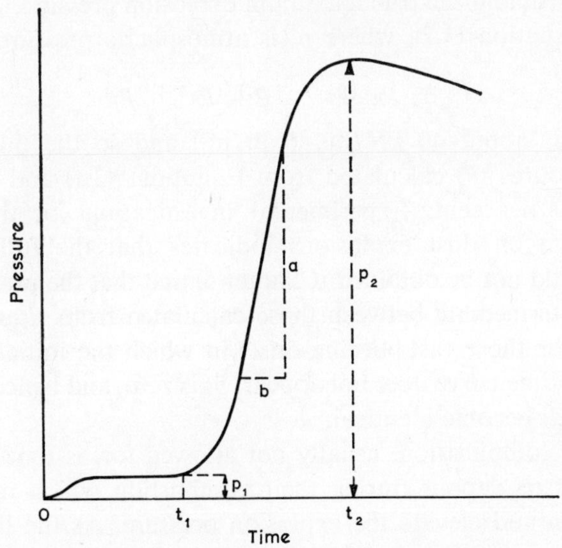

Fig. 3.7 Pressure-time record from explosion pressure test.

explosion of the dust is usually more pronounced and frequently commences only after all the dispersing air has entered the tube, even though the igniting source is present continuously. Occasionally, particularly with dusts which have high maximum explosion pressures, p_2 the plateau in the curve due to the dispersing air is absent and pressure rises continuously from the beginning of the test. The tests are repeated at various concentrations of dust, usually within the range 0.1 to 5.0 g/l, but occasionally to higher values to ensure that maxium values of pressure are obtained under the most severe conditions of test.

The interpretation of the pressure-time records has to be undertaken with care. The usual way of calculating the maximum explosion pressure (*P*) from a pressure-time record such as Fig. 3.7, is given by

$$P = p_2 - p_1. \tag{3.1}$$

Values of maximum explosion pressure (*P*), calculated according to Equation (3.1) are listed in Appendix 2.

The physical meaning of Equation (3.1) is that the dispersing air enters the explosion tube, but does not take part directly in the explosion by supplying air for the combustion. Alternatively, if the dispersing air takes a full part in the explosion, then the vessel is in effect pressurized for combustion. The observed explosion pressures are thus raised, and the true maximum explosion pressure (*P*) would be given by Equation (3.2), where p_0 is atmospheric pressure,

$$P = p_0\ (p_2 - p_1)/(p_0 + p_1). \tag{3.2}$$

Usually p_1 is about 40 kN/m^2 (6 lb/in^2) and so the maximum explosion pressures (*P*) calculated from Equations (3.1) and (3.2) would differ by 40 per cent. Experimental investigations of the effect of pressurization on dust explosions indicates that the full calculated increase would not be obtained (Chapter 5) and that the practical value would be intermediate between those calculated from Equations (3.1) and (3.2). For those fast-burning dusts in which the initial plateau in the pressure-time curve does not appear, p_1 is zero, and hence Equations (3.1) and (3.2) become identical.

A further complication, usually not allowed for, is that some dusts generate gas or vapour during their combustion which remain after the explosion and elevate the explosion pressure. As the listed values of maximum explosion pressures have been calculated according to

Equation (3.1), the procedure is continued in order to obtain direct comparisons. However where maximum explosion pressures are being used as the basis for further calculations, the interpretation of the pressure-time curves should be reassessed.

The maximum rate of pressure rise is calculated from the steepest part of the pressure-time curve, as indicated in Fig. 3.7. The usual procedure is to calculate the gradient directly ($=a/b$), but where considerations applicable to Equation (3.2) would arise the value of a should be multiplied by the appropriate factor in a similar manner to that in Equation (3.2). Values of the maximum rates of pressure rise for dusts are included in Appendix 2.

The average rate of pressure rise is sometimes used for guidance in the design of relief vents; it is calculated from Fig. 3.7 as the pressure $(p_2 - p_1)/(t_2 - t_1)$. Further consideration is given in Chapter 8.

In using the values of maximum explosion pressure and rates of rise in practical situations, the usual method is to assume that the most severe conditions for explosion would apply, irrespective of the dust concentration or the type of ignition source. Values obtained for conditions less than the most severe are thus of very limited interest, and serve mainly to indicate whether the severe conditions are likely to occur under a wide range of dust concentrations, and with different types of ignition source, or whether the most severe condtions are more strictly limited. No quantitative assessment of the explosion hazard can be made on this basis, but if a wide range of dust concentration gives high maximum values, particularly stringent precautions in the design of the plant may be necessary.

A modification of the pressure test has recently been introduced, in which the explosions are in controlled air/nitrogen mixtures. Other inert diluent gases can be used. The apparatus is modified to enable the explosion tube to be filled with the gas mixture; as well as the reservoir for dispersing the dust. The air/nitrogen mixtures are made up, under pressure, in a reservoir using the method of partial pressures. The gas compositions are checked, by direct analysis, in both the reservoir and at the exhaust from the explosion tube. The tube is sealed off before firing the dust. The test is of particular value in measuring the explosion pressures produced in plant with a partially inert atmosphere, so that safety margins can be estimated for plant strength.

UNITED STATES

Tests for evaluating the explosibility of dusts were devised at the laboratories of the Federal Bureau of Mines, US Department of the Interior, Pittsburgh, Pennsylvania. The test methods are extensive and include investigation of dust suspensions and dust layers, with a wide range of ignition sources under a variety of conditions of dispersion. A full account of the equipment and test procedures has been published [3] and a short summary of available test results has appeared [4]. Detailed results have appeared for the explosibility of agricultural [5], carbonaceous [6], metal [7], plastics [8], chemical [9] and miscellaneous [10] dusts. Explosion pressure measurements [11] and prevention of dust explosions by inerting [12], have been the subjects of further reports.

The basic scheme of tests is outlined in Fig. 3.8.

The first step in the preliminary examination of a dust sample is to screen it through a No. 20 U.S. mesh sieve; the fraction not passing the sieve is weighed and discarded. The fraction which passes the seive, referred to as the "as received" dust, is further screened through 100 and 200 U.S. mesh sieves to evaluate the particle size distribution. For a pure substance a through 200 mesh fraction is obtained by sieving, whereas for a mixture a representative sample is ground until it passes the 200 mesh. The complete range of tests is then carried out on the through 200 mesh fraction, with a few tests on the through 20 mesh fraction. The moisture content of the "as received" dust is determined by heating in a drying oven; if the moisture content exceeds 5 per cent the dust is dried before it is tested for explosibility. The dust is also dried if the moisture content is below 5 per cent, but the dispersability is affected by the moisture. The sample is examined under a microscope before testing, and other physical properties are measured when relevant.

Tests on dust suspensions

The minimum ignition temperature of a dust cloud is determined in the Godbert-Greenwald Furnace, similar to that shown in Fig. 3.5. The apparatus and procedure are similar to that given previously, except that the dispersion air reservoir is of 500 cm^3 capacity and is closed by a full port solenoid valve. The dispersion air blows dust from a holder into the Furnace and the criterion for ignition is the appearance of flame below the mouth of the Furnace. Usually 0.1 g

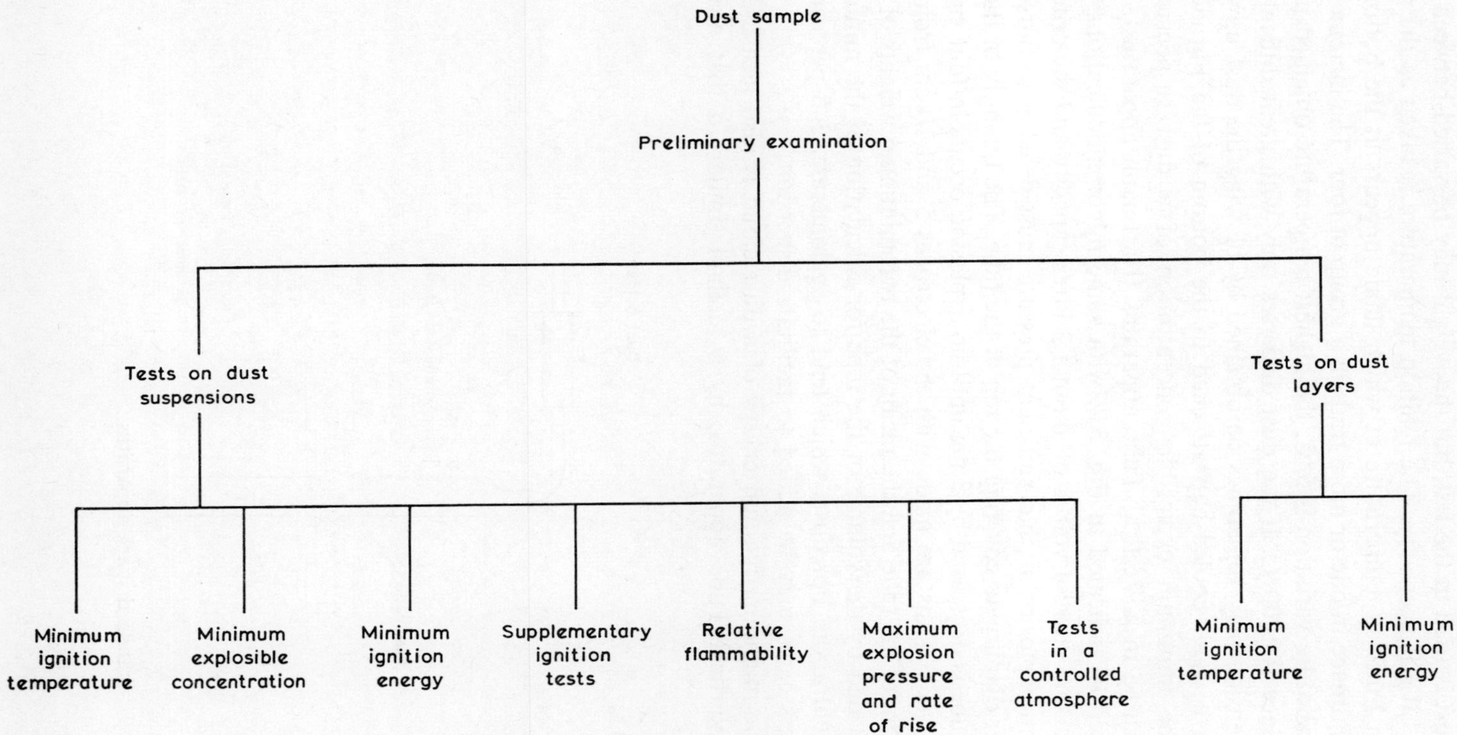

Fig. 3.8 Basic scheme of tests in United States.

of dust is used in the test, but the weight may be varied between 0.05 and 1.0 g if necessary. The ignition temperature is taken as the minimum Furnace temperature at which flame appears at the bottom of the Furnace in one or more trials in a group of four. The increment of temperature variation is 5°C, the highest temperature obtained in the Furnace is 1000°C. If the dust disperses only with great difficulty, a suspension may sometimes be obtained by blowing the dust upwards from a water cooled cup attached to the bottom of the Furnace.

The minimum explosible concentration of a dust suspension is measured in a Vertical Tube apparatus (Hartmann apparatus). This is used as indicated in Fig. 3.2, with some minor modifications. The dispersing air reservoir is of about 1.3 litres capacity and the criterion for an explosion is that sufficient pressure must develop to burst the paper diaphragm covering the top of the tube. The quantity of dust in the tube is decreased by 5 mg until no explosion occurs in four successive trials. Tests are made with the electrodes 5 and 10 cm from the bottom of the tube and the mean of the two minimum weights of dust is divided by the volume of the tube for calculation of the minimum concentration. For dusts which tend to agglomerate, 3–5 per cent of Fullers earth may be added to facilitate dispersion.

The minimum ignition energy of a dust cloud is also measured in the Vertical Tube apparatus, by a method similar to that already

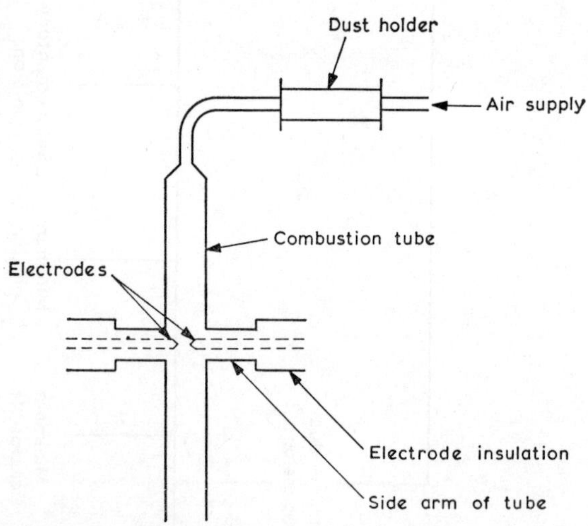

Fig. 3.9 Standard spark apparatus.

described. Sparks are produced across electrodes in the tube by discharging condensers through the primary of a transformer, the electrodes being connected to the secondary winding. The time interval between dispersing the dust and passing the spark is varied and controlled electronically, the optimum time interval being measured in preliminary trials. The criterion for ignition of the dust cloud is the propagation of flame at least 10 cm in the tube; up to 4 trials at each ignition energy are made. The minimum ignition energy measured is approximate because some electrical energy is lost in the transformer circuit and some may remain in the condensers.

Supplementary ignition tests are carried out on dusts which do not ignite with an electric spark in the Vertical Tube apparatus, or in the Furnace apparatus. In the supplementary tests stronger sources of ignition are used, and the dust concentrations are varied to cover the optimum ignition conditions. In one test, the Standard Spark apparatus (Fig. 3.9), a high energy electrical spark is passed through the dust suspension. The apparatus consists of a vertical glass tube, 20 cm in length and 3.8 cm in diameter, down which dust is blown from the dispersion system used in the Godbert-Greenwald Furnace test. Tungsten electrodes are fitted about 12 cm from the bottom of the tube and a continuous spark is obtained from a high voltage transformer. The quantity of dust dispersed is usually 1 g. Other tests using more powerful sources of ignition include the Vertical Tube apparatus provided with an electric arc, gun cotton flame, heated wire coil, or frictional sparking sources of ignition; a flame as ignition source is tested by shaking dust from a dispenser down onto the flame from a horizontal gas burner. In all of these tests the criterion for ignition is the appearance of flame in the dust cloud.

The relative flammability of a dust cloud is defined as the percentage of inert dust that must be mixed with the explosible dust to suppress flame propagation. The property is of particular significance for coal dusts, as it can be used to measure the percentage of non-combustible required to prevent explosion propagation in mines. Measurements are made in the Furnace apparatus at a temperature of 700°C. This value was chosen because it is close to that at which a mixture of 62.5 per cent inert and 37.5 per cent Pittsburgh coal dust would propagate flame, and a mixture of 65 per cent inert and 35 per cent coal dust would not. These two mixtures respectively would and would not propagate explosion in full scale mine tests. Although developed for coal and lignite dusts, the test is used widely for industrial materials. An alternative test for relative flammability is made with the Standard Spark apparatus (Fig. 3.9) operated in the usual way. With either the Furnace or Spark apparatus, 4 trials are made using samples of the explosible/inert dust (Fullers earth) mixtures. Usually 1 g samples of the mixture are used in each test. The criterion for propagation is the appearance of flame at the bottom of the Furnace or Spark tube, and if no propagation occurs the tests are repeated with a lower proportion of inert. The maximum percentage of inert dust which is practicable is 90 per cent; with higher percentages the proportion of combustible dust is too small to give reliable results and the relative flammability is reported as 90 per cent plus.

Maximum explosion pressures and rates of rise are measured in the closed tube apparatus (Fig. 3.6). The dust cloud is usually ignited by an electric spark source, but heated wire coil and gun cotton flame sources are also available. The explosion pressures are measured either with electronic transducers or by a mechanical manometer. Tests are usually carried out at dust concentrations of 0.1, 0.2, 0.5, 1.0 and 2.0 g/l. The maximum explosion pressure, and the maximum and average rates of pressure rise are calculated as described previously for the case where the dispersing air pressure is subtracted directly from the observed pressure.

The tests in a controlled atmosphere provide information on the proportion of inert gas necessary to prevent dust flame propagation. Modified versions of the Furnace Apparatus and the Standard Spark apparatus are used. The inert gas is usually carbon dioxide, but nitrogen and other gases are occasionally employed. The modified Furnace apparatus is as described previously and is maintained at a fixed temperature of 850°C. In the Standard Spark apparatus the tube is increased in length to 48 cm and the inert gas/air mixture is fed both to the dispersing reservoir and also continuously through the manifold containing the dust chamber and through the explosion tube. The criterion for ignition is flame propagation for at least 10 cm from the spark. In other tests with a controlled atmosphere the maximum explosion pressure and rates of pressure rise can be measured in the usual apparatus (Fig. 3.6) but with either reduced or enriched oxygen atmospheres. Slight modification of the explosion tube is necessary for these tests.

Tests on dust layers

The minimum ignition temperature of a dust layer is determined in a modified form of the Furnace apparatus. The fitting at the top of the Furnace is removed entirely (Fig. 3.5) and a plate containing an air inlet orifice is attached to the bottom of the Furnace. A metered airstream is passed upwards through the inlet orifice and the Furnace. The dust under test is placed in a container, 2.5 cm diameter and 1.3 cm deep, made of 40 U.S. mesh stainless steel with a 200 U.S. mesh stainless steel bottom liner. The container is suspended in the centre of the Furnace and the temperature of the dust is measured by a thermocouple embedded in it. Ignition is denoted by an inflection in the time-temperature record. In addition the top surface of the dust is observed in a mirror over the Furnace. In successive trials the temperature of the Furnace is reduced, usually in 5°C decrements, until a minimum temperature is obtained at which ignition of the dust just occurs. The duration of the test is such that the dust attains the set temperature of the Furnace for 5 minutes, unless ignition occurs sooner.

The minimum ignition energy of a dust layer is measured using the spark from charged condensers. A layer of dust, about 0.15 cm thick, is placed in a circular steel vessel 2.5 cm in diameter. The vessel is connected electrically to the condensers. A pointed steel electrode, also connected to the condensers, is lowered towards the dust surface until a spark discharges through the dust. If ignition occurs, the dust is replaced, and the test repeated with lower spark energies. Twenty trials are carried out at the minimum conditions to ensure that the true value has been attained. The ignition energy is calculated from the voltage and capacitance of the condensers, as for minimum ignition energy of dust clouds.

FEDERAL GERMAN REPUBLIC

The principal establishment for the laboratory testing of dusts for explosion and fire properties is the Federal Institute for the Testing of Materials, Dahlem, Berlin. The explosions and fire hazards work does not include investigations of explosives or of toxic hazards, which are undertaken in other sections of the Institute. In addition the hazards associated with coal dust are investigated elsewhere, e.g. at Dortmund.

A detailed account of the test methods used in assessing the ex-

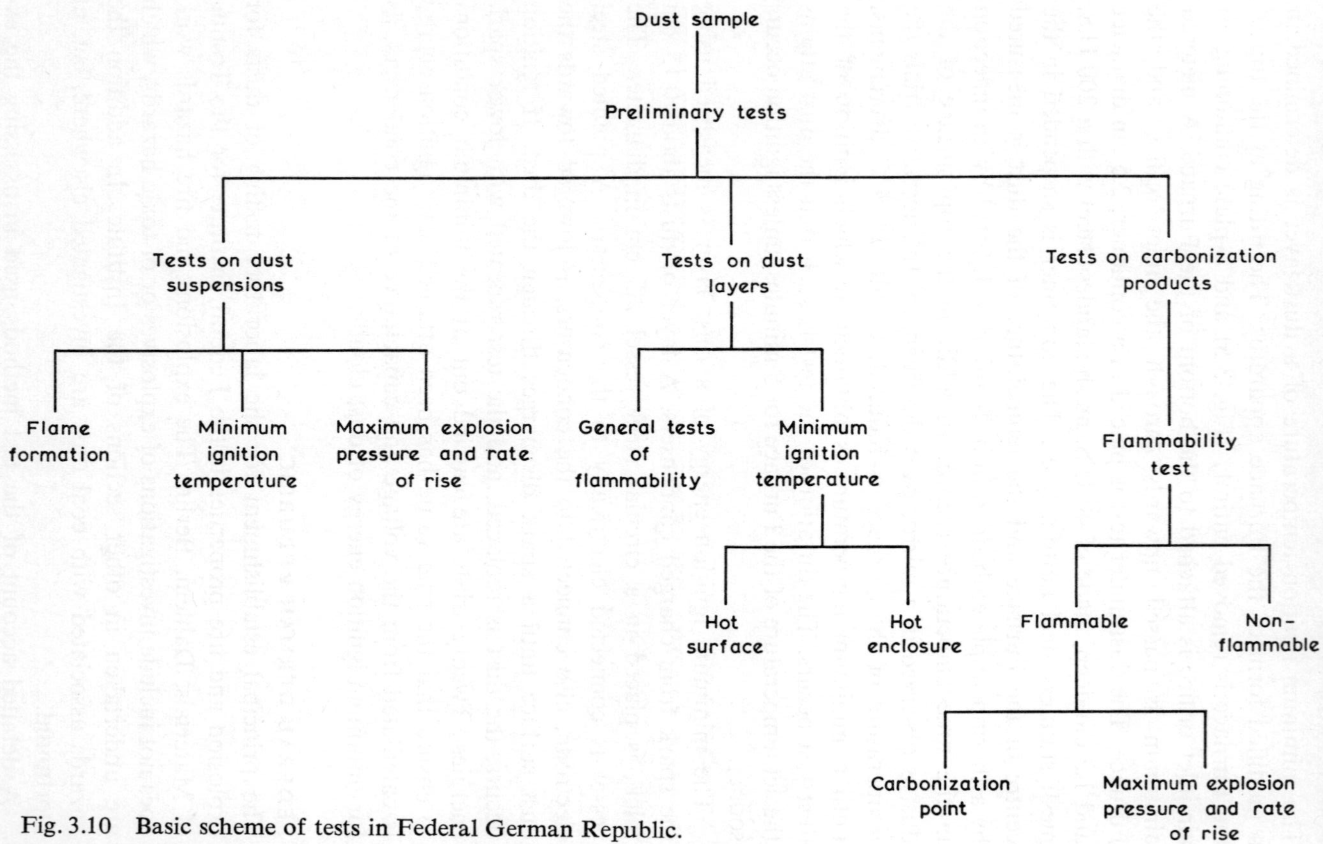

Fig. 3.10 Basic scheme of tests in Federal German Republic.

plosion and fire hazards of dusts at the Federal Institute (BAM) has been published [13]. The tests are wide ranging, and include observations and measurements on dust suspensions, dust layers, and the combustible volatiles produced by the heating of dust layers. The basic scheme of tests is shown in Fig. 3.10. It is recommended that the sample should consist of fine dust that is as fresh and dry as possible, for material in this condition has been found to be the most dangerous. In the preliminary tests the particle size of the dust sample is examined, and when the sample as received contains particles greater than 200 microns the dust is either sieved or ground to yield particles below this maximum size. The dust is then dried for 24 hours at 75°C before any further tests are carried out.

Tests on dust suspensions

The flame formation tests give an approximate idea of the explosibility of the dust when dispersed in air. Usually about 10 cm^3 of the dust is blown into suspension in the presence of a gas flame. The size, type and violence of the explosion are noted, either by direct observation or by cine photography, and by comparison with previous results a rough overall assessment of the explosibility of the dust can be made.

The minimum ignition temperature of a dust suspension is measured in a Furnace apparatus. The apparatus is similar to that in Fig. 3.5 except that the heated inner tube is horizontal and contains a heated surface against which the dust is dispersed manually by means of a squeeze bulb containing air. The criterion for ignition is the emission of flame from the other end of the furnace. The maximum temperature obtainable with the apparatus is about 720°C. The minimum ignition temperatures obtained in the test are taken as the values to be used in the application of V.D.E. Regulation No. 0165, 1957, which details the regulations for the installation of electrical equipment in factories exposed to explosion hazards.

The maximum explosion pressure and rate of rise of pressure are determined in apparatus very similar to that in Fig. 3.6. The ignition source is usually an induction spark, but alternative sources such as a hot wire or a thermite flame are sometimes used. The pressure/time variation is measured with mechanical or electrical pressure gauges and from the records the maximum explosion pressure, the maximum and average rates of pressure rise, are calculated.

The minimum explosible concentration of a dust suspension is not measured directly, but is calculated from the heat of combustion of the dust, its specific heat and that of air, and an arbitrary flame temperature.

Tests on dust layers

Tests on flammability provide preliminary information on the general behaviour of a deposit of the dust as regards the ignition and subsequent burning.

Deposits of the dust are exposed to various ignition sources, such as a bunsen burner flame, a lighted match, a glowing cigarette end, a hot glass rod etc. and observations are made of the behaviour of the dust [14]. Note is taken of whether the dust liquefies and whether glowing or flaming develops. Measurement is also made of the rate of spread of burning across the surface of the dust layer. The tests provide a rapid assessment of the ease of ignition, the types of burning which may develop, and the rate at which combustion can take place on the surface of a dust layer. Such information can be applied directly in the consideration of precautions against fire in layers of the dust, and the tests give direct evidence of the hazards involved with the material. They have a ready practical application in relation to ease of ignition by dropped matches, cigarette ends, etc. and to the subsequent development of fire.

The minimum ignition temperatures of dust layers are measured by two methods. In the first a dust layer is formed on a hot horizontal surface, the upper surface of the dust layer being in contact with air at room temperature. In the other method samples of the dust are heated in hot enclosures, and so the external surfaces of the dust deposit are exposed to air at elevated temperatures. The minimum ignition temperature of a layer on a hot surface is measured by depositing the dust on a hot plate, heated electrically, and with its temperature controlled thermostatically. The maximum temperature attainable by the hot plate is about 450°C. Usually a dust layer of thickness 0.5 cm is used, as this thickness is required for tests under V.D.E. Regulation No. 0165. In the tests observations are made as to whether or not glowing or smouldering combustion is initiated, and the tests are repeated until the minimum ignition temperature is obtained. If the dust should melt, decompose, or otherwise interfere with the test, the dust may be mixed with kieselguhr, after which an improved performance is frequently obtained.

The minimum ignition temperatures of dusts in hot enclosures are measured in two tests, using different sizes of sample. Relatively quick determinations are obtained with small samples (about 30 cm^3) of dust inserted in a metal tube. The internal diameter of the tube is 2.4 cm, and the height of the dust layer is about 5 cm. The tube is vertical and is heated electrically, a metered supply of air or other gas mixture passes through the tube and the dust layer. Ignition of the dust is detected by rise in temperature above that of the tube. Temperatures up to 430°C can be obtained, and because the tests are relatively rapid, prolonged oxidation of the sample during an induction period preceding ignition is avoided. If the minimum ignition temperature is lower than 180°C further tests on larger samples are carried out. About 200 cm^3 of the dust is put in a cylindrical vessel, and a thermocouple is embedded in the dust; the vessel is then placed in a thermostatically controlled oven and the temperature of the dust sample is recorded during the test. By repeating the tests the minimum temperature of the oven required to ignite the dust is found; the values are lower than those in tests with smaller samples or with dust suspensions in air. Because of the lower temperature, the duration of the test is considerably longer than in the other tests at higher temperatures. If the dust sample melts or decomposes, the procedure is again used for mixing kieselguhr in order to contain the sample within the vessel. The results obtained from the test are directly applicable to situations where dusts may be stored under warm conditions, or may be processed for considerable periods of time in warm plant e.g. as in drying ovens.

The subject of the self heating of dust layers, the minimum ignition temperature, and the relation to the dimensions of the dust deposit has been the subject of experimental and theoretical investigations. The technical background to the small scale tests on dust layers is considered in more detail later, see Chapter 9.

Tests on carbonization products

When many combustible materials are heated, they pyrolyse and produce flammable vapours and gases which can form explosible mixtures with air. In this way an explosion may be produced from the dust, or a massive solid, which has been heated but not dispersed in the air. The hazard is particularly serious when the carbonization point i.e. the minimum temperature at which flammable vapours etc. are involved, is low or when the explosibility of the gas/air mixture produced is greater than that of the mixture produced by dispersing the dust itself in air. The experimental investigation of explosibility of gases produced by the carbonization of solids is often difficult because of the number of factors involved. It can be foreseen that the amount and composition of flammable gas and vapour would depend on the rate at which the dust was heated, as well as the final temperature obtained. The time for which the flammable vapours are maintained at the elevated temperatures is important, because of the possibility of decomposition and oxidation of the vapours before testing for explosibility. In addition some of the vapours may be flammable, whereas others may not be and the proportion of the two would depend upon their relative rates of production. Consequently, the results obtained from the tests are sometimes ambiguous but where they are clear the hazard of the dust may be compared with that of other materials tested previously whose properties are already familiar. The first test on carbonization products is that of flammability; it is carried out by heating a small part of the dust in a test tube and attempting to ignite the gases evolved by means of a flame. Flammable gas is deemed to have been evolved if flames are still observable when the igniting flame has been removed. If the carbonization products are shown to be flammable, further tests are carried out.

The carbonization point indicates the temperature to which the dust must be raised in order to produce sufficient vapours to give rise to an explosion hazard, and bears some resemblance to the flash point of liquids. A small quantity of the dust is placed in a cylindrical steel vessel, provided with a lid, which is heated externally. As the temperature increases, and the production of gas and vapour commences, a small flame is introduced into the air space above the dust within the vessel and observation is made as to whether the gas/air mixture can be ignited.

The maximum explosion pressure and the rate of pressure rise in explosions of carbonization gas/air are measured using a spherical pressure vessel. The volume of the vessel is about 1.5 l, and a small quantity of the dust sample is heated within it and the resulting carbonization gas ignited after allowing it to mix with air. The pressure produced by an explosion deflects a diaphragm fitted in the top of the sphere. The deflection is recorded from an external mirror and a camera and the pressure/time curve can be derived. The maximum rate of pressure rise and the average rate of pressure rise are calculated in the same way as with pressure tests on dust suspensions. Direct comparison of the results obtained with carbonization gas and with dust suspensions should be made with caution as the explosion vessels differ in volume and shape, and the concentration of combustible in the air is likely to differ.

Hazards arising from other ignition sources

In addition to the basic scheme of tests shown in Fig. 3.10 some supplementary tests are available. Alternative sources of ignition which are considered in these tests include friction sparks, electrostatic sparks, and intense light radiation.

Friction sparks may arise in industrial processes when two surfaces either rub or impact together. Tests are carried out in which a stream of sparks is produced by the action of a grinding wheel on steel; the sparks are directed either on to a dust layer 0.5 cm thick or a dust/air suspension is blown into the stream of sparks from the grinding wheel. For tests with dust layers, the following types of behaviour have been reported for different dusts: [14, 15]

1. Smouldering produced after short exposure to sparks without initial change in the dust.
2. Smouldering after change in the dust such as melting, swelling, or carbonization.
3. Melting, discolouration, generation of odour without smouldering.
4. No noticeable change in the dust.

   For dust/air suspensions observation is made of whether ignition is easy, difficult, or not obtained.

Intense light radiation is used as an ignition source in tests with dust layers and suspensions in air. The light sources are flash bulbs of various powers and in tests with dust layers these are mounted from 1 to 6 cm away from the surface of the dust. The thickness of the dust layer is 0.5 cm and its area is 10 cm^2. For dust suspensions the lamp is provided with a reflector and separated from the dust by a glass window; parallel tests can be carried out in which the ignition source is a burning dust/air suspension. In this case both the dust under test and the dust suspension used as ignition source, are contained in glass tubes [16].

OTHER COUNTRIES

In the USSR a test is used in which the minimum ignition energy of dust suspensions is defined in terms of the discharge energy of electrical condensers capable of igniting the optimum concentration of dust, with the optimum characteristic of the discharge circuit. The gap between the spark electrodes is such as to exclude any flame quenching effects [17]. The apparatus consists of a steel box containing a dust hopper with a perforated base; on vibrating the hopper a column of dust falls through the box to a dust collector in its base. The electrodes, which are pointed, are carried in insulators on opposite sides of the box; the number of spark discharges is counted automatically. A control unit synchronizes the operation of the hopper, the sparking circuit, and the counter. The voltage of the condenser is measured before and after the discharge.

The test procedure is to measure firstly the optimum mixture concentration, then the optimum inductance and resistance of the circuit, and finally the minimum gap between the electrodes to prevent quenching. In each case the probability of ignition is defined as the number of ignitions divided by the total number of discharges, the number of ignitions must be not less than 10. With each of the variables the maximum ignition probability is measured and the corresponding experimental conditions used in the test. The minimum ignition energy is measured by plotting the ignition probability as a function of ignition energy, and the minimum is taken as that as which the probability is 10^{-2}. In this test the particle diameters must be less than 75 microns.

The minimum explosible concentration, or lower limit of explosibility, of magnesium powder in air has been measured using a 2 l closed steel sphere [18]. The dust is placed in a small bowl at the centre and dispersed by air from a jet blowing down upon the powder, and spark electrodes are nearby. The presence of flame is detected by observation through ports in the wall of the sphere. The minimum explosible concentration was that below which explosion did not occur when the experiment was repeated five times. Dust/air mixtures with a minimum explosible concentration of 0.065 g/l or less are considered explosible, and those with a higher minimum concentration are flammable.

Special methods

In addition to test methods recognized nationally within different countries, special methods have been developed by industries, authorities etc in various countries. The purpose of the special methods has usually been to provide *ad hoc* information for specific industries, processes, or materials, which enables comparisons to be made between materials more rapidly than by the national tests. They are particularly useful for rapid comparisons within a series of similar materials, e.g. a range of dyestuffs, where the *ad hoc* approach is often helpful.

An outstanding example of special test methods is the series adopted by a group of chemical manufacturers in the Federal German Republic [19]. The tests were aimed at ascertaining those dusts which were particularly hazardous during drying or grinding operations, and since coming into use have led to a substantial decrease in the incidence of decompositions and fires. There are five tests in the series, as follows:

1. A flammability test in which a platinum wire at 900°C is applied to a small heap of the dust.
2. An ignition test in which about 100 mg of dust is deposited on an aluminium block, thermostatically controlled, at various temperatures.
3. A decomposition test in which the dust, contained in a tube, is inserted in an aluminium block, the temperature of which is raised in stages, and signs of exothermic reaction are sought.
4. A drop-hammer test in which a 5 Kg weight is dropped about 75 cm onto a steel punch, of area 1 cm^2, resting on the dust which is deposited on a steel anvil. Observation is made for flames, sparks or explosion.
5. A dust explosion test in which the Vertical Tube apparatus (Fig. 3.2) is fitted with a hinged aluminium lid, and the opening angle of the lid is taken as a measure of explosibility.

In each test, except for the decomposition test, the response of the dust is converted to a numerical index. The indices are then added as follows:

Flammability + Ignition to give assessment of drying hazard.

Ignition + Drop hammer + Dust explosion to give assessment of grinding hazard.

The Decomposition test gives an indication of maximum safe working temperatures in plant.

References

[1] Raftery, Monica M. (1968), Fire Research Technical Paper, 21, London, HMSO.

[2] Dust explosions in factories. Classified list of dusts that have been tested for explosibility in the form of a dust cloud. Dept. of Employment and Productivity HM Factory Inspectorate SHW 830. London, HMSO (1968).

[3] Dorsett, H. G., Jacobson, M., Nagy, J. and Williams, R. P. (1960), US Bureau of Mines RI 5624.

[4] Marks, L. S. (1958), *Marks Mechanical Engineers Handbook*, 6th Edn, McGraw Hill, New York.
[5] Jacobson, M., Nagy, J., Cooper, A. R. and Ball, F. J. (1961), US Bureau of Mines RI 5753.
[6] Nagy, J., Dorsett, H. G. and Cooper, A. R. (1965), US Bureau of Mines RI 6597.
[7] Jacobson, M., Cooper, A. R. and Nagy, J. (1964), US Bureau of Mines RI 6516.
[8] Jacobson, M., Nagy, J. and Cooper, A. R. (1962), US Bureau of Mines RI 5971.
[9] Dorsett, H. G. and Nagy, J. (1968), US Bureau of Mines RI 7132.
[10] Nagy, J., Cooper, A. R. and Dorsett, H. G. (1968), US Bureau of Mines RI 7208.
[11] Nagy, J., Cooper, A. R. and Stupar, J. M. (1964), US Bureau of Mines RI 6561.
[12] Nagy, J., Dorsett, H. G. and Jacobson, M. (1964), US Bureau of Mines RI 6543.
[13] Leuschke, G. (1966), *Staub*, **26**, 2, 49–57.
[14] Peters, H. (1965), *VFDB Zeit*, **14**, 2, 59–64.
[15] Leuschke, G. and Zehr, J. (1962), *Arbeitsschutz*, 6, 146–7.
[16] Leuschke, G. and Zehr, J. (1962), *Arbeitsschutz*, 10, 249–51.
[17] Smelkov, G. I., Fetisov, P. A. and Verevkin, V. N. (1968), *Tsentralnyi nauchno-issledovatelskii institut protivopozharnoi oborony MOOP. SSSR*, 22, 1–18 ; (1969), Nat. Lend. Lib. Boston Spa, Yorks. RTS 5035.
[18] Popov, E. I. (1968), *Poroshkovaya Metallurgiya*, **72**, 12, 91–4.
[19] Lutolf, J. (1971), *Staub-Reinhalt* (English version), **31**, 3, 1–6.

4 *Application of Results of Tests*

In interpreting the results of dust explosibility tests for industrial application, problems arising from the scaling-up and the reproducibility of the results are met continuously. The information available is considered in some detail, because of its basic importance. A generalized method of indexing the explosion hazard of a dust is then described, and is followed by assessments of the merits and limitations of the individual dust explosibility tests.

Scaling

All the routine tests for measurement of the explosibility of dust suspensions, described in Chapter 3, use laboratory scale equipment as a practical necessity. In many cases the results of the tests need to be applied to full industrial scale plant. It is important to establish that the results of the laboratory scale tests are realistic in relation to the industrial problem.

This necessity is particularly acute in the United Kingdom, where considerable importance is placed on the classification tests, which decide whether a dust should be regarded as an explosion hazard or not. The traditional approach has been that dusts which exploded vigorously in the test apparatus should be definitely regarded as a dust explosion hazard, whereas those dusts which gave a nil response were not. Some of these latter dusts were however combustible, with relatively- high calorific values, and were still regarded as a fire hazard. Practical experience over many years showed that the interpretation of the results of the tests, when as clear-cut as described, was realistic and many of the vigorously explosible dusts did in fact cause explosions in industry. In contrast none of the dusts which did not respond in the small scale tests gave any explosions. In fact a number of dusts which were vigorously explosible in small scale tests also did not give explosions in industry; there are a number of reasons for this, principal ones being the type of industrial operation, the quantity of dust involved, and the energy of potential ignition sources.

Difficulty arose in considering dusts which responded only feebly in the small scale tests. It was not clear whether, on a larger scale, the explosion would be much more vigorous and hazardous, or whether the flame would die away a short distance from the source of ignition, so that pressure effects through the plant would be negligible and the dusts would in practice be non-explosible. Experience in industry indicated that vigorous explosions did not occur, but whether this was due to the dust being unable to generate such explosions, or to the

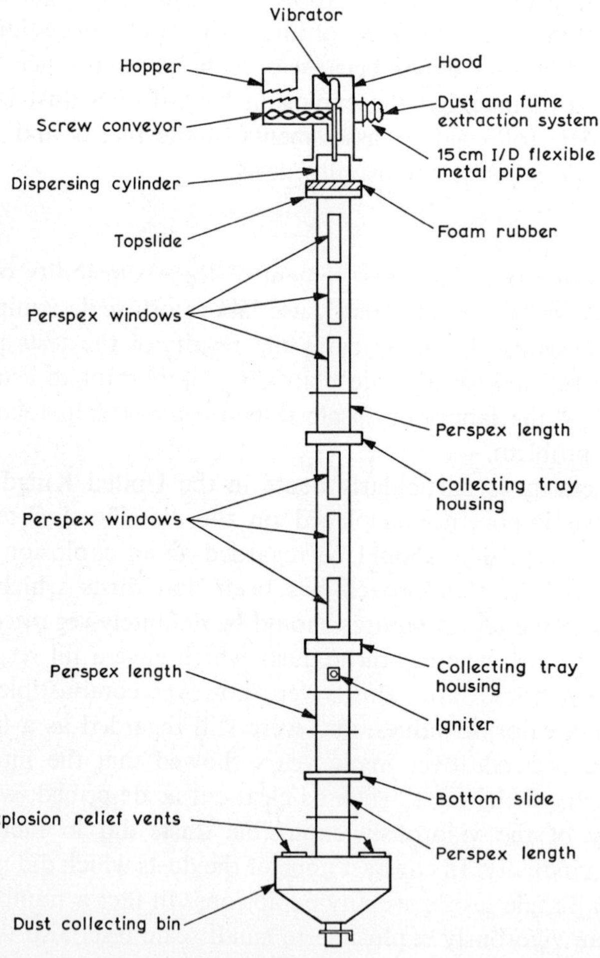

Fig. 4.1 Large scale dust explosion tube [1].

processes and quantities of dust involved to date being unfavourable for the generation of vigorous explosions, was not clear. As the dilemma was unlikely to be solved in the short-term by observation of the behaviour of dusts in industrial plant, an investigation was carried out at the Fire Research Station, to establish whether the classification system, then in use, was realistic, and to examine the explosion hazard associated with marginally explosible dusts, i.e. those which gave a feeble but positive response in the small scale tests.

The experiments were carried out using a large scale tube apparatus (not to be confused with the small Vertical Tube test apparatus). Large scale tubes have been used in a number of investigations and will be considered in detail later (Chapter 5). In the present case the tube, which was 25 cm in diameter and 5.2 m in length, was erected vertically (Fig. 4.1). Dust was fed in at a constant rate, to the top of the tube, and allowed to fall, under gravity, into a collecting bin attached to the bottom. One end of the tube was closed to prevent a stream of air being drawn through. When the dust was falling at a steady rate its concentration was measured, and ignition was attempted by means of a propane gas flame injected at a point towards the bottom of the tube. Observation was then made as to the extent and velocity of the flame propagation. A fuller description of the experimental technique has been published [1]. Because the tube was of light and relatively weak construction, the dust concentrations were restricted to those near the explosibility limits. This condition was fully adequate for the situation, but much more vigorous explosions could have been generated if necessary. Three types of explosion behaviour were observed:

Flame propagated the whole length of the tube
Flame propagated part way along the tube
Flame did not propagate away from the ignition source.

A dust suspension into which flame propagated for part way along the tube was recorded as explosible because the extent of flame propagation was not affected by a substantial increase in the size of the ignition source. It was unlikely that the extent of partial propagation was governed by energy derived from the igniting source, and a more likely explanation is that as the flame was propagating relatively slowly, and its composition was near the explosibility limit, it was readily affected by random variation of concentration in the dust suspension.

In order to obtain dusts of controlled and graded explosibilities, some of the experiments were carried out with dust mixtures containing an explosible dust (phenol formaldehyde resin), and a non-explosible dust (magnesium oxide) premixed in various proportions. As the proportion of non-explosible dust in the mixture was increased, the explosibility of the mixture diminished. Each of the dust mixtures was dispersed in the large scale apparatus, and the concentration range over which flame propagation occurred was measured. As the proportion of non-explosible dust in the mixture was increased, the range narrowed until eventually sufficient non-explosible dust was present in the mixture to ensure that propagation did not occur at any concentration of the suspension. For this to occur the mixture had to contain at least 80 per cent by weight of magnesium oxide. Further tests were carried out with individual industrial dusts to ascertain whether they could propagate flame in the large scale apparatus. A summary of the results is given in Table 4.1. All the dusts used in the large scale tube were also tested in the small scale apparatus and classified (Table 4.1) according to the present system of grouping. Further experiments were carried out with coal dust, with which comparisons could be made with results available elsewhere on coal mine galleries of the order 1 m in diameter and 100 m in length. The results supported the earlier findings [2].

Comparison of the results from the large scale tube and the small scale tests showed that there was a broad correlation, with a tendency

Table 4.1 Explosions in large scale tube [1]

Dust	*Flame propagation in large scale tube*	*Small scale classification tests giving flame propagation*	*Classification*
Phenol formaldehyde—magnesium oxide (25:75)	Yes	Vertical tube (spark or coil igniter)	Group (a)
Phenol formaldehyde—magnesium oxide (20:80)	No	Vertical tube (coil igniter only)	Group (a)
Methyl cellulose	Yes	All	Group (a)
Manioc	Yes	Inflammator and vertical tube (spark or coil igniter)	Group (a)
Sodium carboxy methyl cellulose	No	Inflammator and vertical tube (coil igniter only)	Group (a)
Polyvinylidene chloride	No	None	Group (b)
Calcium citrate	No	None	Group (b)

for the small scale tests to slightly over-estimate the explosibility of the dusts. As the large scale tube was considerably closer to industrial plant proportions than are the laboratory scale test apparatus it is reasonable to conclude that any dust giving flame propagation in the small scale tests, i.e. as a Group (a) dust, should be regarded as explosible in full scale industrial plant whereas those of Group (b) were not. It is of interest that all dusts and dust mixtures listed in Table 4.1. ignited in the Furnace apparatus (Fig. 3.5), which seriously over-estimated the explosibility hazard of dusts dispersed in plant at ordinary temperatures. The Furnace apparatus should not be used in isolation to assess the explosibility hazard of dusts but should be regarded mainly as a means of estimating minimum ignition temperatures of dust suspensions.

It is reasonable to assume that dusts which ignite and propagate flame in the other small scale tests, excluding those in which the dust is dispersed in a furnace or similar heated enclosure, should be regarded as explosible when dispersed in full scale industrial plant at ordinary temperatures.

Finally, it should again be emphasized that classification tests and other tests in which the appearance of flames in dust suspensions is observed, should be used only for assessing whether or not a dust explosion hazard can arise. These tests give very little evidence as to the degree of hazard other than that of qualitative observations of explosion severity and a full assessment of dust explosibility can only be made if additional tests are done.

Hazard index

A method of indexing dusts according to their explosibility has been put forward by the U.S. Bureau of Mines [3]. Ratios called the Ignition Sensitivity and the Explosion Severity are calculated for each dust and the Index of Explosibility for the dust is obtained as the product of the two ratios. The Ignition Sensitivity and Explosion Severity for a dust are obtained by comparing various explosibility measurements of the dust with the values obtained in the same tests using a standard dust, for which a Pittsburgh coal dust is used. The reason for selecting the coal dust is that large scale explosions in an experimental mine have been carried out over many years and considerable data on full scale explosions for this dust have accumulated. The explosibility of a dust may thus be compared with that of the known material, and in this way previous practical experience can be used in assessing the explosibility hazards of new materials. The two ratios are calculated as follows:

$$\text{Ignition Sensitivity} = \frac{(\text{Min.ign.temp.} \times \text{Min.ign.energy} \times \text{Min.expl.conc.})\ \text{Pittsburgh coal dust}}{(\text{Min.ign.temp.} \times \text{Min.ign.energy} \times \text{Min.expl.conc.})\ \text{Sample dust}}$$

$$\text{Explosion Severity} = \frac{(\text{Max.expl.pressure} \times \text{Max.rate pressure rise})\ \text{Sample dust}}{(\text{Max.expl.pressure} \times \text{Max.rate pressure rise})\ \text{Pittsburgh coal dust}}$$

$$\text{Index of Explosibility} = \text{Ignition Sensitivity} \times \text{Explosion Severity}$$

Values for the minimum ignition temperature, minimum ignition energy, and minimum explosible concentration of dust suspensions are measured by the usual methods. The maximum explosion pressure and maximum rate of pressure rise are also measured by the usual techniques, at a dust concentration of 0.5 g/l. Values of these quantities for the Pittsburgh coal dust are available [3 and Appendix 2].

The Index of Explosibility has been correlated with the explosion hazard rating on an empirical basis as shown in Table 4.2. A dust with an Index of Explosibility greater than 1.0 indicates a hazard greater than that of the coal dust, the hazard rating of which is taken to be on the border line between moderate and strong.

Table 4.2 Relation between hazard rating and index of explosibility

Relative explosion hazard rating	*Ignition sensitivity*	*Explosion severity*	*Index of explosibility*
Weak	<0.2	<0.5	<0.1
Moderate	0.2—1.0	0.5—1.0	0.1—1.0
Strong	1.0—5.0	1.0—2.0	1.0—10
Severe	>5.0	>2.0	>10

The Index of Explosibility and the relative hazard rating give a useful guide, particularly when new dusts are being considered. A dust which yields a rating well above that of coal dust is clearly likely to be a major dust explosion hazard and may need special precautions in its handling; conversely, a dust having an Index of Explosibility well below 0.1 can usually be regarded as primarily a fire hazard. An exception would be necessary if a low value for one of the terms in the ratio was due to limitations of the test apparatus rather than to inherent low explosibility of the dust. If, for any reason, ignition cannot be obtained in one or more of the relevant tests then the Index cannot be applied, and it would not be permissible to insert a zero value in either of the ratios.

The Index is on an empirical basis and is dimensionless because each of the quantities involved is compared with that for the Pittsburgh coal dust. Equal weight is given in the ratios to each of the quantities and this may not be completely justifiable. For instance, minimum ignition energies and minimum explosible concentrations can vary over relatively wide ranges, whereas minimum ignition temperatures are well above room temperature and due to limitations of the apparatus cannot exceed 1000°C. The Ignition Sensitivity is more heavily dependent on minimum ignition energy and minimum explosible concentration than on minimum ignition temperature. The values for the standard coal dust are constant and would not counteract this weighting.

The Index of Explosibility is an attempt to put on a numerical basis the explosion hazard rating of dust, and as such it is useful, although it should be used with caution because of the limitations of an empirical relationship. It is likely to continue to be a useful notion until the theoretical background to dust explosions has been strengthened and an Index based on explosion properties, with their correct interactions, can be derived.

The Index of Explosibility and the hazard rating of course apply only to the dust itself. In any practical situation consideration would also have to be given to the processing the dust was likely to receive and the mass involved. These factors introduce further problems which at present have to be dealt with on an empirical basis as will be discussed later (Chapter 10).

Reproducibility of tests

Because of the impossibility of obtaining perfect dispersion of dust, apart from variation in the apparatus itself, very accurate reproducibility is not obtained in dust explosibility tests. In carrying out tests it is usual for a number of repetitions to be made, and providing the scatter of the results is not too large, which is a subjective decision in most cases, the most hazardous value of the quantity being measured is taken. Generally it is not practicable for economic reasons to undertake a large enough number of tests for a full statistical treatment to be applied to the results, and hence, the less satisfactory procedure of taking extreme values is used. The reproducibility obtainable under relatively favourable conditions has been published for a range of tests on corn starch (through 200 US mesh) [4]. Ten samples of the dust were put through a range of tests and the values obtained, together with the standard deviation are shown in Table 4.3. If a range of

Table 4.3 Reproducibility of laboratory tests at US Bureau of Mines [4] (Cornstarch dust, through a 200-US mesh sieve)

Test	*Mean*	*Standard deviation*	*Range of values measured in tests*
Ignition temperature of dust cloud °C	403	4.8	400—410
Relative flammability, standard spark apparatus, per cent	86.3	1.3	85.0—87.5
Minimum ignition energy of dust cloud mj	22.5	2.6	20—25
Minimum explosible concentration g/l	0.044	0.002	0.043—0.047
Limiting percentage of oxygen in air–CO_2 mixture to prevent ignition by electric spark	12.3	0.5	11.8—12.8
Maximum pressure (lb/in^2) and rates of pressure rise (lb/in^{2}s) at concentrations of:			
0.10 g/l:			
Maximum pressure	42.0	4.4	34.8—48.3
Average rate	1053	191	720—1320
Maximum rate	2842	602	1860—3890
0.20 g/l:			
Maximum pressure	62.6	1.7	60.4—65.5
Average rate	1478	107	1350—1620
Maximum rate	4382	417	3580—4690
0.50 g/l:			
Maximum pressure	88.0	5.9	81.1—98.0
Average rate	1781	254	1460—2210
Maximum rate	5645	927	4260—6740
1.0 g/l:			
Maximum pressure	101.1	6.5	88.7—111.0
Average rate	1617	245	1270—2100
Maximum rate	4798	964	3730—6470
2.0 g/l:			
Maximum pressure	110.5	5.9	102.8—120.7
Average rate	1266	284	830—1820
Maximum rate	3428	973	1860—4690

variation equivalent to two standard deviations is assumed to be likely, substantial variation in test results on repeat tests would be expected. Even greater variation would be expected with dusts less readily dispersed than finely divided corn starch.

The reproducibility shown in Table 4.3 is likely to be as good as that obtained with the general run of dust samples, hence attempts at higher accuracy in determination of explosion properties than those shown in Table 4.3 are not warranted. In applying the results of tests sufficient safety margin should be included in any calculations to allow for the likely error.

All the results in Table 4.3 refer to the ignition of dust suspensions. In tests on dust layers, the variation in the results arising from chance fluctuations in dispersion would not arise, and so a higher level of reproducibility would be expected. In general this is found to be the case, and particularly so where the source of ignition can be accurately controlled e.g. the temperature of a furnace. The ignition of dust layers in heated enclosures, or on hot surfaces, is considered in more detail later (Chapter 9).

Greater variation than that shown in Table 4.3 is likely to arise in tests where the production of the explosible material itself is subject to chance. This is the case in tests on carbonization products (Fig. 3.10) where the quantity of volatile combustibles produced, and their rate of production can both vary. In carrying out explosibility experiments with these products further variations arise in the extent to which they are mixed with air and also in the ignition process. Interpretation of the results of these tests should take account of these variations.

Interpretation of individual tests

MINIMUM IGNITION TEMPERATURE

The values usually quoted in the literature for the minimum ignition temperature of dust suspensions are those obtained from the Furnace apparatus. The principle of this test is that the dust suspension passes through an enclosure which is an extended heat source. The analogy with industrial practice is with a relatively large heated area in the plant, such as a failed bearing or a furnace wall, or a dryer. There are limitations on the extent to which the ignition temperatures, obtained in the tests, can be directly applied to industrial situations. These limitations arise from two sources, the design of the test and variation of the conditions on the plant.

If the test apparatus and procedure are considered in detail a number of limitations are apparent. In the test the Furnace apparatus is allowed to reach equilibrium at a pre-set temperature and the air within the Furnace tube attains a temperature close to that of the tube itself, which is the recorded temperature. However the dispersing air reservoir and the dust holder are at room temperature and the glass adaptor connecting them to the Furnace (Fig. 3.5) would be heated by conduction from the Furnace to a temperature above that of the room. The volume of air used in dispersing the dust, and which passes along the connection tube to the Furnace, is usually more than half the total volume of heated air initially within the Furnace tube. This dispersing air is cold and on entering the heated part of the apparatus would expand, probably approaching the original volume of air in the heated Furnace. The extent to which the dispersing air and the heated air, within the Furnace, mix during the performance of the test is not measured. If perfect mixing occurred, the temperature of the air within the Furnace at the time of entry of dust would be well below the Furnace wall temperature; if mixing were poor, part of the air temperature would again be below that of the Furnace wall. The air temperature would rise as it was heated by the Furnace wall, the rate depending on the heat transfer from the wall to the air. The situation during the test, immediately following the dispersion of the dust, is thus that of a dust suspension in air at an unknown temperature, lower than that of the Furnace wall. If ignition is governed by the air temperature the true minimum ignition temperature would not be that measured, which is of the Furnace wall.

When the dust dispersion enters the Furnace tube some of the particles are likely to be deposited on the wall, whilst the others remain in suspension. The test does not indicate whether dust on the wall ignites first, or whether the dust suspension ignites, or whether the controlling factor is the production of carbonization volatiles from dust deposited on the wall and ignition of these gases on mixing with the heated air. If ignition of the dust whilst it is on the Furnace wall is the usual mechanism then the ignition temperature of the dust is likely to be close to that of the quoted values because the solid surface temperature would be near to that actually measured. Ignition of the dust suspension would require a higher Furnace temperature than the true minimum because of the cooling by the dispersion air. If production of carbonization volatiles is necessary then ignition is governed by two criteria, namely that the wall temperature should be sufficiently high to produce the carbonization, and the air temperature sufficiently high to ignite the volatiles. For these circumstances the minimum ignition temperatures quoted would not be directly linked to full scale plant, and would depend substantially on the design of the test apparatus and the procedures used with it.

A detailed investigation of the processes occurring in the Furnace apparatus has not been carried out. A number of superficial observations are made in carrying out routine tests, some of which are as follows:

(a) At temperatures well above the minimum ignition temperature, production of flame usually occurs very quickly after dispersion of the dust into the apparatus and the flame is relatively large.
(b) At temperatures near the minimum for ignition the flames are smaller and may be produced after a delay as long as several seconds.
(c) Unburnt and apparently unaffected dust can pass through the Furnace apparatus at temperatures above the minimum for ignition.
(d) If ignition does not occur, considerable quantities of vapour may be evident, particularly with carbonaceous dusts.
(e) Even for dusts which are readily dispersible, the volume of air used for dispersion affects the probability of ignition at temperatures near the minimum.
(f) The probability of ignition also depends on the amount of dust dispersed.
(g) Some dusts produce sparks instead of flame.

These observations make it unlikely that any of the individual mechanisms for ignition suggested above occurs predominantly with all dusts, but that a combination of mechanisms may act with a single dust.

The second limitation of application of results arises from the plant design. A hot surface within plant is unlikely to have the configuration and dimensions of the tube in the Furnace apparatus and may be much larger; the volume of heated dust suspension may be correspondingly greater. In addition the residence time of the dust suspension near the heated surface may be greatly different, this time is not controlled or measured in the small scale test. Both these factors will affect the ignition temperature.

Minimum ignition temperatures determined in the Furnace apparatus have two main uses. Firstly they demonstrate the minimum temperatures at which the ignition of a dust suspension can occur. This temperature represents a maximum which should not be exceeded in industrial plant, but under industrial conditions ignition may in fact occur at a lower temperature than the minimum obtained from the test. The minimum ignition temperatures are basically those which should never be exceeded, and plant surfaces may require keeping well below the minimum. As some Group (b) dusts, which would not be explosible under ordinary temperatures, propagate explosion in a heated enclosure the need for explosion precautions should be investigated, where these dusts may be dispersed in the presence of extended heat sources.

The second use for the minimum ignition temperatures determined by small scale tests is in the comparison of the ease of ignition of different dusts by hot surfaces. The comparison is particularly useful when new materials are considered, because of the large number of values already reported for dusts dispersed in air and for which considerable practical experience has accumulated.

Usually the values of minimum ignition temperatures exceed 300°C, and any material having a minimum value near this temperature should be regarded as easily ignited. At values below 250°C the dust should be regarded as particularly vulnerable and strict precautions taken to prevent it being heated during manufacture or processing. The VDE regulations in the Federal German Republic stipulate that the maximum permissible temperature of surfaces in factories having a dust explosion hazard shall not exceed two-thirds of the minimum ignition temperature.

The minimum ignition temperatures obtained for dust layers or dust deposits are of more direct application. The two types of test involve either ignition of dust on a heated surface, with the upper surface of the dust layer exposed to atmospheric temperature, or the ignition of the dust deposit in a hot enclosure in which case all surfaces are at elevated temperature. Which test is selected depends upon the industrial conditions. Where dust may deposit itself on hot pipes or on a hot outer casing, with the upper surface of the dust layer exposed to atmospheric temperature, even with very thick dust deposits, the minimum ignition temperature determined by a hot plate technique is appropriate. Where dust is being dried in an oven, or where a dust layer may be deposited inside a heated enclosure, even with very thin layers, then the minimum ignition temperature measured by a technique involving a hot enclosure would be appropriate. In either the minimum ignition temperature will depend upon the thickness, depth, or other characteristic dimension of the dust deposit.

With a dust layer on a hot plate, the minimum ignition temperature decreases sharply as the thickness of the deposit is increased [5] as shown in Fig. 4.2. Under routine test conditions the minimum ignition temperatures are measured with a dust deposit of thickness 5 mm, but if under industrial conditions thicker deposits can occur a lower minimum ignition temperature should be used. The VDE regulations lay down that the maximum permissible surface temperatures in factories with a dust explosion hazard must be 75°C below the minimum ignition temperature of a 5 mm layer of the dust. As the minimum ignition temperature is reduced, because the thickness of the layer is increased, the time required for ignition also increases. This is considered in more detail later (Chapter 9) but assessments of industrial hazards should take account of the fact that the time for ignition may vary between seconds and days depending on the dust, its thickness, and the temperature.

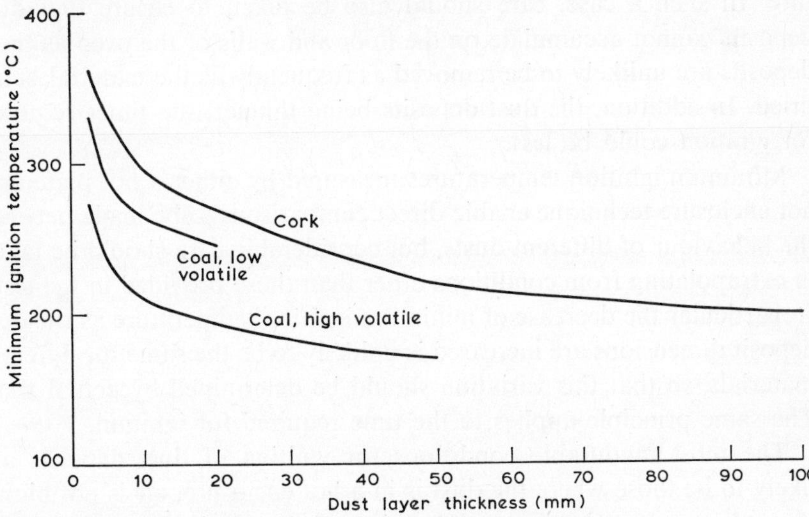

Fig. 4.2 Minimum ignition temperatures of dust layers [5].

The minimum ignition temperatures obtained in a heated enclosure also depend on the dimensions of the deposit. The larger the deposit the lower the minimum ignition temperatures. The minimum values will also be affected by the aerodynamic conditions of the test; in particular whether air is passed through the dust during the test or whether the interior of the specimen is subject only to natural convection. Vigorous movement of the air in the furnace or oven will also affect heat transfer from the dust surface to the air, and during the ignition process the dust surface temperature is likely to rise above that of its surroundings and vigorous stirring of the air will cool the specimen.

If the industrial situation for which the tests are being carried out is such that a relatively large amount of dust is exposed to the heating, then tests may be inconvenient on full scale. A short series of tests with a limited range of thicknesses of deposit should be done. Extrapolation from the test scale to full industrial scale may then be carried out with reasonable confidence because the theoretical background is available, at least for simple cases (Chapter 9).

As with dust layers, the time required for ignition increases with the dimension of the deposit and may be a controlling factor under industrial conditions. For instance, where a material is being dried in an oven, the time for which the material is heated may be far shorter than the time required for ignition of the material at the oven temperature. In such a case, care should also be taken to ensure that dust deposits cannot accumulate on the floor and walls of the oven as these deposits are unlikely to be removed as frequently as the material being dried. In addition, the dust deposits being thinner, the times required for ignition could be less.

Minimum ignition temperatures measured by either a hot plate or a hot enclosure technique enable direct comparisons to be made between the behaviour of different dusts, but considerable care should be taken in extrapolating from conditions other than those provided in the tests. In particular the decrease of minimum ignition temperature as the dust deposit dimensions are increased is unlikely to be the same for different materials, so that this variation should be determined by actual tests. The same principle applies to the time required for ignition.

The most favourable conditions for ignition of dust deposits are likely to be those where the dust is in a hot enclosure, air is not blown through the dust deposit, and stirring of the air in the enclosure is gentle. Tests under these conditions give values of minimum ignition temperatures lower than those of other methods, after allowing for the effect of deposit thickness.

Occasionally a dust does not ignite, or ignites only with great difficulty in hot surface and hot enclosure tests; because of liquefaction or decomposition. Ambiguous results may thus be given by small scale tests. However, on full industrial scale, where deposits of thickness of the order of 1 m may be present, ignition may occur more readily. This is due to the products of liquefaction or decomposition not being able to escape readily from the centre of the heated deposit, whereas in small scale tests escape can occur before the process leading to ignition has time to develop. Dusts which fail to ignite in small scale tests because of liquefaction or decomposition should therefore be considered very carefully, particularly if there are chemical reasons for supposing that the products lost in small scale tests could ignite under conditions of full scale heating.

INERTING WITH GAS OR DUST

The tests concerned with the prevention of explosion by inerting involve either the Furnace apparatus (Fig. 3.5) or the Standard Spark apparatus (Fig. 3.9). In industrial practice use may be made either of inert gas, to prevent explosion, or of an inert dust which is mixed with the explosible dust. The use of inert dust is principally confined to the coal industry, but relative flammability tests with inert dust are carried out in the United States for industrial dusts generally (Fig. 3.8).

The use of the Furnace apparatus for assessing the percentage of inert gas to be added to air to prevent dust explosions is subject to disadvantages, some of which are similar to those arising in the measurement of minimum ignition temperature. For determining the inert gas requirements, the Furnace apparatus is modified as described previously and maintained at a temperature of 850°C. As in the measurement of minimum ignition temperature the air/inert gas mixture used to disperse the dust and carry it into the Furnace is initially at room temperature and so the temperature of the dust dispersion in the Furnace is uncertain.

If the minimum ignition temperature of the dust in air/inert gas mixture were higher than the temperature of the suspension, and certainly if it were higher than 850°C, then ignition of the suspension in the test would not occur. Ignition might have been possible with a hotter source, such as an electric spark.

The results from the Furnace apparatus could err on the side of danger by underestimating the percentage of inert gas required to prevent ignition, particularly with those dusts having a high minimum ignition temperature in air. Results obtained for dust having a relatively high ignition temperature in air, say over 600°C, must be interpreted with caution. This proviso only applies to explosible dusts, i.e. those in Group (a). It would not apply to those Group (b) dusts which produce flame in the Furnace apparatus, providing they are not used in heated environments.

Conversely a dust could explode in an air/inert gas mixture in the Furnace apparatus but would not subsequently propagate flame in surroudings at atmospheric temperature. The test would then overestimate the percentage of inert gas required to prevent explosion, and the permissible oxygen percentage would be unnecessarily low. This is an error on the side of safety, and if it were small it would be acceptable in practice. However if the error were large it could make the technique of using inert gas to prevent explosion appear to be uneconomic, and this would be undesirable because the technique is sound in principle, providing adequate monitoring is installed to ensure that the inert gas percentage is correctly maintained (Chapter 8).

The extent to which substantial overestimates of the required percentage of inert gas can influence the economics of the technique may be illustrated by an example. If the true maximum permissible oxygen concentration were 10 per cent, the air/inert gas mixture would contain approximately equal proportions because atmospheric air contains about 21 per cent of oxygen by volume. If the test indicated that the oxygen percentage should be 5 per cent, because the test substantially overestimated the necessary extent of inerting, then the air/inert gas mixture would need to contain about three volumes of inert gas to each volume of air. The inert gas requirement is therefore trebled. In this example no safety margin has been provided.

In practice the position may not be as difficult as in the example because the total volume of supply of the air/inert gas mixture may be fixed. If the required delivery is 100 volumes per hour, then in the actual case an inert gas delivery of 50 vol/hr would be required whereas the test results would indicate a delivery of 75 vol/hr. Nevertheless because of the uncertainty introduced by the use of the

Furnace apparatus, and in particular because it may err on the side of hazard or of safety, the overall position is unsatisfactory.

The extent to which the present test method is realistic has been assessed by measuring the maximum permissible concentration of oxygen in an air/nitrogen mixture, to avoid ignition of terephthalic acid dust in the Furnace apparatus at 850°C and at 1000°C [6]. The maximum permissible concentration was also measured in a larger scale tube apparatus, 15 cm diameter and 150 cm long approximately, at room temperature and at 150°C. In the larger scale tube apparatus the maximum permissible concentration was substantially higher than in the Furnace apparatus; the results are summarized in Fig. 4.3. The Furnace apparatus indicated a maximum permissible oxygen level of 4 per cent when tested at 850°C and only 2.4 per cent when the temperature was 1000°C, whereas with the large scale apparatus at room temperature the maximum permissible oxygen concentration was about 12.5 per cent. The permissible oxygen level decreased steadily as the temperature of the apparatus rose (Fig. 4.3). A subsequent investigation at the Fire Research Station, with a modified large scale dust explosion tube (Fig. 4.1), extended these findings with a wider range of dusts and inert gases. A modified test, based on the Vertical Tube apparatus, is being developed.

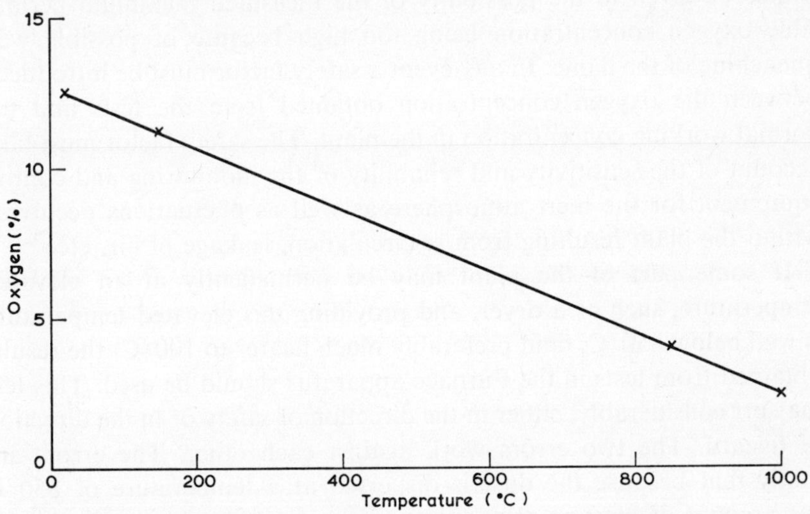

Fig. 4.3 Effect of temperature on maximum permissible oxygen concentration for terephthalic acid dust [6].

The maximum permissible oxygen concentrations measured with the Standard Spark apparatus are considerably higher than with the Furnace apparatus. But the Standard Spark apparatus has a tube of diameter only 3.8 cm (Fig. 3.9), which could possibly cause wall quenching of flames propagating from the spark. If quenching should occur it would assist the inerting action, and might lead to an underestimate of the percentage of inert gas. There is evidence that, at least with coal dust, wall quenching effects are present in tubes of diameter as high as 7 cm (Chapter 5). Some contribution from the wall assisting the inert gas in quenching the flame in the Standard Spark apparatus cannot be ruled out, and requires investigation.

The results obtained from the Furnace apparatus and the Standard Spark apparatus for the percentage of inert gas required to prevent dust explosions will differ and should be interpreted with care. In deciding which of the tests is the more relevant, the industrial application should be the starting point. If all the plant is run at about atmospheric temperature so that no part of the dust suspension is in a permanently heated confined volume, then the results from tests in the Standard Spark apparatus could be used as a basis for estimation of inerting requirements even though it is envisaged that the ignition source might be relatively large, such as an overheated bearing. In using results from the Standard Spark apparatus account must of course be given to the possibility of the measured maximum permissible oxygen concentration being too high because of possible wall quenching of the flame. In any event a safety factor must be introduced between the oxygen concentration obtained from the tests and the normal working concentration in the plant. The safety factor must take account of the sensitivity and reliability of the monitoring and control equipment for the inert atmosphere as well as fluctuations occurring within the plant resulting from re-circulation, leakage of air, etc.

If some part of the plant may be permanently at an elevated temperature, such as a dryer, and providing this elevated temperature is well below 850°C, (and preferably much nearer to 100°C) the results obtained from tests in the Furnace apparatus should be used. This test may err considerably, either in the direction of safety or in the direction of hazard. The two errors work against each other. The errors are firstly that because the dust is dispersed at a temperature of 850°C the amount of inerting required may be a considerable overestimate. This is an error in the direction of safety, and may be expected to predominate with dusts having an ignition temperature in air of below 600°C. The second error may arise if the ignition temperature of the dust dispersed in the air/inert gas mixture is above 850°C. This is an error in the direction of hazard because ignition will not occur although the mixture might be ignited by an alternative source of ignition. It is unlikely to arise with dusts having a minimum ignition temperature in air below 600°C, but if the ignition temperature is above this value and particularly if it approaches 850°C, then this error would be important. Where the uncertainty due to error is unacceptably large the use of small scale testing may need to be abandoned, and direct experimentation on an apparatus nearer in scale and design to the industrial plant may be necessary. In the long term it is to be hoped that the uncertainties arising from small scale testing will be overcome, using a modification of the test apparatus and procedure.

Two of the inert gases likely to be available are nitrogen and carbon dioxide. Because the gases differ in density and specific heat their inerting capabilities are slightly different, carbon dioxide generally being more effective than nitrogen. The difference is relatively small and may be covered in the general case by a rule of thumb stating that one per cent more oxygen is permissible with carbon dioxide than with nitrogen e.g. if the maximum permissible oxygen concentration is 12 per cent, by volume, using nitrogen, then with carbon dioxide the value would be 13 per cent. This rule is only applicable where there is no chemical interaction between the dust and the inert gas; notable exceptions are metals such as magnesium which reacts with carbon dioxide, so that a lower value is obtained than with nitrogen. In severe cases, other inert gases may be necessary, such as helium or argon. Of these two helium is usually the more effective, probably because of its higher thermal conductivity, but may be costly.

In applying the results of tests in which an inert dust is mixed with the explosible dust to prevent propagation of flame much of the comment relevant to inert gases is applicable. The same test methods, using the Furnace apparatus and the Spark apparatus, are applied in determinations of relative flammability and similar difficulties arise in application of the results. In practice the difficulties of interpretation are less severe than with inert gases because the principal application of the technique of adding inert dust is in the coal industry, where relatively few inert dusts are involved and experience on full scale has accumulated over many years. The technique of using dusts for inerting is relatively little used in surface industries, because of contamination of the product or because alternative explosion prevention or protection techniques are more economic.

The use of inert materials arises indirectly when the explosibilities of mixtures of dusts, some of which are combustible and some not, are evaluated. In these cases the proportion of inert dust in the mixture is usually governed by external factors and no attempt is made to ensure that the proportion is above the minimum value for prevention of explosion. In such cases direct assessment of explosibility and explosion properties is the usual course taken, providing information on the dust explosion hazard in the plant.

Just as dust explosions may be prevented in inerting, so ignition of dust layers can also be avoided by these techniques. There are no routine tests at present commonly used. Fires in dust layers, particularly in systems inerted with gas, can continue at oxygen percentages below those necessary for explosion. Thus although a plant may run successfully with an oxygen concentration sufficiently low to prevent explosion, this does not mean that fires are also automatically prevented in any dust deposits which may be present within the plant.

MINIMUM EXPLOSIBLE CONCENTRATION

The reasons for measuring the minimum explosible concentration of a dust are firstly to obtain an indication of hazardous concentration in plant, particularly in dust extraction systems, and secondly to enable the Ignition Sensitivity and the Index of Explosibility of the dust to be calculated.

The method of measuring the minimum concentration was originally developed to give comparable results to those from larger scale experiments. It is convenient, but cannot be expected to give high accuracy. The main source of error is the non-uniform distribution of dust when dispersed, giving high concentrations in part of the apparatus. These may ignite and expand whilst burning to give a visible flame and calculations based on the total amount of dust present and the volume of the apparatus will give a low estimate. Also the suspension is turbulent, due to the dispersion mechanism and a different value for the minimum explosible concentration might have been obtained if the dust were allowed to fall gently through stationary air. Photographic evidence of the dispersion of dust in the test apparatus [7] has shown that the dust dispersion is not uniform, and that greater uniformity can be obtained by modifying the dispersion cup in the standard apparatus, Table 4.4. Under standard conditions of test, the values obtained are unlikely to be highly accurate but enable comparisons to be made between dusts as in calculation of Ignition Sensitivity. Practical experience indicates that the data obtained are realistic for industrial processes.

With industrial plant such as dust extraction systems, calculations in which the total throughput of dust is divided by the throughput of air during the same time, giving a mean value for dust concentration, are unlikely to be realistic. The dust and the air may not move at a constant rate through the plant so that variation in the concentration occurs. A simple example is the delivery of a dust/air mixture from a pipe into a larger enclosure; substantial change in dust concentration would be likely. Secondly dust may settle out within the plant, and then be redispersed, and would result in concentrations quite different from the calculated mean. By working with a mean dust concentration below the minimum explosible concentration, a measure of safety is obtained, but it may not be complete. The position with dusts is different to that with gases, where a flammable gas which has been diluted with air to a concentration below the lower flammable limit will not reconcentrate itself. Even if dust does not fall out of suspension, into deposits within the plant, concentration of suspensions can occur in transit. An obvious example is a dust suspension passing into a cyclone; although the dust particles do not become stationary within the cyclone, the concentration changes drastically and the suspension could be explosible at some point.

With many dusts the minimum explosible concentration is sufficiently high for the dust cloud to be opaque, a thickness of the cloud of order 1 m being too dense to see through. Operatives could not work within a cloud of this concentration for any substantial length of time. This criterion is not invariable, and with some dusts of which polyethylene is a notable example, suspensions are explosible which are transparent over 1 m or more.

MINIMUM IGNITION ENERGY

Measurements of minimum ignition energies can be made on dust suspensions and on layers. Comparison between dusts, suspended in air, can be made directly or by incorporation in the ratio for Ignition Sensitivity. There are limitations on the accuracy of the values

Table 4.4 Minimum explosible concentrations (g/l) for various dispersion arrangements in Vertical Tube apparatus [7]

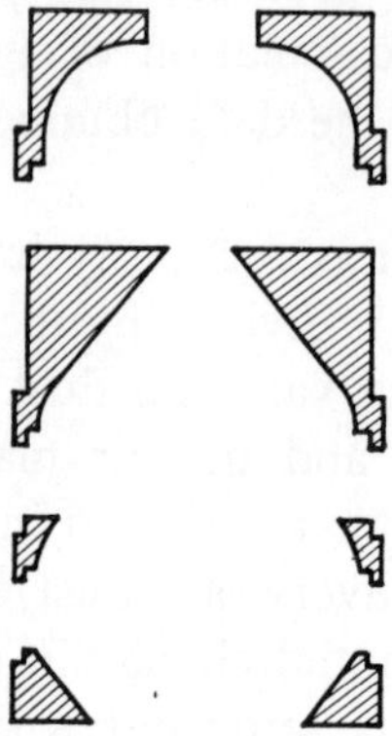

Dust	Standard dispersion	Small ring	Conical ring	Curved ring
Coal	0.090	0.070	0.115	0.065
Polypropylene	0.015	0.015	0.010	0.008
Cornstarch	0.050	0.030	0.030	0.030
Aluminium	0.040	0.025	0.025	0.025

measured, due to variation in the dust dispersion during test and also to the characteristics of the electrical circuit used for generating the spark. As described earlier, the spark is usually produced by the discharge of condensers at a known voltage through a transformer. In calculating the energy of the spark no allowance is made for loss in the transformer, and complete discharge of the condensers is also assumed. The values for minimum ignition energy are thus likely to be too high. The extent of the error introduced has not been investigated in detail, but in some tests where high voltage condensers were discharged directly without a transformer, comparable minimum ignition energies were obtained for several dusts [4].

Apart from energy losses in the electrical circuit, the characteristics of the circuit may also be important. For several dusts the minimum ignition energy depends upon the electrical resistance of the circuit generating the spark [8], but not with coal dust [9]. Increase in the inductance of the discharge circuit did not reduce the minimum ignition energy of coal although it visibly changed the character of the spark. The spark voltage is also a factor.

Increasing the temperature of the dust suspension would be expected to lead to lower energies, and at temperatures near to the minimum ignition temperature of the dust suspension, the electrical energy required for ignition might be very low. An investigation using coal dusts of various volatile contents [9] showed that the minimum ignition energy decreased as the temperature rose but that a minimum value was obtained, and at higher temperatures the minimum ignition energy rose sharply, Fig. 4.4. The existence of the minima in Fig. 4.4 was attributed to reaction of the dust with the oxygen of the air before dispersion, and the extent to which this occurred would depend upon the quantity of dust present and the design of the apparatus. In this investigation the apparatus was non-standard, and high values were obtained for the minimum ignition temperatures of the coal dusts; the results were not comparable with those from the standard test apparatus and should be applied with caution to full scale plant. But the effect observed, namely the reaction of the dust with the oxygen in the air, would be applicable to industrial conditions.

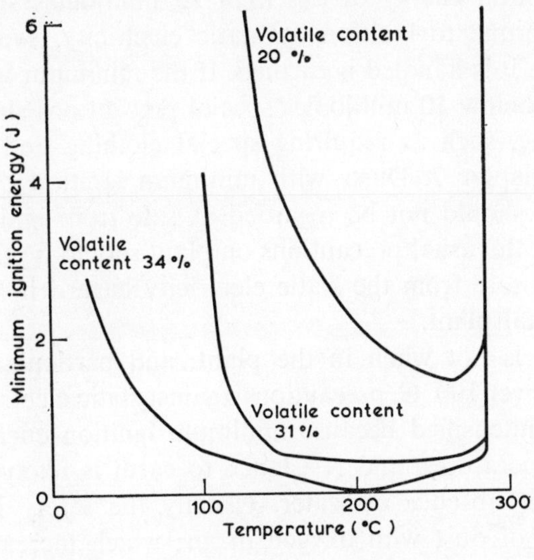

Fig. 4.4 Minimum ignition energies of coals at various temperatures [9]. Air in the apparatus is not flushed.

The principal application of minimum ignition energies to industrial plant is in consideration of the hazard of static electricity. In the movement of dust in plant its generation is to be expected, and precautions necessary in the construction of the plant itself are considered in Chapter 7. If adequate precautions are taken, the accumulation of a hazardous charge on the plant is prevented, but the dust itself may retain some charge which may then become transferred to operatives. Although the resistance to earth of operatives can be minimized, e.g. by the use of special footwear, it is still high compared with the resistance of a plant properly bonded to earth. There is thus a risk that a spark would pass from a charged operative to earth, in the presence of dust and under these conditions the minimum ignition energy of the dust is an important criterion. The maximum spark energy that could be obtained from a charged operative is indeterminate, but a rule of thumb value of 25 millijoules is often taken. It depends on the electrical capacity of the operative and the resistances of the charging and earthing circuits. This value should not be regarded as an absolute maximum, but it is probable that higher values could only be obtained under very unfavourable circumstances. It is based on a generous value of 5×10^{-4} μF for the electrical capacity of an operative [8], charged to a potential of 10^4 volts. A dust which has a minimum ignition energy of less than 25 millijoules should thus be regarded as prone to ignition by static electricity, even though the plant in which it is handled is earthed. If the minimum ignition energy of the dust is below 10 millijoules special precautions should be taken in its handling, such as requiring special clothing and footwear for operatives, Chapter 7. Dusts with minimum ignition energies above 25 millijoules should not be regarded as safe from ignition by static electricity and the usual precautions on plant should be taken. No dust is likely to be safe from the static electricity hazard if handled in an unearthed metal plant.

If the dust is hot when in the plant, and particularly if it is at temperatures over 100°C, precautions against static electricity (Chapter 7) should be intensified because minimum ignition energies decrease at higher temperatures, and resistance to earth is likely to be higher because of the absence of water. Usually the safety factor arising from reaction of dust with oxygen in air, which increases minimum ignition energies, will not apply because oxidation would lead to deterioration of the dust. Working temperatures are likely to be kept lower than that at which oxidation proceeds rapidly, and higher working temperatures should be regarded as increasing the hazard from static electricity.

With dust layers the minimum ignition energy is again applied mainly to the problem of ignition by static electricity. Similar criteria to those for dust suspensions should be applied in assessing whether a hazard is likely from charged operatives.

The minimum ignition energy, measured by means of electric spark discharges cannot be applied to the problem of ignition of dusts by friction sparks. The two forms of spark are quite different in character, a friction spark being basically incandescent particles expelled from a solid surface. The energy of such particles is supplied mechanically by the impact or rubbing, and sometimes by chemical reaction between the displaced particle and oxygen in the air. As the size of the particle cannot be controlled its energy cannot be measured before it enters a dust suspension and causes ignition. Where information is required on the ignition of dusts by friction sparks, the tests in which friction is the source of the sparks should be used. These test are basically non-quantitative, but a material which was particularly susceptible to ignition by friction sparks would become evident. All dusts should be regarded as prone to ignition when exposed to badly maintained machinery, or when they become contaminated with tramp metal, flints etc.

MAXIMUM EXPLOSION PRESSURE AND RATE OF PRESSURE RISE

Values of maximum explosion pressure are required if the plant is to be sufficiently strong to withstand the full explosion pressure. In most applications this approach is not practicable, and weak plant is frequently protected by explosion relief venting. The maximum rate of pressure rise is then required, and it is also useful if automatic explosion suppression is to be installed (Chapter 8).

The apparatus in which maximum pressures are measured are closed vessels, so that no gas or dust escapes during the measurement. For this reason the test does not give direct information on the venting requirements for the dust, and these have to be assessed subsequently. This point is considered in more detail later in Chapter 8.

Experience with the test indicates that the results can be applied directly to full scale plant. The principal sources of uncertainty are cooling by the walls of the test vessel. thereby reducing maximum pressure, and the effect of vessel volume on the maximum rate of pressure rise. In the highly turbulent suspension generated in the test, burning dust is likely to be present throughout the whole volume. This appears to counteract any loss of pressure due to wall cooling, as compared with larger vessels when burning throughout the whole volume is more difficult to obtain. The relationship between the maximum rate of pressure rise and vessel volume, with the highly turbulent suspension, is considered further under explosion relief venting.

It may readily be shown that for a non-turbulent suspension in a closed spherical vessel, and with a flame propagating at a constant velocity from the centre, the maximum rate of pressure rise is inversely proportional to the diameter of the vessel. In turbulent conditions the proportionality breaks down and should not be used indiscriminately when being applied to industrial plant, see also Chapter 8.

The maximum rates of pressure rise, as measured in the test, vary considerably between different dusts (Appendix 2). The higher the maximum rate the more severe is the explosion hazard and it can be used as a guide in assessing risk with new materials. If the maximum rate is below 35 000 kN/m^2s (5000 lb/in^2s) the dust may be regarded as behaving similarly to many carbonaceous and agricultural dusts, whereas if the maximum rate is above 70 000 kN/m^2s (10 000 lb/in^2s) the explosion risk is as severe as with reactive metals such as magnesium and aluminium (Chapter 8).

Both the maximum explosion pressure and the maximum rate of rise are required for the calculation of the Index of Explosibility. When the product is divided by the corresponding product for Pittsburgh coal dust, the Explosion Severity ratio is obtained.

The average rates of pressure rise are sometimes also used as a rough guide to the explosion hazard of the dust. There is usually an approximate correlation between the average and maximum rates of pressure rise for a dust, and so either may in fact be a basis for venting calculations, although on theoretical grounds the use of the maximum rates is preferable. The average rate of pressure rise would be influenced by a slow development of the explosion, whereas the maximum rate, and the venting requirements, are more concerned with the fully developed explosion. Usually the average rates are one-half to one-third the values for the maximum rates. Numerous exceptions occur and a check should be made for the particular dust if it is desired to use values for the average rather than the maximum rate of pressure rise. The limitations listed above for the interpretation of maximum rates also apply to that for average rates of pressure rise.

Pressure measurements obtained with carbonization gas/air mixtures, and not with dust suspensions, require a different interpretation. The errors involved in the method of test are likely to be greater than with dust suspensions, as discussed above, and the explosions are more akin to gas flames than to those in dust suspensions. In applying the results of the tests with carbonization gas it is preferable to use the same criteria as for combustible gases and vapours. For the venting of the explosions two comprehensive guides are available [10, 11]. The properties of the carbonization gas can be related to those for common gases and vapours by comparing the relative maximum and average rates of pressure rise. In this way explosion venting data for gases and vapours of the appropriate explosion pressure characteristics may be used, bearing in mind that the maximum and average rates of pressure rise in gas and vapour explosions are inversely proportional to the linear dimension of the explosion test vessel.

IGNITION BY FRICTION SPARKS

Interpretation of the results of tests in which dust suspensions or layers are exposed to friction sparks is essentially qualitative, but has important industrial applications if the dusts are easily ignited. Under these circumstances particular care should be taken in the design of industrial plant to prevent the generation of friction sparks, the hazard being greater with impacts which are produced by mechanical means rather than by manual action. Precautionary measures, which may be taken with industrial plant to prevent generation of friction sparks, are considered in Chapter 7. In the tests the sparks are produced by the action of stone on steel, as these materials are most likely to be present in industrial plant generally. The hazard from friction sparks is greater if light metals, such as aluminium or magnesium are involved in the frictional process, because their sparks are more incendive.

The results of tests involving friction sparks cannot be used to assess the hazard arising from electric sparks, either electrostatic or from the mains supply. In considering the explosion hazard arising from sparks, the appropriate tests for the particular type of spark must be used.

References

[1] Palmer, K. N. and Tonkin, P. S. (1968), Third Symposium on Chemical Process Hazards with special reference to plant design. Inst. of Chemical Engineers, 25. London, 66–75.

[2] Palmer, K. N. and Tonkin, P. S. (1971), *Comb. and Flame*, **17**, 159–70.
[3] Jacobson, M., Nagy, J. and Cooper, A. R. (1962), US Bureau of Mines RI 5971.
[4] Dorsett, H. G., Jacobson, M., Nagy, J. and Williams, R. P. (1960), US Bureau of Mines RI 5624.
[5] Selle, H. and Zehr, J. (1957), *VDI Berichte*, **19**, 73–87.
[6] Craven, A. D. and Foster, M. G. (1967), *Comb. and Flame*, **11**, 408–14.
[7] Eggleston, L. A. and Pryor, A. J. (1967), *Fire Tech.*, **3**, 2, 77–89.
[8] Boyle, A. R. and Llewellyn, F. J. (1950), *J. Soc. Chem. Ind.*, **69**, 173–81.
[9] Carpenter, D. L. and Davies, D. R. (1958), *Comb. and Flame*, **2**, 35–54.
[10] National Fire Protection Association. Guide for Explosion Venting No. 68 Boston, Mass. (1954).
[11] Ministry of Labour. Guide to the use of flame arrestors and explosion reliefs. Safety, Health and Welfare New Series No. 34, (1965), HMSO, London.

5 *Experimental Data on the Propagation of Dust Flames*

Introduction

Numerous factors govern the initiation and subsequent propagation of dust flames; they influence all stages of an explosion, from the formation of the dust cloud before ignition to the completion of the combustion. Study of the factors involved, and their interaction, is necessary if a theoretical basis is to be established which will enable the explosion processes to be thoroughly understood. Knowledge of these processes could then be applied to practical situations, ensuring that the necessary steps for the protection of plant are taken with the maximum economy and reducing the extent to which *ad hoc* measures are taken.

There have not yet been comprehensive investigations in which all the various factors have been studied, and their interactions measured in a controlled series of experiments. Such work would involve considerable difficulties of experimentation arising from the number of factors and their control and measurement. Many investigations have been made of individual factors affecting flame propagation and much useful information is available, but because the investigations have been on an individual basis, the resulting picture is fragmentary. The position is sufficiently confused for generalization to be made only with difficulty; this confusion has undoubtedly hindered the development of unifying ideas and theories of dust flame propagation, the need for which is acute.

In this Chapter a survey of available information is given, selected from the large number of investigations involved.

The experimental evidence is considered under the following headings:

Formation of dust suspensions
Influence of ignition source

Dust composition
Particle diameter and specific surface
Dust concentration
Ambient temperature
Ambient pressure
Presence of inert gas
Presence of diluent dust
Presence of water
Presence of flammable gas
Turbulence
Detonation

The evidence provides data on the propagation of dust flames, and is complementary to theoretical considerations of the mechanism of propagation discussed in Chapter 6.

Formation of dust suspensions

By definition, before a dust explosion can occur, a suspension must be present. In industry the dust could be dispersed either before or after it was ignited, and the formation of the suspension may be either by independent mechanical action or by an earlier, smaller, explosion.

In industries other than coal mining, dust suspensions will in any case be generated by independent mechanical action. such as handling or transportation of the dust. Secondary explosions can occur when deposits on horizontal surfaces etc., are disturbed and ignited by the primary explosion. The raising of dusts into suspension, in industries other than coal mining, has not been studied in detail in relation to dust explosions and little information is available.

In the special circustances of coal mining the raising of dust suspensions during an explosion is of considerable importance. The usual method of preventing dust explosions in mines is to introduce deposits of non-combustible dust, which become mixed with coal from the atmosphere during normal working. If dispersion of the deposits should occur, the non-combustible dust will be raised as well as the coal. Studies of the dispersion of dust layers, in a wind tunnel, have shown two mechanisms [1, 2]. The dust is either eroded, i.e. removed particle by particle, or denuded, i.e. the entire dust layer is removed suddenly. The mode and ease of the dispersion depends on the properties of the dust and of the underlying surface. The diameter, density and cohesion of the dust particles are important, and the properties of the underlying surface such as its smoothness and whether it is vibrating also affect dispersion. A dust layer that is removed by erosion is likely to be more uniformly dispersed than one that is denuded, because subsequent break-up of the denuded suspension into individual particles could require substantial energy. This aspect has not been studied to the same extent as the initial dispersion of the dust layers.

The initial stages of the dispersion have been studied by a shock tube method [3] in which the dust was dispersed from a horizontal surface. The type of dust and the thickness of the deposit were varied, together with the Mach number of the shock wave. Measurement of the height of the dust cloud above the surface, as a function of the distance from the origin of the cloud, showed that for a given dust the cloud profile was of constant shape and was independent of the thickness of the dust deposit and the strength of the shock wave.

A comprehensive investigation of the dispersion of coal and stone dust deposits in a wind tunnel has shown that the dust pile geometry and pulsations of the airstream also influence dispersion [4]. Relative humidities above 75 per cent, and additives to the dust, cause significant effects.

The formation of dust suspensions has been investigated in greater detail from the point of view of obtaining uniform suspensions, with which to undertake subsequent experimental investigations. A review [5] of the methods up to 1959 showed that the available techniques for the dispersion of dust for experimental purposes could be divided into three groups.

(1) Dispersion by an air-stream passing over or through a bed of dust or by direction of a jet of air at a dust deposit. (Fig. 5.1)

(2) Dispersion by mechanical disturbance or stirring of the dust within a vessel. (Fig. 5.2)

(3) Dispersion by mechanical feeding of dust into still air, or into a metered air-stream. (Fig. 5.3)

Subsequently there has been no widespread introduction of new methods, although existing methods have been refined and the relative frequency of use of the methods has changed.

Method 1 has been widely used in routine test methods (see Chapter 3). Its main advantage is simplicity and for many applications adequate dispersion is obtained. The disadvantages of the method are that no steady state may be obtained, which would enable the properties of the dust suspension to be measured before the explosion is initiated,

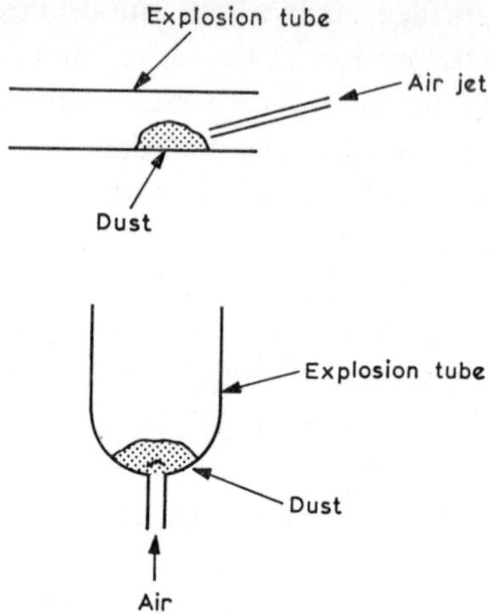

Fig. 5.1 Dispersion of dust by airstream.

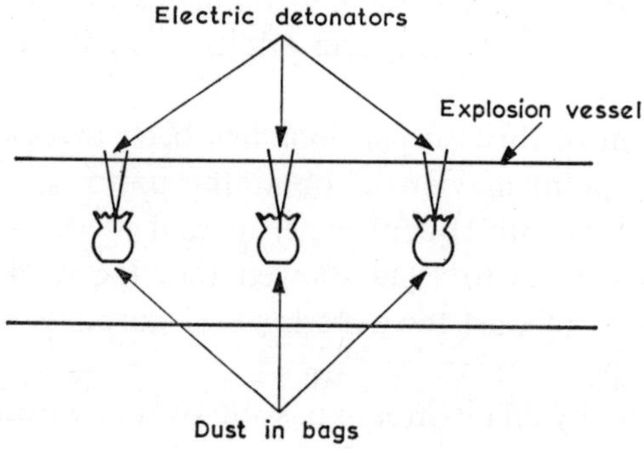

Fig. 5.2 Dispersion of dust by mechanical disturbance.

and that a turbulent suspension is almost inevitable. Because the energy on dispersion is adjustable, a separation of the suspension into individual particles is possible, at least theoretically. The actual dispersion may be difficult to assess.

Method 2 is also used in routine test methods (Chapter 3), particularly where a closed vessel is involved. The method has also been applied to open vessels [6, 7, 8] which were relatively small and for which the method is convenient. Where mechanical stirring is applied, attrition of the dust particles may result in change in the particle size distribution. With mechanical stirring in a closed system a steady state

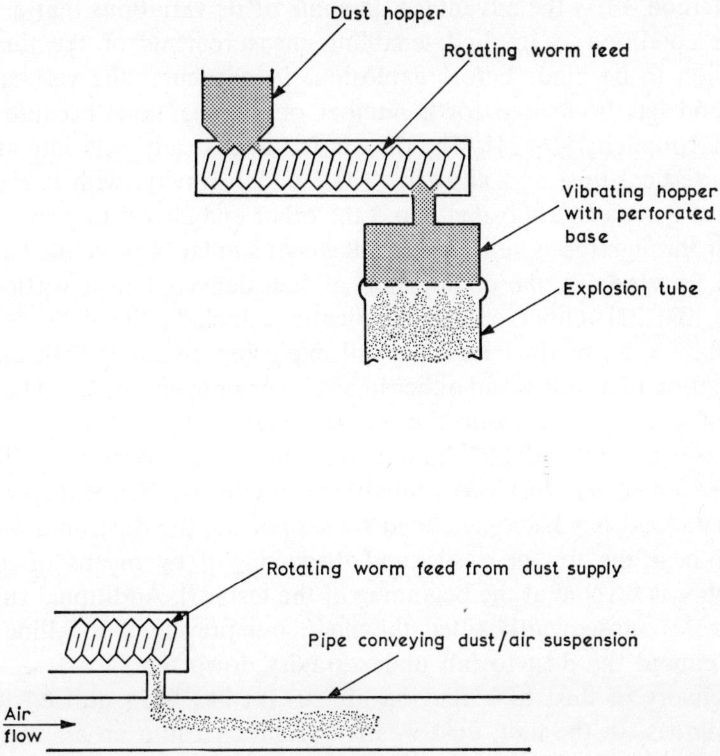

Fig. 5.3 Dispersion of dust by continuous feed.

condition is obtainable, apart from attrition of the dust particles, permitting measurement of the suspension properties, uniformity of suspension may not however be obtained [9]. In coal mine gallery experiments it is frequent practice for the ignition source to give sufficient energy to raise dust in its vicinity and to commence the propagation of an explosion. The energy released by the explosion then continues to raise further dust [10, 11, 12]. An alternative means of dispersion, particularly useful for experiments in large scale industrial plant, is by the use of detonators exploded inside bags containing the dust [13, 14]. This method gives rapid and efficient dispersion, but without steady state conditions. In all the variations of method 2, turbulence of the dust suspension is usual, and may be of high intensity, also the extent to which the suspension consists of individual particles and not clusters is not usually assessed because of experimental difficulties.

Method 3 has the advantage, in some of its variations that a steady state condition is implicit, enabling measurements of the dust suspension to be made before explosion. In particular the vertical tube method has been used for a number of investigations because of its basic simplicity [16–21]. The dust is fed at a steady rate into the top of a vertical tube and allowed to fall under gravity, with one end of the tube open to atmosphere and the other end closed to prevent flow of air through the system. An arrangement similar in principle has been used for studying the combustion of fuel delivered to a vertical furnace [22, 23]. Other similar applications include the delivery, at a constant rate, of dust into a small explosion chamber [24], and the formation of a dust cloud either in a tube or in open air [25]. The blowing of a mass of dust into the top of a vertical tube, down which the dust consequently fell [26] did not give a steady state condition although dispersion of the dust was probably very effective. On a larger scale, a dust cloud has been generated by supporting the dust on a wooden shelf near the top of a silo and dispersing it by means of a cord which was severed at the beginning of the test [27]. Additional supporting wires subsequently tilted the shelf, but prevented it falling over, and caused the dust to fall under gravity down the silo.

Delivery of dust to a moving air stream has been utilized in two techniques. In the first, dust was fed from a jet into an air blast and the subsequent cloud explored with a flame for signs of flammability [28]. Alternatively, and more usually, the dust suspension has been delivered to a burner, frequently equipped with a stabilizing pilot flame for measurement of flame velocity or investigation of flame structure [29–35]. In some applications of method 3, particularly the vertical tube method, turbulence of the dust suspension is controllable to some extent, and under suitable conditions may be prevented.

The extent to which the suspension consists of individual particles, rather than clusters, would vary between dusts because relatively little energy is put into the separation of the particles. However, because of the steady state condition, samples can be taken from the dust suspension and the efficiency of dispersion measured, even though it may not be easily controlled.

The introduction of dust at a measured rate into a metered air stream is particularly suitable for experiments in which dust flames are stabilized on burners. For this purpose either the dust-laden air stream can be passed directly to the burner, or a portion of the stream may be tapped off to the burner. A further advantage is that sometimes, usually with dilute suspensions, the tracks of individual particles may be observed and this is difficult with most alternative methods. In using a metered dust suspension care should be taken to ensure that the concentration is calculated or measured at the appropriate part of the system, due to differential movement of the dust and air. For instance, erroneous values may be obtained if the concentration is measured for the suspension in a pipe, but the explosion experiment takes place in a vessel of different diameter fed by the pipe.

The methods selected for the generating of dust suspensions will depend upon the type of experiment to be carried out, the quantity of the suspension required, the time for which a steady state is required, and the properties of the dust itself. Many dusts disperse only with difficulty, and the energy and time required to separate the particles may control the method of dispersion selected. In some extreme cases, dusts collected from an industrial plant cannot be redispersed because of matting or cohesion of the particles, and in order to undertake experiments it may be necessary to actually generate the dust from the parent material in the experimental apparatus, rather than attempt to redisperse dust formed previously. The dispersability of a dust cannot be predicted with confidence and direct measurements are necessary.

Ignition source

The influence of the ignition source on subsequent dust flame propagation is conveniently dealt with under the following headings.

(1) Efficiency of dust dispersion.

(2) Existence of minimum ignition energy, or minimum ignition temperature

(3) Ignition by large sources

(4) Ignition by explosives

The efficiency of dispersion is an important factor in situations where the dust is not in suspension when the ignition source is initially present. As explained previously under most industrial conditions a suspension is liable to be formed in advance of the presence of an ignition source due to mechanical movement of the dust. The disturbance caused by the ignition source may be then only local and a serious effect on the extent of dispersion would not be produced. Where the dust is initially at rest the extent to which the ignition source can raise a suspension may be an essential factor in the initiation of an explosion. Formation of the suspension depends upon the size of the igniting source and, if it is a gas mixture, upon its concentration and uniformity. A layer of combustible gas is less effective than a mixed volume, and if the expansion of gas in the ignition source can be relieved in directions other than towards the dust deposit the effectiveness is reduced [36].

Although the effect of the explosion properties of the gas on the dispersion of dust has been investigated for mine conditions, a detailed analysis is not available. It is not at present possible to calculate the extent of dust dispersion, as a function of the properties of the gas explosion.

For an explosion to propagate, the ignition source must have more than a minimum energy or temperature. For ignition by electric spark or heated surface routine tests are available, and discussion of them has been given in Chapters 3 and 4. The values obtained are for dust suspensions in air, as is usually the case in practice. Sometimes other gases are involved and ignition temperatures and energies are then quite different. For instance, the minimum ignition temperature of coal dust in oxygen may be as much as 300°C lower than in air, (Fig. 5.4), depending upon the volatile content of the coal [37]. Similar considerations apply to other ignition sources; the ease of ignition of

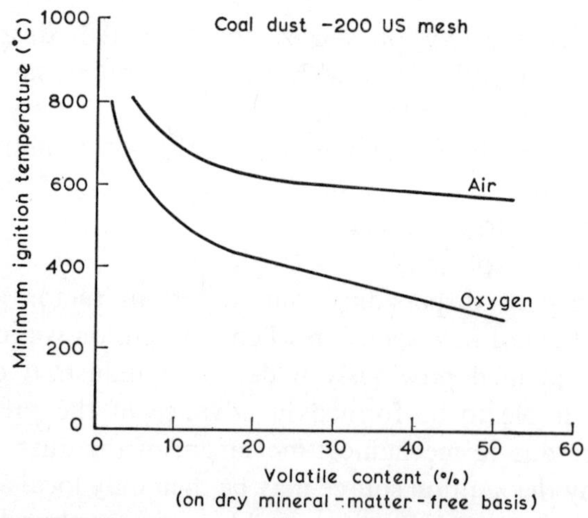

Fig. 5.4 Ignition temperatures of coal dust in air and oxygen using Furnace apparatus [37].

dusts must always be considered in relation to the gas in which they are suspended. With electric spark ignition, the minimum energy may depend on the resistance and inductance of the electric circuit [38]. Using lycopodium dust there were optimum values for the electrical properties of the circuit and also the minimum energy was smaller for a dust suspension in a tube than for the same dust falling freely.

A flame, whether of dust or gas, is an effective method of initiating explosions. A minimum volume, temperature, and duration of the flame is probably necessary for the successful initiation of an explosion but an adequate theory has not yet been developed. In addition there is no recognized routine test for their measurement. The most detailed information is in relation to the ignition of coal dust explosions by small transient gas jets [39]. The practical problem which led to these investigations is specialized, being concerned with ignition by flames or hot gases during the firing of explosives in coal mines. These conditions differ in rate of energy release from the general case in industry where a suspension might be exposed for a short period to a small flame arising from the failure of plant. Conversely, large flames also have an effect on dust explosion propagation; this point is considered in more detail below.

The ignition of dust clouds by friction or impact sparks is a process about which relatively little is known. A test has been described (Chapter 3) which illustrates whether the dust can be ignited, but no measurements are made of the energies of the sparks. It is likely that a minimum frictional spark energy for ignition of dusts exists, but values are not available. Ignition of grain dust by the impact of magnesium alloy on rusted steel has been demonstrated [40], but insufficient experiments were done to establish the minimum impact energy required. No attempt was made to measure the energies of individual sparks. By analogy with electrical sparking, the influence of friction sparks on the subsequent flame propagation in dusts is likely to be small, once ignition has been established.

Certain dusts are pyrophoric, that is they ignite on dispersion without requiring the presence of an ignition source. Pyrophoric dusts thus have zero ignition energy. These dusts are usually reactive metals, such as zirconium, uranium and Raney nickel [41, 42], or some of their compounds such as hydrides, nitrides, and carbides. For pyrophoric behaviour to occur the dust must be of small particle size, often about 1 micron diameter. On dispersing, a cloud of pyrophoric dust is likely to ignite simultaneously at a number of points and, as the dusts are reactive materials, flame propagation is usually rapid. Theoretically, if the dust could be uniformly and rapidly dispersed, ignition should occur throughout the cloud simultaneously and the rate of flame propagation would then be infinite. However because of variations in cloud density simultaneous ignition is not obtained, and a flame propagates rapidly through the cloud.

If a dust cloud is ignited by a large source, the subsequent propagation of the explosion can be affected by the energy of the source. The process has been demonstrated in experiments using coal dust, in a mine roadway, a gallery, and in a pulverizing mill. In the roadway tests [36] one end of the roadway was filled with a natural gas/air mixture extending over distances of 25 or 50 ft. The volumes of gas mixture were 1300 and 2600 ft^3 respectively. The length of flame developed in exploding the gas, in the absence of coal dust, is shown in Fig. 5.5; the flame could extend up to seven times the length of the zone of the original gas, depending on the composition of the mixture. The effect of the igniting source on subsequent coal dust flame propagation is illustrated in Table 5.1, where the percentage of inert dust needed to be mixed with the coal to prevent explosion increased with the energy of the ignition source. The increased severity of the

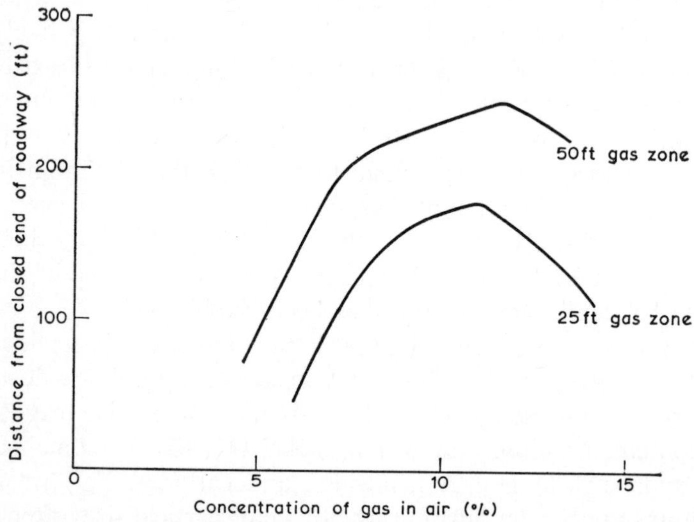

Fig. 5.5 Length of natural gas/air flames in an experimental roadway [36].

conditions with more powerful ignition sources may have led to increased uniformity of dust dispersion. In the gallery tests [43] pyrotechnic igniters were used for explosions with coal dust/stone dust mixtures. Two types of explosion were obtained, weak explosions in which the pressure rises due to the flame were less than those due to the ignition and strong explosions in which the pressures generated by the flame were eventually greater.

Table 5.1 Diluent dust requirement to prevent explosion in experimental coal mine as function of natural gas/air ignition source [36]

Percentage gas in air	*Length of gas mixture zone ft*	*Minimum percentage diluent dust to prevent coal dust explosion*
6.0	25	40
9.5	25	60
9.5	30	65

The procedure with the coal pulverizing mill was to disperse the dust before ignition by means of detonators [15]. Alternative ignition sources were then used, either a further two detonators or the introduction of a dust explosion from an external auxiliary vessel into the mill. The volume of the external ignition source was 110 ft^3 and that of the mill was 183 ft^3. The mill was protected by means of explosion relief vents, and measurements were made of maximum explosion pressures and maximum rates of pressure rise. Considerably higher pressures and rates of rise were obtained with external ignition (Table 5.2) even though the area of vent was four times that with the smaller internal source. The higher pressures could be attributed to the larger flame which ignited the dust in the mill.

Table 5.2 Vented explosions in coal mill, with two sizes of ignition source [15]

Ignition source	*Vent ratio $ft^2/100\ ft^3$*	*Maximum explosion pressure lb/in^2*	*Maximum rate of pressure rise $lb/in^2.s$*
Small, internal (two detonators)	0.84	3.1	14
Large, external (dust explosion in attached vessel)	3.2	8	290

The initiation of a dust explosion by explosives is likely to arise in coal mines. Gun powder has been used, in conjunction with coal dust, as an igniting source for tests in which the amount of diluent dust (limestone or clay/slate) or of water needed to be added to coal dust to prevent propagation of explosion was measured [44]. The required percentage of diluent dust or water increased with the strength of the igniting source, as measured by its heat output (Fig. 5.6). The heat output was varied by increasing the weight of coal dust in the ignition source from nil to 75 kg, the weight of gun powder being constant at 0.75 kg. The gun powder was placed at the closed end of the mine roadway, and the coal in the ignition source was spread out along the roadway over distances between 4 and 30 m depending on the quantity.

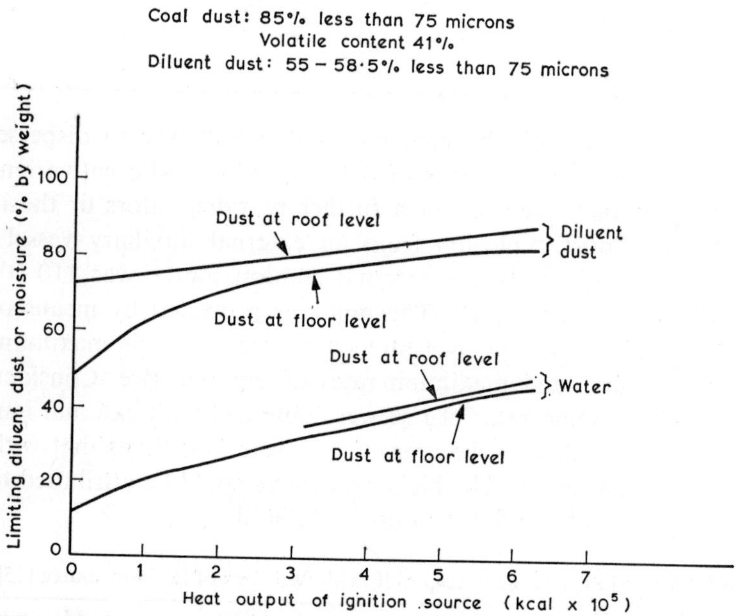

Fig. 5.6 Limiting diluent dust or water content to prevent explosion in coal dust in a mine roadway [44].

The moistened or diluted coal dust was distributed along the remainder of the roadway either on the floor or on an overhead structure. It is unlikely that all the thermal energy was effective, because an igniting source using 0.75 kg gunpowder and 50 m³ of a 9 per cent methane/air mixture gave conditions comparable to a source of potentially ten times the energy using gunpowder/coal. As with other large sources such as flames, in experiments with mine roadways the extent to which the energy of the source accelerated the combustion and the extent to which it increased the dispersion of dust, thereby promoting flame propagation, could not be separated. Some further consideration is given in Chapter 6.

Finally, the propagation of flame in a dust cloud is influenced by the type of ignition source as well as its energy. In the testing of dusts for explosibility [45] both electric spark and heated wire coil igniters may be used in the same apparatus. In general, dusts of organic or vegetable origin are ignited more readily with a heated coil, whereas metal dusts tend to ignite more readily with an electric spark. With coal dust, and electric spark igniter, explosion was not obtained in routine tests when the volatile content of the coal was below 20 per cent [46], although coal of lower volatile content could be ignited by a heated coil.

Dust composition

The chemical composition of the dust particles has a fundamental influence on the propagation of flame in an explosion. For an explosion to propagate there must be a chemical reaction between the dust particles and the oxygen in the air, in addition to any chemical reaction between constituents of the particles themselves, and hence the dust must be combustible. However not all combustible dusts can cause a dust explosion. The heats of combustion of a representative selection of dusts, together with those of the stoichiometric mixtures in air are given in Table 5.3, together with the explosibility Classification (Chapter 3). There is no simple correlation between the heat released by a stoichiometric mixture of the dust in air and the explosibility Classification, i.e. whether flame propagation can continue away from the source of ignition through a suspension on the dust in air at atmospheric temperature. The characteristics of flame propagation, such as the flame velocity and the pressure developed, are found in practice also to depend upon the chemical composition of the dust but the dependence in the general case cannot at present be put on a quantitative basis.

By taking a series of dusts of related chemical compositions a gradation of explosibility can be measured, as has been done with coal dust, for which there is a dependence of the explosion properties on

Table 5.3 Heats of combustion and explosibility classifications of dust/air dispersions

Dust	*Heat of combustion (approximate) cal/g*	*Heat of combustion of stoichiometric mixture in air (approximate) cal/l*	*Explosibility classification*
Cellulose	4 200	1 060	Group (a)
Sugar (sucrose)	3 950	1 060	Group (a)
Polyethylene	11 200	980	Group (a)
Polyvinylchloride	5 150	1 210	Group (a)
Urea	2 550	960	Group (b)
Coal (35 per cent volatiles)	8 500	950	Group (a)
Anthracite (5 per cent volatiles)	8 500	940	Group (b)
Aluminium	7 400	2 500	Group (a)
Sulphur	2 200	660	Group (a)

the volatile content of the coal. This was shown clearly in full scale mine gallery tests in which the minimum percentage of diluent dust needed to prevent explosion when mixed with the coal was determined for various volatile contents. The diluent dust percentage included the moisture and ash content of the coal; the volatile content was calculated for the coals on a dry, ash-free, basis. The required percentage of diluent depended markedly on the volatile content of the coal (Fig. 5.7) and there were also systematic differences between the various investigations. Differences in the properties of the coal dust, diluent

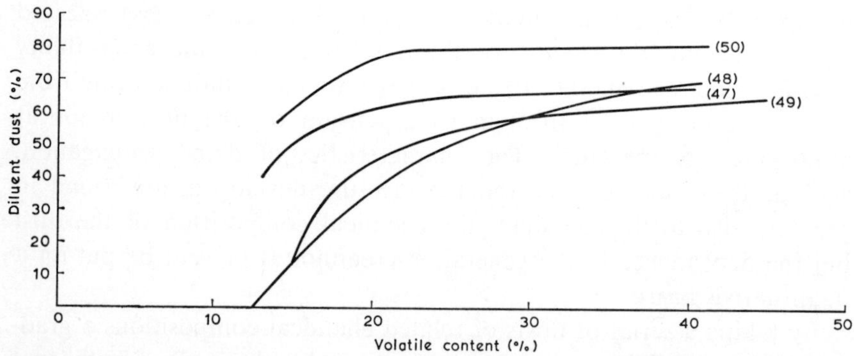

Fig. 5.7 The relation between the diluent dust requirement for explosion suppression and the volatile content of coal (see Table 5.4).

Table 5.4 Experimental conditions in Fig. 5.7

Reference	*Coal dust, particle size*	*Diluent dust, and particle size*	*Ignition source*	*Other information*
47	95 per cent through 63 micron sieve	Boiler flue dust, 95 per cent through 63 micron sieve	240 g gelignite with 11.5 Kg coal dust	Mine gallery 325 m long and 2.8 m^2 cross-section
48	85 per cent through 63 micron sieve	Fullers earth, 85 per cent through 63 micron sieve	20 oz. gunpowder with 1 lb coal dust	Mine gallery 325 ft long and 4 ft diameter
49	20 per cent through 74 micron sieve	Fullers earth, 33 per cent through 74 micron sieve	4 lb black powder with 100 lb coal dust	Two parallel inter-connected mine galleries, each 350 ft long and 55 ft^2 cross-section
50	85 per cent below 75 microns	Limestone, 64 per cent below 75 microns	750 g gunpowder with 75 Kg coal dust	Mine roadway 200 m long and 5 m^2 cross-section

dust, and the strength of the igniting source (Table 5.4) made the results of the various investigations not strictly comparable.

The results of the investigations had certain broad features in common; there was a volatile content of coal below which explosion did not take place, and at high values of volatile content the diluent dust requirement changed little. In one investigation [48] the following empirical relationship was established between the percentage volatile content (V) and the minimum percentage (S) of diluent dust (Fullers earth) required to prevent explosion

$$S = 100 - \frac{1250}{V} \tag{5.1}$$

The reason for the dependence of diluent dust requirement on volatile content is discussed in Chapter 6.

For values of (V) of 12.5, and less, the diluent dust requirement (S) was nil, because the coal dusts of volatile content less than 12.5 per cent were non-explosible. Comparable empirical relationships for the other results represented in Fig. 5.7 were not obtained but there was again experimental evidence that coals of low volatile content would be non-explosible. Even in experiments with an intense ignition source [50], coals of volatile content up to 11.5 per cent did not propagate explosion, but the minimum volatile content at which explosion did occur was not established with accuracy.

One instance has been reported [51] of a coal dust of volatile content around 12.5 per cent which was able to propagate dust explosion. This coal, whose volatile content was stated to be in the range 6–14 per cent, contained a substantial proportion of methane, 40 ml being released from each gram of coal, and the methane would of course assist the flame propagation. To prevent explosion 40 per cent of slate dust was needed with the coal. This percentage was sufficiently low to indicate that even with the methane the coal dust was only moderately explosible (Fig. 5.7).

Other explosion properties are also affected by variation of the volatile content of coals. The minimum ignition temperature in air decreased regularly as the volatile content increased (Fig. 5.4), although other investigators using a similar furnace apparatus did not find the uniform variation and found that temperatures about 800°C were required for ignition [52]. Greater variation with volatile content was found for the minimum ignition energy and minimum explosible concentration [37] as shown in Fig. 5.8. The ignition source was an electric spark and flame could not be initiated at volatile contents below 26 per cent.

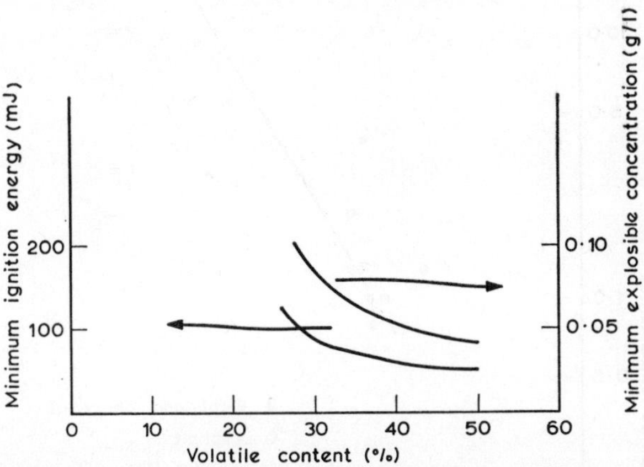

Fig. 5.8 Variation of minimum ignition energy and explosible concentration of coal dust with volatile content [37].

The variation of explosion properties with composition, fairly marked with coal, has been extended to a wider range of carbonaceous dusts. Thus an approximate relationship has been found between the Index of Explosibility (Chapter 4) and volatile content for the series of materials: charcoal, coal, lignite, asphalt and pitch [46] (Fig. 5.9). The Index of Explosibility is compounded of a number of explosion properties, and as an approximate relationship exists between the Index and volatile content, there are probably underlying technical reasons. These have not yet been put on a quantitative basis and the dependence of flame propagation on the chemical composition of dusts still needs further elucidation.

Although the volatile content of coal, to some extent, defines the composition it is not a fundamental property of the fuel. Volatile contents as quoted in the literature are determined in various standard ways, usually involving the insertion of a weighed sample of the coal in a furnace at a fixed temperature (usually in the range 800–1000°C) for a specified time, normally of the order of a few minutes [53]. The volatile content is then calculated on the basis of the weight loss after allowing for the presence of moisture and ash. The rate of temperature

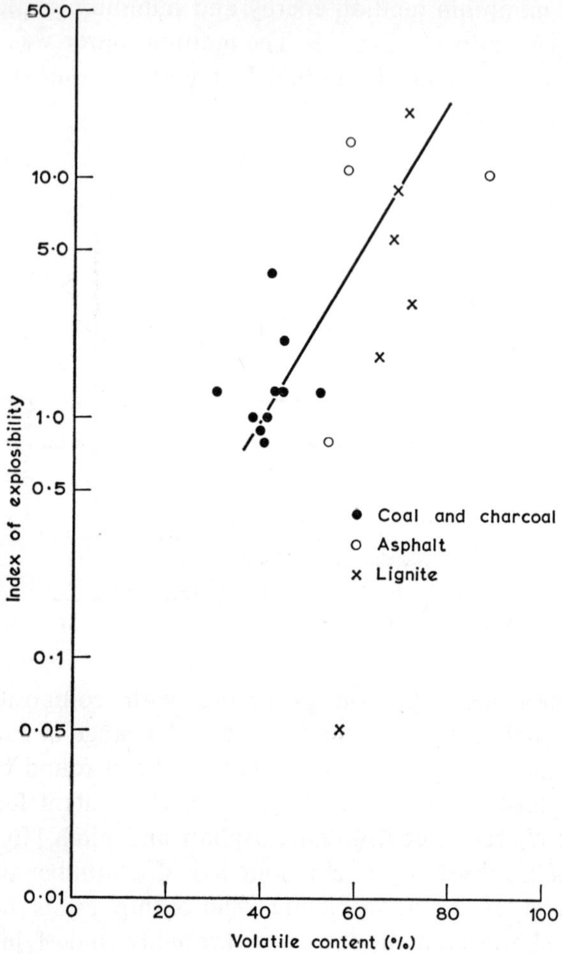

Fig. 5.9 The dependence of the Index of Explosibility on the volatile content [46].

rise of the coal after insertion in the furnace is not measured, but is likely to be relatively slow and probably would not exceed 10^2°C/s. In dust explosion flames the rates of temperature rise are likely to be far quicker and probably in the order 10^4 to 10^5°C/s. The volatile content is increased for these high rates of heating, whether these are obtained with a furnace [54] or by the use of laser or flash tube irradiation [55]. When individual coal particles were dropped through a vertical furnace in nitrogen the loss of volatiles differed for coals that swelled on heating and those that did not [56]. The following equations were obtained for heating rates of the order of 10^5°C/s to temperatures above 1000°C.

$$\text{Swelling coals} \quad W = QV \quad , \tag{5.2}$$

$$\text{Non-swelling coals} \quad W = QV(1-C), \tag{5.3}$$

where the symbols are defined at the end of the chapter.

For swelling coals Q is approximately 1.7. For non-swelling coals Q is approximately 1.5, and C is 0.14. Thus for the two coal types the total loss of volatiles would be greater by factors of 1.7 and 1.3 respectively than for slow rates of heating.

An increased production of volatiles has also been reported for the rapid heating of cellulose. On exposing cellulose sheet to high intensity thermal radiation (10^2 to 10^3 cal/cm^2 s) complete volatilization without charring appeared to take place [57]. At lower levels of intensity (11.6 cal/cm^2 s) the proportion of charred residue was again small, and was not more than 4 per cent of the original cellulose. At still lower irradiation levels the charred residue was about 20–35 per cent of the cellulose. Because the specimens were heated by radiation, a temperature gradient existed within the sample, but the rate of tmperature rise of the exposed face was likely to have been 10^2 to 10^3 °C/s. For comparison, the volatile content of wood measured at a relatively slow rate of heating (probably not more than 10^2 °C/s) was about 80 per cent indicating that about 20 per cent remained as charred residue [58]. This value is in reasonable agreement with that observed from the slow rates of heating by thermal radiation. At the high rates of heating that occur in dust explosion flames (10^4 to 10^5 °C/s) small particles of cellulosic materials are likely to volatilise completely.

Particle diameter

Most investigations of the effect of particle size on the propagation of flames in dust clouds have been concerned with particles having three approximately equal dimensions, i.e. they approximated to spheres, and the particle diameters were characterized in terms of standard sieve apertures. On this basis, there is a general tendency for the explosibility of dusts to increase as the particle sizes become smaller. The relationship between Index of Explosibility and mean particle diameter, for atomized aluminium [41], is shown in Fig. 5.10. The Index of Explosibility ranged from severe to weak explosions as the particle diameter was increased from about 10 to 120 microns. The minimum ignition energy and the minimum explosible concentration increased with particle size, as exemplified in Fig. 5.11 for corn starch [59]. The particle diameters ranged from 20 to 180 microns. The dependence of minimum explosible concentration of xanthate dust on its particle size has been investigated [8].

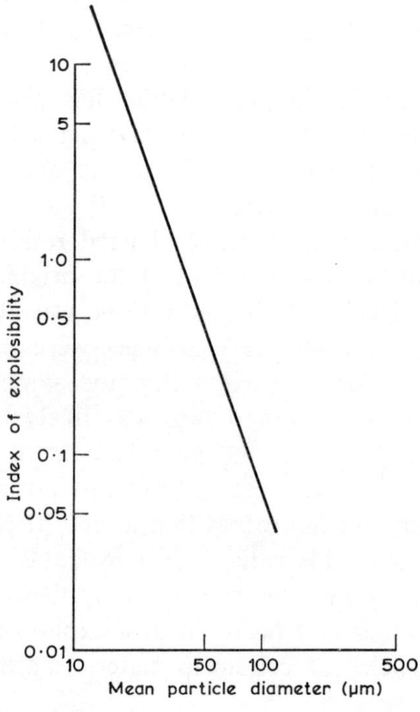

Fig. 5.10 Index of Explosibility of aluminium dust as a function of the particle diameter [41].

The particle diameters stated, of 120 and 180 microns respectively, do not necessarily imply that coarser particles of these dusts could not propogate flame and insufficient information is available for predicting the maximum particle size at which an explosion could be propagated. It would of course be expected that the maximum size would depend upon the composition of the dust and its combustion characteristics, but no detailed ruling can be given at present. From the point of view of deciding whether a dust explosion hazard is present with a particular dust of uniform size, the best procedure would be to subject the dust to routine explosibility tests (Chapter 3) if the particle diameter is less than 12 B.S. mesh.

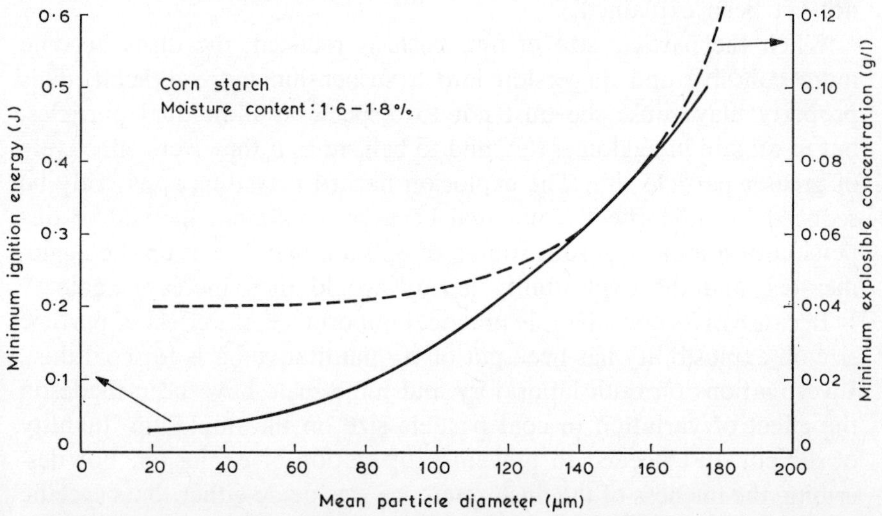

Fig. 5.11 Variation of minimum ignition energy and explosible concentration with particle diameter [59].

It is most unlikely that all particles in a given sample of dust would have a uniform diameter; even materials which are manufactured to a constant particle size are likely to suffer abrasion on movement or transportation so that a proportion of fines is generated. In a dust explosion propagating through these fines, a contribution to the combustion can be made by coarse particles, even though the latter may be of too great a diameter to sustain a flame themselves. If the proportions of fine and coarse dust are reversed, so that small amounts of coarse material are present in the fine powder, little change is caused in the explosibility from that of the fine dust alone. With aluminium pellets of diameters ranging from 590 to 2000 microns mixed with a fine atomised aluminium powder, the maximum pressure and rates of pressure rise were similar with or without the aluminium pellets [41].

With fine dusts, as the particle diameter is further reduced below 50 microns, the explosibility generally tends to increase, but not markedly (Fig. 5.11) with most materials. Exceptions have been reported for certain metals, and silicon, where very small particles showed a markedly increased explosibility. A chromium dust (100 per cent finer than 74 microns) was not ignited by an electric spark whereas a sample of mean particle diameter 3 microns was ignited and presented a moderate explosion hazard [41]. The reasons for this behaviour have not yet been explained.

When the particle size of fine dusts is reduced, the dusts become more cohesive and dispersion into a suspension more difficult. This property may cause the dust not to disperse as individual particles, but to remain in agglomerates, and to behave as if they were effectively of greater particle size. The explosion hazard may thus apparently be reduced, but the effect is not real because a different method of dispersion or a more vigorous source of ignition may break up the agglomerates, and the explosibility hazard would then increase again.

Because of its considerable practical importance, the effect of particle size on explosibility has been put on a quantitative basis for coal dust. Investigations on both laboratory and mine scale have been made on the effect of variation in coal particle size on the minimum quantity of diluent dust needed to prevent propagation of explosion. For describing the fineness of the dust, reference was made either to its specific surface, (surface area of particles per unit mass of dust) or to a mean particle diameter calculated from the dimensions of sieve meshes from which the particular dust sample was prepared. As a result of a large number of full scale mine experiments [12], the following empirical relationship was found.

$$Z = S/(100 - S) = AF + B \quad , \qquad (5.4)$$

where Z is explosibility coefficient

S is percentage of diluent dust required to prevent explosion

F is specific surface of dust (cm²/g)

A is a constant ($= 5.8 \times 10^{-4}$ for a number of coals)

B is a constant (= either $-$ 0.24 or + 0.78 depending on dispersion conditions).

In these experiments the particle size of the diluent dust was held constant. The value of B, and hence of Z, depended upon the power of the ignition source and it was pointed out that during the explosion the pure coal dust used in the ignition zone mixed with the coal diluent dust mixture under investigation. The extent of the mixing would depend upon the ignition source.

From Equation (5.4) it follows that the explosibility of the coal dust should increase progressively with the specific surface of the dust particle provided that no other effects become important. The behaviour could, however, be modified by the following:

1. Change in dust dispersability. It was confirmed that the dispersability changed with increasing fineness, and dusts containing a high percentage of particles below 10 microns had much greater cohesion than dust with a smaller proportion of this fraction. The effects of diminished dispersability were more marked in laboratory scale tests, where the dispersion forces were relatively weak, than in full scale mine tests where powerful sources of ignition were used and the blast from the explosion was of high intensity.
2. The volatile content of the coal might vary with particle size. An unexpectedly vigorous explosion in the experimental mine with a coal of nominal volatile content 11.5 per cent was attributed partly to the fine dust used having an actual volatile content of 17.3 per cent. In addition the particle size of the dust was smaller than had been intended.
3. The oxidation of the coal before its use could be more severe for fine dusts. No evidence was obtained that the explosibility of a freshly prepared dust differed markedly from that of dust which had been able to oxidise through prolonged exposure to air.

A similar relationship between explosibility and particle size had been reported earlier [10]. The explosibility was defined as $100/(100-S)$, i.e. $Z+1$, and the fineness factor F' was inversely proportional to particle diameter. F' was taken as a unity for a diameter of 63 microns and the following empirical equation was devised.

$$100/(100-S) = 1.1F'+0.7 \quad . \qquad (5.5)$$

An arbitrary value of 2 was taken for F' when the dust was finer than 63 microns and this assumption limits the use of the relationship for the fine coal. The diluent dust used in the experiments was again of constant particle size.

Variation of the diluent dust particle size has been taken into account in one empirical derivation also based on experiments in mines [49]. The explosibility coefficient Z was related to the fineness factor for the coal (F_c) and a similar factor for the diluent dust (F_i) by an equation of the form

$$Z = (aF_c+b)/F_i^n \quad , \qquad (5.6)$$

where a, b, and n were constants.

The fineness factors were based on the specific surfaces of the dusts, but as these were not of uniform diameters a method of averaging based on sieve analyses was used. The equation was in reasonable agreement with the experimental results to which it was applied.

A further empirical treatment of the dependence of explosibility on fineness and volatile content gave the following equation derived using the above notation [60].

$$Z = m(V - V_0)(F' - F'_0) \quad , \tag{5.7}$$

where V and F' are volatile content and fineness factor (based on specific surface) respectively

and m, V_0 and F'_0 were constants

The equation fitted the result for a number of coals of constant fineness, and other coals of varying fineness. The particle size of the diluent dust was not taken into account.

An alternative approach to the problem of relating explosibility on a quantitative basis to the particle size considered the explosion properties of coal alone and did not involve the addition of diluent materials [61]. The method was to measure the rate of pressure change in explosions of the coal in a closed vessel of volume 40 1 in which uniform dispersion of the dust was claimed. The explosion pressure was recorded with a spring-loaded gauge. From the pressure/time curves the maximum rates and average rates of pressure rise were calculated and an 'explosion number' (k) was obtained as the product of the maximum and average rate. The explosion number represented the ability of the explosion to raise a dust cloud as well as the maximum severity of the explosion. The explosion number was also related to the behaviour of explosions in a 200 m gallery, a high k value always corresponding to a strong explosion. It was shown that the transition from non-propagation to propagation of explosion in the gallery, occurred when $k = 7$ (units probably $atm^2.s^{-2}.10^{-3}$.). Information obtained with the 40 l vessel could thus be related quantitatively, by means of k, to behaviour in a full scale gallery. The small vessel was designed to enable experiments to be carried out in rapid succession.

The relation between dust concentration and k, for various particle sizes, was measured for a number of coals. Typical results are shown in Fig. 5.12 (volatile content of coal not stated) and the coal of smallest particle diameter (5.8 microns) gave a lower value of k than another

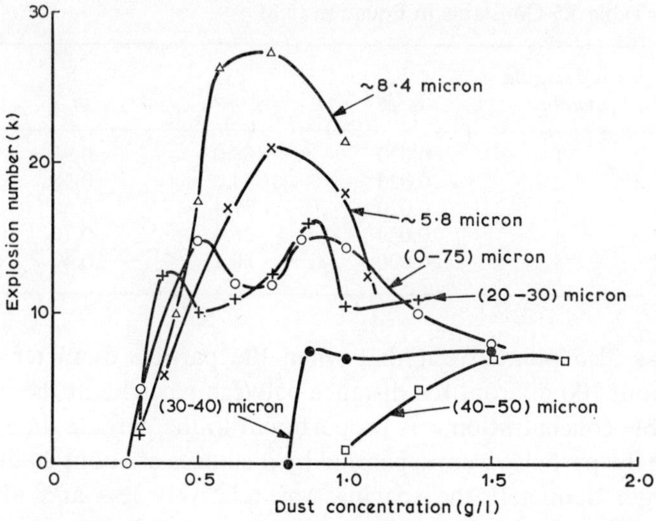

Fig. 5.12 Relation between the explosion number (k) and the coal dust concentration [61].

fraction of particle diameter about 8.4 microns. For fractions coarser than 20 microns, the value of k had decreased sharply and this characteristic continued as the particle diameter was increased further.

In another investigation involving coal dust alone, the dependence of the minimum explosible concentration on the specific surface of the coal dust particles was studied[24]. The specific surface was calculated assuming the particles were spherical, from the sieve analysis; the volatile contents of the fractions of a given coal varied by up to 5 per cent but no correction was made. The relation between the specific surface (F) and minimum explosible concentration (c) was of hyperbolic

$$(F-F_0)(c-c_0) = K \quad , \tag{5.8}$$

form where F_0 and c_0 were constants relating to the coal sample, and K was another constant. Typical values are given in Table 5.5 and indicate a limiting specific surface of 0.02 m²/g for explosion, corresponding to a particle diameter of about 300 microns. For particle diameters below about 60 microns the minimum explosible concentration varied little. This conclusion was in agreement with previous findings elsewhere.

Table 5.5 Constants in Equation (5.8)

Coal sample number	F_0 m^2/g	c_0 g/m^3	K m^{-1}
1	0.020	35.0	0.90
2	0.025	54.0	0.50
3	0.020	39.4	0.60
4	0.020	26.4	0.70
5	0.020	20.1	0.40

It was also pointed out that when the particle diameter was less than about 100 microns the distance between particles at the minimum explosible concentration was proportional to the particle diameter. On average the particles were separated by distances of about 24 diameters. For larger diameters the spacing was relatively less and ultimately, with particles greater than 150 microns, the separation distance actually decreased as the particle diameter increased.

The effect of particle size on flame propagation has also been studied in terms of the burning velocity of a flame on a burner. By means of a stabilizing ring above the burner tube, a coal dust flame was obtained which enabled the variation of burning velocity with particle size to be measured [33]. The results, combined with those of a previous investigation [31] covered a range of particle diameters from 2.6 to 120 microns for coals of volatile contents 34–36 per cent. For experimental convenience, the tests with small particle diameters were carried out with air enriched with oxygen [33]. When the variation of burning velocity with oxygen content was taken into account, values could be calculated for the burning velocity of the coal in air. At a dust concentration of 0.3 g/l in air, the burning velocity was approximately inversely proportional to the square root of the particle diameter over the range tested. The results were used in discussions on possible mechanisms of flame propagation (Chapter 6).

Little evidence is available on the effect of particle shape on flame propagation in dust. For experimental convenience particles are usually selected to have spherical, or three approximately equal dimensions. However, in practice, dusts may have one dimension considerably longer than others, as in fibres or needles, or considerably smaller, as in flakes. Occasionally the three dimensions all differ considerably, as in shredded thin sheets. Evidence with aluminium showed clearly that the maximum explosion pressure and rate of rise were lower for atomized than for flaked aluminium, at the same dust concentration. The difference could be as great as 50 per cent [41]. Secondary effects were likely to have been present because the atomized dust was stated to have an oxide coating, whereas the flaked aluminium had a protective coating of stearic acid. Other factors which could influence the effective particle surface area, include the porosity of individual particles. If a gas/solid reaction is necessary for the flame propagation, the surface area in the interior of particles could be relevant. The particle diameters of dusts are measured before they enter the flame and apart from the decrease in diameter as combustion proceeds, which would be expected, other effects occur. These include melting and vaporization, expansion to form hollow spheres (e.g. cenospheres) and fragmentation. In each case the diameter of the particle undergoing combustion would differ from that initially measured.

Dust concentration

The effect on flame propagation of variation in concentration has similar characteristics for many dusts, including the existence of lower and upper explosibility limits. The measurement of the lower limit by standard techniques, and the interpretation of results obtained, has already been considered in Chapters 3 and 4.

At the dust concentration is increased above the lower explosibility limit, keeping other conditions the same, the flame speed usually increases to a broad maximum and subsequently decreases as the upper limit is approached. Although the concentration at which the maximum flame speed is obtained cannot be measured precisely, because the peak is wide, it is usually well above the stoichiometric concentration. This general pattern of behaviour is exemplified by cork dust, dispersed in a vertical tube [16], for which dust concentrations and flame speeds were measured directly (Fig. 5.13). In addition to variation in flame speed, visible changes in the structure of the flame were also apparent. Between the lower explosibility limit concentration (0.05 g/l) and about 0.1 g/l the flames were described as relatively thin, having a thickness of about 10 cm. At higher concentrations the flame thickness was 50–100 cm. The upper explosibility limit was at a concentration of 1.8 g/l and the stoichiometric concentration was 0.15 g/l. Such a wide range of explosible concentrations is common with dust flames in air.

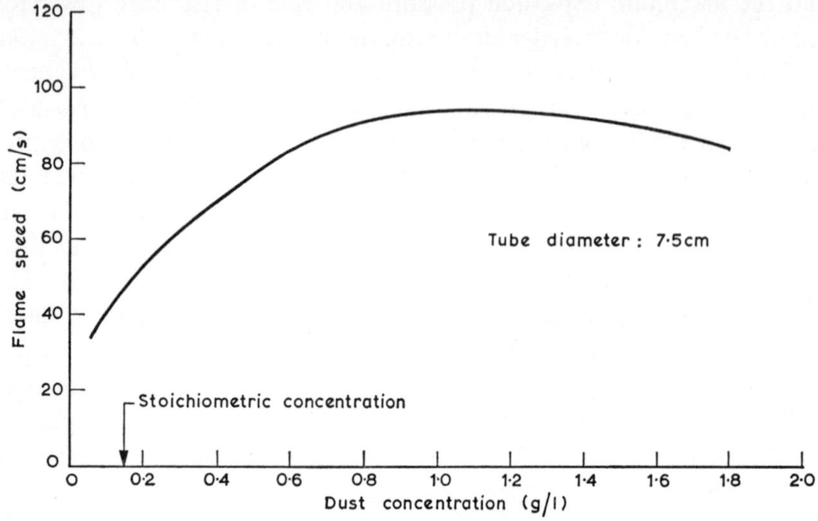

Fig. 5.13 Variation of the flame speed with the cork dust concentration [16].

The maximum explosion pressure and rate of rise also vary with dust concentration, following a similar pattern to that for flame speed. Typical results for corn starch [59] and aluminium [41] showed that a broad peak was again obtained, and vigorous explosions occurred at concentrations several times the stoichiometric. A similar pattern was obtained with bituminous coal dust [37].

The burning velocities of dust flames have been studied using burner methods; the area of the approximate cone which the flame resembles is required, and this has been obtained by calculations from photographs [33]. Some difficulty is encountered in measuring areas from photographs, because of the brush-like flames, and only approximate values can be expected. The relationship between burning velocity and dust concentrations, Fig. 5.14, included results from earlier work [31] and shows the effect of ambient oxygen concentration. Because of the variation in oxygen concentration the stoichiometric mixture varied from that in air (0.11 g/l approximately) in proportion to the increase in oxygen. The characteristic broad maximum for the burning velocity, at concentrations above the stoichiometric, was still plain. The range of dust concentrations investigated did not extend to the upper explosibility limit, and the reduction in burning velocity for concentrations approaching this limit, as would be expected by comparison with the results for cork dust (Fig. 5.13), was not found. An alternative technique for flame stabilization was to have an acetylene diffusion flame on the axis of the burner, which permitted a stable coal/air flame around it [62]. The maximum burning velocity occurred with an air/coal ratio by weight of 3, whereas the stoichiometric ratio for total combustion was 10 and for the volatiles only was 5.

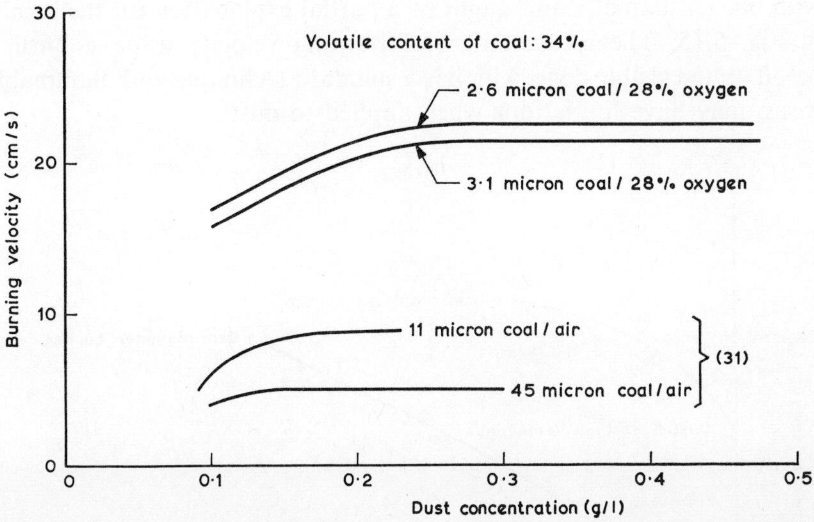

Fig. 5.14 The burning velocity of coal as a function of the dust concentration [33].

In experiments with finely divided aluminium dust on apparatus similar to bunsen burners, the effect of dust concentration was again clearly defined. The burning velocity increased steadily with concentration over the range 0.1 to 0.27 g/l, but as the stoichiometric concentration is about 0.33 g/l the maximum burning velocity was probably not obtained [34]. The dust finer than 10 microns had an appreciably higher burning velocity than that finer than 30 microns (Fig. 5.15); surprisingly the burning velocities on a burner diameter 1.9 cm were lower than with a burner diameter only 1.4 cm. The reverse would have been expected because of the reduced heat conservation and the smaller contribution of thermal radiation to the flame on a narrower burner. A possible explanation was that the flame on the smaller diameter burner was of greater curvature and relative surface area; in the calculation of the burning velocity the flame areas were estimated from photographs but as the flames were diffuse and brush-like great accuracy could not be expected. In addition the typical bunsen flame was obtained, with an inner cone on which measurements were made, although the aluminium/air mixture was leaner than stoichiometric. The outer flames on the burner must have been supported by aluminium passing through the inner conical flame, the effective concentration in which was less than that calculated. The effect could vary with burner diameter and might be a partial explanation for the results in Fig. 5.15. The measurement of burning velocity using a burner, based on the visible cone, which is a standard technique with flammable gases, may have limitations when applied to dusts.

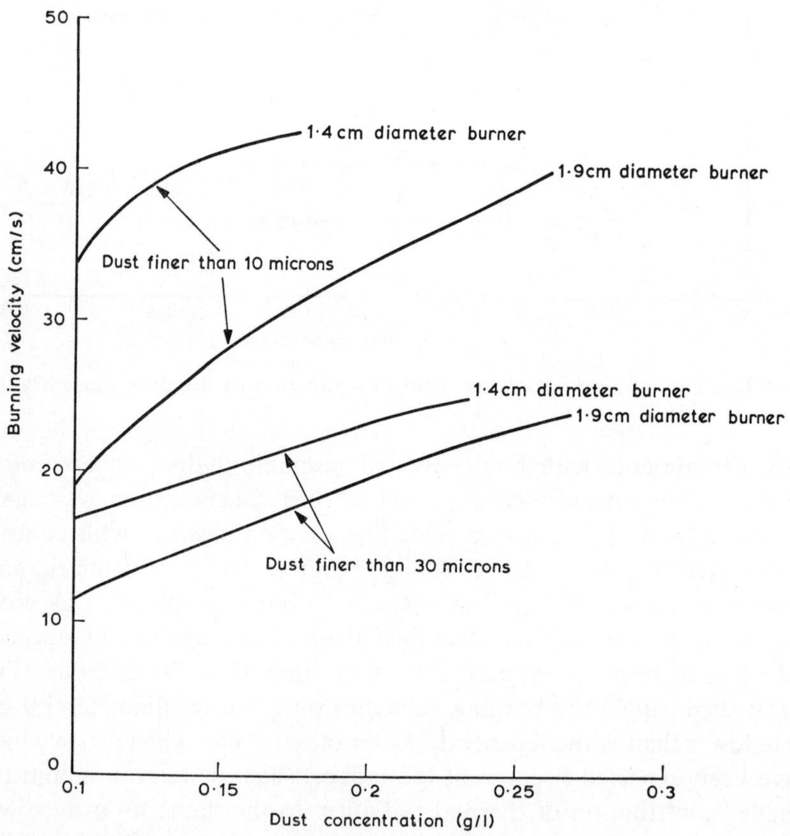

Fig. 5.15 Burning velocites of aluminium/air flames [34].

A more detailed study of a burner flame [35], using lycopodium/air mixtures, showed that the dust concentration near the tube wall was substantially lower than on the axis and that the burning velocity fell substantially towards the edge of the flame. The burning velocity was roughly constant over the central region of the flame front only, and the average for the whole flame was much less than the central value. In addition to the decrease in dust concentration, quenching effects from the walls were also found. In the lycopodium flames the concentration was always well above stoichiometric and a continuous flame front was obtained. Flames were stabilized at concentrations from 0.125 to 0.19 g/l and their visual characteristics could be changed by varying the flow rate of suspension. In this way flat or saucer shaped flames could be obtained and with the latter the blue flame front was surmounted by a conical region, reddish yellow in colour with luminous streaks. The lycopodium flames were of considerably different structure to those of aluminium [34] and generalizations as to the structure of dust flames must be made with caution.

The dependence of burning velocity or flame speed on dust concentration is sufficiently well defined for graphs to be obtained between these quantities, but to date there is no theory available to support the observed relationships, which necessarily remain empirical. Further consideration is given in the discussion on mechanisms of flame propagation in Chapter 6.

Whereas relatively few investigations have been concerned with studying the flame speed or burning velocity of dusts as a function of concentration over a substantial part of the explosible range, particular emphasis has been given to the measurement of the lower and upper explosibility limits. More attention has been given to the lower explosibility limit than to the upper limit, probably because of difficulties in the experimentation. The results of standard tests (Chapter 3) enable direct comparisons between dusts to be made. Other methods of measurement of explosibility limits have been used, not suitable for routine test work, in attempts to obtain greater accuracy. Some of these investigations will be considered here in order to emphasize particular points of interest.

Because of its industrial importance much work has been done with coal dust and the variation of minimum explosible concentration with volatile content, using a standard test procedure, has already been considered (Fig. 5.8). In an attempt to obtain steady conditions, so that the dust concentration could be measured and the explosion carried out before the suspension altered, a vertical tube was used [19] together

Table 5.6 Explosibility limits of coal dusts by various methods

Coal		*Method of measurement*	*Lower explosibility limit g/l*	*Upper explosibility limit g/l*	*Reference*
Volatile content per cent	*Particle diameter*				
36	30μ	Vertical tube, 14 cm diameter, with ignition near open end.	0.2	2.1	19
36	30μ	Vertical tube, 25 cm diameter, with ignition near open end.	0.09	—	63
36	30μ	Vertical tube, 25 cm diameter, with ignition near closed end.	0.05	—	63
50	—200 mesh	Small explosion chamber.	0.058	—	24
41.5	85 per cent less than 75μ	Mine gallery, as ref. 50 of Table 5.4.	0.041	0.68	65
Not stated	85 per cent less than 63μ	Mine gallery, as ref. 48 of Table 5.4.	0.07	—	10

with a coal of high volatile content (36 per cent) and small particle size (mean diameter 30 microns). The tube diameter was 14 cm and it was reported that flame propagation could not be obtained in a tube of diameter only 7 cm. The results were erratic and were considered on a probability basis. For a probability of propagation of 0.5, and for the flame to propagate at least 0.5 m from the lower open end of the tube the lower and upper explosibility limits were respectively about 0.2 and 2.1 g/l. Subsequent work with a 25 cm diameter [63] tube, using the same dust and a similar distance criterion for propagation, gave a value of 0.09 g/l for the lower explosibility limit. For conditions where the ignition was at the closed end of the tube so that during the explosion the dust/air mixture would be propelled along the tube and the concentration of dust would change, a lower limit of about 0.05 g/l was found, based on dust concentration before initiation of the explosion [63]. The conditions in this experiment would be comparable with those in mine gallery experiments in which the ignition source is usually near a closed end.

A comparison of the lower and upper explosibility limits for coal, for both laboratory and full-scale experiments, is shown in Table 5.6. Despite the variation in volatile content, the values of lower limits were generally in the range 0.04 to 0.09 g/l, apart from the higher value already noted from experiments in the 14 cm diameter tube. In the laboratory scale tests the dust clouds were raised mechanically, by independent means, whereas in the large scale gallery tests, dispersion of dust was by the explosion itself. This difference in behaviour is of interest in considering the reason for the existence of the lower explosibility limit, particularly whether it is related to the ability of the explosion to disperse further dust. The explosibility limits of coal do in fact vary with volatile content, and in mine scale experiments the lower limit increased as volatile content decreased [64].

The lower explosibility limits of a number of other dusts have been measured in various investigations using apparatus incorporating a vertical tube open at the lower end. A summary of the results is given in Table 5.7, and the limits are those measured directly in the experiments. The correct limits are, however, larger because of the relative movement of dust and flame. Thus for polyethylene, at the lower limit concentration, the velocity of fall of the dust due to gravity was 40 cm/s and the upward velocity of the flame relative to the tube was 90 cm/s. The concentration of dust entering the flame assuming it all vapourised, was therefore 0.010 × 130/90, i.e. 0.015 g/l approximately. Similarly the corrected value for the lower explosibility limit of phenol formaldehyde dust was 0.045 g/l [20]. Lower concentrations of dust support flame propagation, if ignited at the closed end of the tube, but the concentrations would vary during the experiment because of movement of the suspension ahead of the flame. Accurate measurement is then difficult.

Table 5.7 Lower explosibility limits of dusts falling down a vertical tube

Dust	*Lower explosibility limit g/l*	*Tube diameter cm*	*Reference*
Charcoal (81 per cent C)	0.200	27	66
60 per cent charcoal: 40 per cent sulphur	0.066	27	66
30 per cent charcoal: 70 per cent sulphur	0.030	27	66
Lycopodium	0.022	7	18
Manioc	0.12	25	20
Methyl cellulose	0.035	25	20
Phenol formaldehyde resin	0.030	25	20
Polyethylene	0.010	25	20
Sulphur	0.0075	27	66
Wood meal	0.032	27	66

The upper explosibility limit is also difficult to measure and few investigations have been attempted. The most complete [68] employed a technique whereby the dust was deposited in a horizontal tube, internal diameter 5 cm, and which was then pressurized with air until a flame could be made to propagate through the dust. The dust concentration was then calculated (g/l of air at atmospheric pressure) and it was necessary to assume that the upper limit under conditions of high pressure would be the same as at atmospheric pressure. Typical values for the upper explosibility limit were

Cork: 1.4 to 2.5 g/l
Polyurethane: about 1 g/l
Polyvinyl alcohol: 2.0 to 2.3 g/l

The values for cork dust may be compared with 1.8 g/l obtained in a vertical tube apparatus [16], and which was in reasonable agreement. Upper explosibility limits were also obtained, at high pressures for a number of dusts mixed with kieselguhr. Although possible variation of upper explosibility with pressure has to be taken into account, the method has the advantage that because of the high pressures involved the influence of wall quenching on the propagation would be much reduced when compared with similar experiments at atmospheric pressure. There is scope for further development of the technique.

Ambient temperature

The propagation of a flame through a dust suspension would be expected to become more vigorous if the temperature of the dust and its surroundings were above atmospheric and if the chemical composition and physical properties were unaltered. Such circumstances could arise in industrial plant such as dryers, and, as an extreme example, in furnaces. The reasons for this more rapid flame propagation include:

(1) Chemical reaction rates increase with temperature

(2) The quenching effect of any moisture in the dust would be reduced at high temperatures

(3) Dispersability of the dust might be improved as the moisture content diminished.

On the other hand as regards the development of pressure during the explosion, the maximum temperature reached by the combustion products could be limited by molecular dissociation of the product gases. The ratio of the final to initial temperatures would decrease as the initial temperature is raised and the ratio of final to initial pressures would also diminish. The maximum explosion pressure obtained in a closed vessel would be lower as the initial temperature of the vessel was raised, other factors such as initial pressure being kept constant. Because the burning of the dust is likely to be more rapid if its initial temperature is raised the rate of pressure rise may be higher than for a cold mixture, with important implications as regards explosion relief venting of plant.

The above comments are based on technical considerations and, surprisingly, no information is available describing actual measurements of dust explosion behaviour in heated enclosures of the scale of industrial plant. With the increasing size of plant such as dryers, the need for such information is becoming pressing.

Tests on the dependence of the minimum ignition energy of coal on the temperature has already been considered (Fig. 4.4). As expected the minimum ignition energy decreased as the temperature rose but only over a limited range, beyond which it increased; the increase was attributed to reaction of the dust with the oxygen of the air before dispersion. Similar behaviour might be expected with other explosion properties but the temperature at which reaction of the dust with the oxygen becomes apparent would depend not only on the dust but also upon any previous exposure to high temperatures.

A substantial amount of work clearly needs to be done before the effect of temperature of dust flame propagation can be put on a quantitative basis. Such investigations would have both theoretical and practical importance.

Ambient pressure

For most practical applications and in many laboratory investigations of dust flame propagation the pressure of the gas, usually air, supporting the combustion was atmospheric. In a few investigations the initial pressure was not atmospheric and in such cases it is preferable to consider dust concentrations in terms of mass of dust per unit mass of air, rather than the mass per unit volume. Rises in pressure during the course of explosions are not considered as changes in ambient, if the pressure at the initiation of the explosion was atmospheric. For instance, in the routine tests for measurement of maximum explosion pressure (Chapter 3) increases above atmospheric occur, but the rises are the result of the explosion.

Coal dust explosions have been generated in the Vertical Tube apparatus (as described in Chapter 3) for initial pressures up to 30 lb/in^2 gauge. The coal had a volatile content of about 38 per cent and 65 per cent of the dust passed a 53 micron sieve [69]. The pressure rise in the apparatus due to the air used in dispersing dust was subtracted from the observed maximum pressure, as in the routine test. The relation between the maximum explosion pressure P (lb/in^2 gauge) and the initial pressure P_i, for a given fuel/oxygen ratio per weight was approximately.

$$P = P_0(1+0.75\, p_i/p_a) \quad , \tag{5.9}$$

where P_0 is maximum explosion pressure (lb/in^2 gauge) with the tube initially at atmospheric pressure p_a (lb/in^2 absolute).

Thus a 100 per cent increase in initial pressure produced only a 75 per cent increase in the maximum explosion pressure of the coal dust. The maximum rate of pressure rise was also measured and varied as follows:

$$(\mathrm{d}P/\mathrm{d}t)_{max} = A + Bp_{\mathrm{i}}^{0.5} \quad , \tag{5.10}$$

where A and B are constant.

In comparative tests with methane/air mixtures the maximum explosion pressure increased proportionately to the initial pressure, and contrasted with the smaller increases obtained with coal dust. The results for methane would be expected on theoretical grounds, but the reason why the dust explosion results followed a different pattern was not explored. Further results for coal dust have confirmed the general conclusions [70].

In addition to affecting the explosion pressures in a closed vessel, change of initial pressure also varied the explosibility limits of a coal dust [71]. Over a range of pressures 0.5 to 2.5 atmosphere absolute (7 to 37 lb/in^2) the upper explosibility limit was over 2 g/l for an initial pressure of atmospheric, but was only 1.7 g/l for an initial pressure of 0.5 atmosphere, and was even lower for pressures above atmospheric. The dust concentrations were measured in terms of mass of dust per unit volume of explosion vessel, which had a total internal volume of 40 litres. The reason for atmospheric pressure particularly favouring high upper explosibility limits was not explored; the lower explosibility limits were little changed with pressure.

In coal dust explosions in mine galleries the time required for pressure waves to travel the length of the gallery may be sufficiently great to have important effects on the movement of flame along the gallery. Although the initial pressure in the explosion is atmospheric, periodic pressure variations due to passage of pressure waves and their reflections can markedly effect the flame propagation [43]. The theory of shock wave propagation is well developed and enables the pressure behaviour in dust explosions in very long enclosures to be predicted and correlated with some of the sudden changes in flame movement observed in explosions. For practical purposes the phenomenon is of importance principally in coal mine explosions and the available theoretical treatments reflect this [72].

Inert gas

In dust explosions the gas in which the dust is dispersed, and which supports the explosion, is usually air, which contains approximately

21 per cent by volume of oxygen. If an inert gas is mixed with air, so that the oxygen concentration is correspondingly reduced, the propagation of the flame in a dust explosion becomes progressively more difficult and usually a limit is reached at which flame no longer propagates. This limit corresponds to the maximum permissible oxygen concentration to prevent dust explosions and routine tests have been developed for measuring this maximum concentration. These tests have been described and discussed in Chapters 3 and 4. Consideration must be given to the fact that a gas may not be inert to all dusts and the selection of the gas must be made specifically for the dust to be protected. For instance, carbon dioxide would be satisfactory for many carbonaceous dusts but not for fine magnesium powder which can burn in it. Thus no air-carbon dioxide mixture would be safe with this dust. An indication of possibly satisfactory combinations of dust and inert gas may often be obtained by considering the chemistry of the materials; if the dust is known to react with the inert gas with the evolution of heat then that gas may not be satisfactory. As an example magnesium is known to react with carbon dioxide, and steam, and so neither of these agents can be used, and a gas such as argon would be required. A test should be carried out if there is any doubt as to the suitability of an inert gas for a particular material.

Mixtures of inert gases may be used if convenient [73]. Thus the maximum permissible oxygen concentration for one aluminium dust in air/nitrogen was 9 per cent, in air/carbon dioxide was 7 per cent, and was 7.5 per cent in air/flue gas. The flue gas contained about 88 per cent nitrogen and 12 per cent carbon dioxide.

At oxygen concentrations above the maximum permissible for preventing explosions flame propagation is possible. Explosibility limits exist, the values for which depend upon the oxygen concentration. Lower and upper explosibility limits of terephthalic acid dispersed in a 15 cm diameter vertical tube [21] were measured using nitrogen as inert gas, and showed that for oxygen cencentrations less than 15 per cent by volume the lower and upper limits converged rapidly. The range of flammable dust concentrations narrowed sharply and when the oxygen concentration had been reduced to 12.4 per cent, flame would no longer propagate. Raising the temperature of the tube from ambient to 150°C, but keeping the same ignition source, caused the maximum permissible oxygen concentration to diminish to 11.1 per cent, demonstrating that the conditions are more stringent for inerting at temperatures above ambient. This point has already been discussed. (Fig. 4.3). The inerting of brown coal dust explosions, also in a vertical tube, showed that at ambient temperatures the oxygen concentration should not exceed 14 or 19 per cent depending upon whether nitrogen or carbon dioxide was used as the inert gas [25]. For tube temperatures of 110°C the required oxygen levels were reduced to 7 and 12 per cent respectively for nitrogen and carbon dioxide, and was also 12 per cent for steam. With both terephthalic acid and brown coal the quoted concentrations are experimental values, with no allowance for a safety factor.

The narrowing of the range of explosible concentrations, as the oxygen concentration was reduced, was followed in experiments using a large-scale vertical tube 25 cm in diameter (Fig. 4.1). Results for diphenylolpropane (Bisphenol-A) in air, inerted with nitrogen, clearly showed that the explosibility limits converged to a value of oxygen concentration that could be regarded as the maximum permissible to prevent ignition (Fig. 5.16). With phenolformaldehyde resin, using nitrogen, the results were of a similar form, but converging on a higher oxygen concentration and with carbon dioxide as inert the maximum permissible oxygen concentration was about 2 per cent higher still.

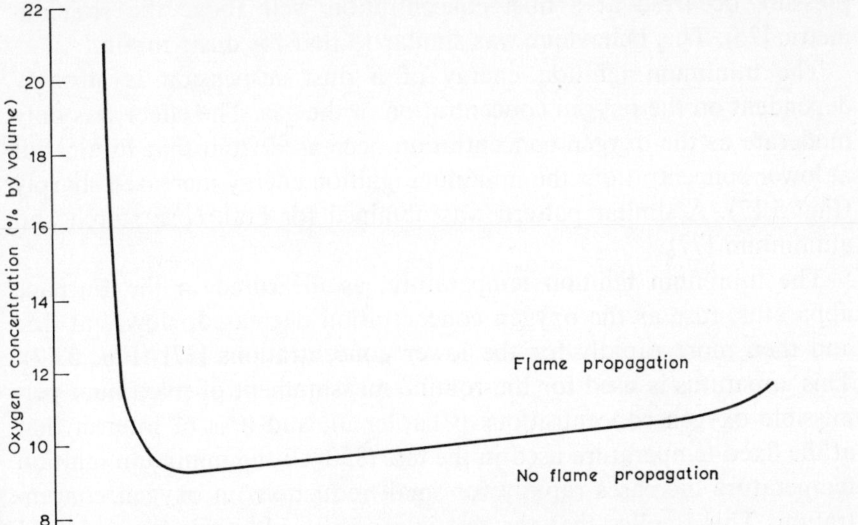

Fig. 5.16 Explosibility of diphenylol propane (Bisphenol A) in air/nitrogen mixtures.

As the oxygen concentration is decreased from that in air, the maximum explosion pressure and rate of rise of pressure both diminish. For example, the maximum explosion pressure of a sample of corn starch, at a concentration of 0.5 g/l decreased slowly from 97 lb/in^2 in air to 92 lb/in^2 at an oxygen concentration of 18 per cent by volume and then dropped sharply to nil for a concentration of 12 per cent [74]. The inert gas was carbon dioxide. With wood dust (spruce) a maximum pressure of 4.2 atmospheres in air decreased to 3.7 atmospheres at 17 per cent oxygen, and sharply to 1.8 atmospheres at 13 per cent and to nil at 11 per cent oxygen. Similar results were obtained with beech wood; in both cases the inert gas was nitrogen. This pattern of behaviour should not be taken as applicable in general, because with sulphur dust the maximum explosion pressure diminished steadily from 3.7 to 1.5 atmospheres as the gas mixture was changed from air to one in which the oxygen concentration was 10.5 per cent by volume. The inert gas was carbon dioxide. When the oxygen concentration was reduced further to 8.4 per cent no ignitions were obtained [76].

As would be expected, for a given oxygen level the maximum explosion pressure depended on the dust concentration. Results for brown coal, using nitrogen as diluent, indicated that the maximum explosion pressure occurred at a dust concentration well above the stoichiometric [25]. This behaviour was similar to that for dusts in air.

The minimum ignition energy of a dust suspension is strongly dependent on the oxygen concentration of the gas. The effect was only moderate as the oxygen concentration decreased from that in air, but at lower concentrations the minimum ignition energy increased sharply (Fig. 5.17). A similar pattern was obtained for both corn starch and aluminium [77].

The minimum ignition temperature, as measured in the Furnace apparatus, rose as the oxygen concentration decreased, slowly at first and then more rapidly for the lower concentrations [77] (Fig. 5.17). This apparatus is used for the routine measurement of maximum permissible oxygen concentrations (Chapter 3), and it is of interest that at the fixed temperature used in the test (850°C) the minimum ignition temperature increases rapidly for small reductions in oxygen concentration. This implies that the test is sensitive, although for technical reasons there may be doubts about its applicability to many practical situations (Chapter 4).

There has been insufficient development of the theory to enable

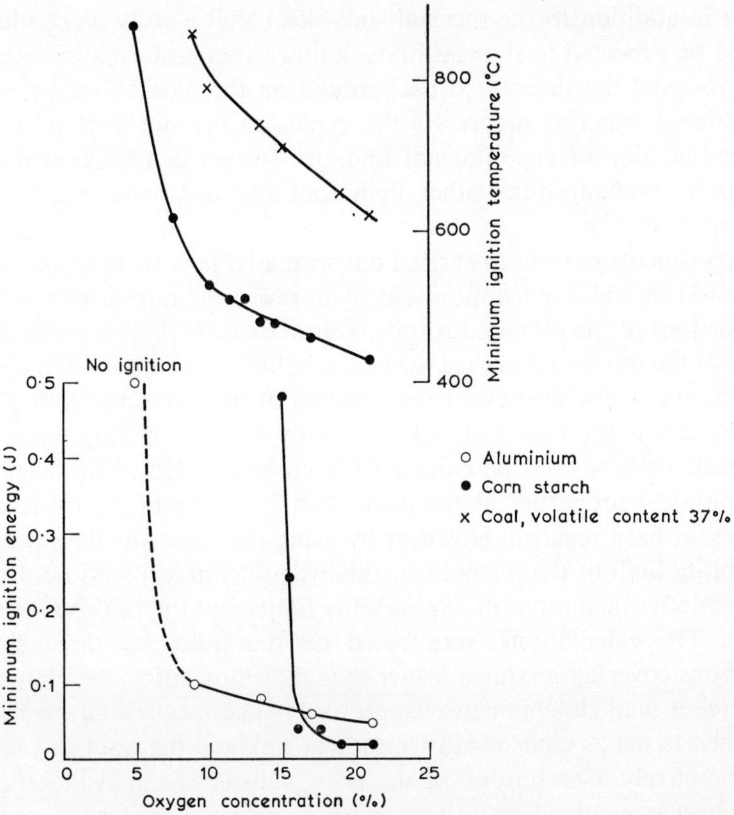

Fig. 5.17 Variation of minimum ignition energy and temperature with oxygen concentration of air/CO_2 mixtures [77].

the effect of inert gases on the propagation of dust flames to be fully understood. Assuming that the inert gas acts as a thermal sink, for which its heat capacity properties would be relevant, it is not yet possible to calculate the observed effect of the addition of inert gas on explosion properties such as maximum pressure and minimum ignition temperature.

Diluent dust

The only industry where diluent dust is used regularly to prevent a dust explosion hazard is coal mining and a large amount of information is available on the properties of various dusts in this respect and their use in mines. The term 'diluent' is preferred to 'inert' dust because although non-combustible, some materials appear to exert a chemical effect in addition to the thermal sink effect that a truly inert additive would be expected to show. Although much more information is available for coal the theoretical background on the mechanism by which the diluent material suppresses the explosion has not kept pace with the publication of experimental findings. Insight can be gained from research involving dusts other than coal and this work will be considered first.

Experiments have been carried out with a large scale tube apparatus (Fig. 4.1) in which phenolformaldehyde resin was mixed with various proportions of the diluent dust magnesium oxide [20]. The explosibility limits of the mixtures were measured, including the proportion of diluent dust required just to prevent propagation at any concentration of the dispersed mixture (the peak value). Ideally the explosibility limits and the peak value should be calculable from knowledge of the chemical and physical properties of the dusts, but this stage of theory has not at present been reached. However by using the value for the lower explosibility limit of the phenolformaldehyde dust in air, with no diluent dust added, calculations of explosibility limits and the peak value were made. The calculations were based on the following three simple equations covering mixtures leaner than stoichiometric, stoichiometric, and richer than stoichiometric respectively. The magnesium oxide was assumed to act as a chemically inert heat sink and the resin to vapourize completely. Combustion of the resin, subject to the availability of oxygen, was assumed to be to carbon dioxide and water.

Leaner than stoichiometric

$$y = \frac{H(x-x_1)}{r\,c_2(T_1-T_0)} \quad . \tag{5.11}$$

Stoichiometric (peak value)

$$y = \frac{H(x_2-x_1)}{r\,c_2(T_1-T_0)} \quad . \tag{5.12}$$

Richer than stoichiometric

$$y = \frac{H(x_2-x_1)}{r\,c_2(T_1-T_0)} - \frac{c_1}{r\,c_2}(x-x_2) \quad , \tag{5.13}$$

where the symbols are defined at the end of the Chapter.

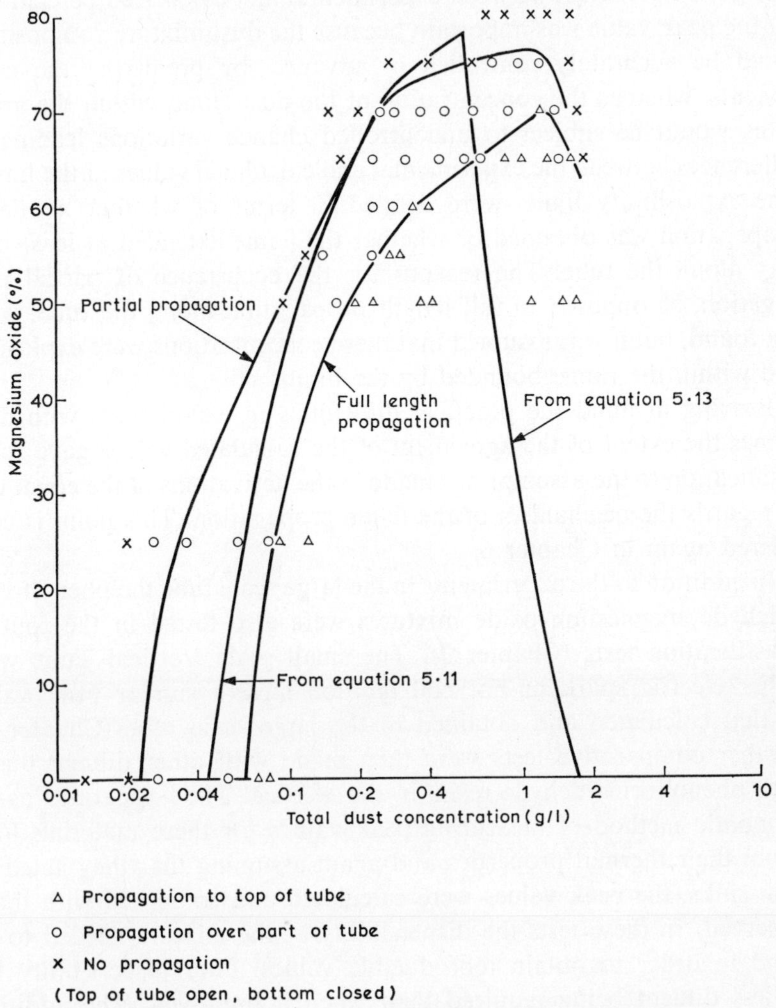

Fig. 5.18 Explosibility limits of phenolformaldehyde/magnesium oxide mixtures [20].

From the known properties of the dusts, including the experimentally determined value of x_1, the two lines shown in Fig. 5.18 were constructed. These lines met at the peak value, the calculated percentage of magnesium oxide being 78, and the experimentally determined value being between 75 and 80 per cent. Agreement with the lower explosibility limits of the various mixtures was fair, but for the upper explosibility limits the calculated values were considerably lower than experimental.

The good agreement between experimental and calculated percentages for the peak value was important because the dust mixture composition could be accurately controlled in advance, by premixing the components, whereas the concentration of the dust cloud within the apparatus would be subject to uncontrolled chance variations leading to differences between the experimental and calculated values of the limits. The explosibility limits were defined in terms of whether no flame propagation was obtained or whether the flame extended at least part way along the tube. The reasons for the occurrence of partial propagation, as opposed to full length propagation along the tube, were not found, but it was assumed that these concentrations were explosible and within the range bounded by the limits.

Bearing in mind the practical difficulties in experiments with dust flames the extent of the agreement of the calculated values gave some justification to the assumptions made in the derivations of the equations as regards the mechanism of the flame propagation. This point is considered again in Chapter 6.

In addition to the experiments in the large scale tube the phenolformaldehyde/magnesium oxide mixtures were also tested in the routine Classification tests (Chapter 3). The small scale Vertical Tube with either electric spark or hot coil ignition gave a similar peak value to that calculated and obtained in the large scale tube (Chapter 4). Further comparative tests were then made with other diluent dusts, with phenolformaldehyde resin, in the Vertical Tube apparatus as an economic method of measuring peak values for these materials [67]. From their thermal properties and again assuming that they acted as heat sinks, the peak values were calculated and compared with those observed. In these tests the dispersibility of the mixture needed to be good in order to obtain reproducible values. Poor dispersibility led to less diluent being required than when it improved. The addition of one per cent of magnesium stearate to the diluent gave good dispersibility, and as magnesium stearate is a readily explosible dust, it was taken as part of the fuel in the mixtures. The results are shown in Fig. 5.19 and 5.20 where the observed compositions are plotted against peak values calculated from the equation 5.12. The major systematic discrepency in the results was for sodium iodide, although titanium dioxide was substantially less effective than predicted and calcium sulphate more effective in one case only. Figures 5.19 and 5.20 include the results for magnesium oxide mixtures given above. In general the agreement between the observed and the calculated values was sufficiently close to give additional support to the assumptions on which the equations were based and which are considered in detail in Chapter 6.

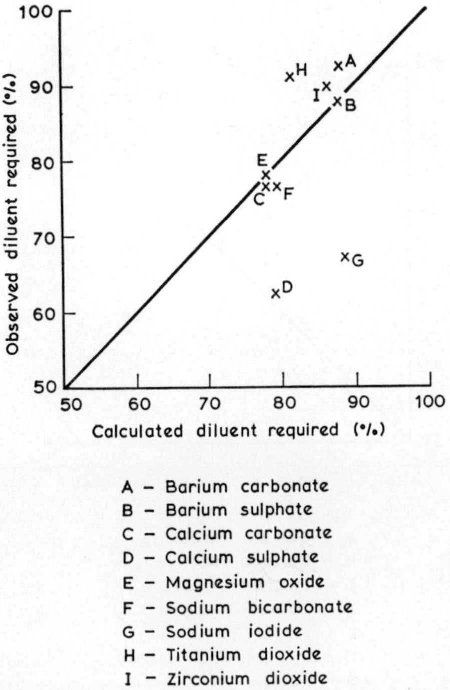

Fig. 5.19 Percentages of diluent dusts to prevent explosion of phenolformaldehyde resin using vertical tube apparatus with electric spark ignition.

The explosibility limits of lycopodium dust with limestone have been measured by dropping the mixed dusts down a tube 7 cm in diameter and 4 m long, open at the lower end [18]. The 5, 50 and 95 per cent probabilities of propagation of flame were measured and, when the ratio of limestone to lycopodium was about 4 to1, small changes in composition gave large variation in the probability of flame propagation. The peak value was in the region of 80 per cent limestone in the mixture.

The suppression of explosion in wheat flour and cork dust by a range of diluent dusts has been investigated by a method in which the dust was discharged from a jet into air and the dust cloud explored for ignition with a meker burner [28]. The proportion of diluent required to suppress flame propagation was greater with the cork dust than with the flour, even though the former was coarser (Table 5.8). Comparative

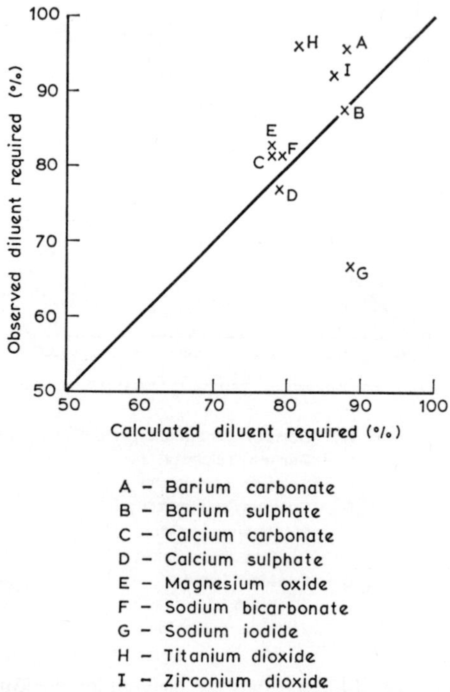

Fig. 5.20 Percentages of diluent dusts to prevent explosion of phenolformaldehyde resin using vertical tube apparatus with hot coil ignition.

tests with the cork dust showed that the alkali metal salts were particularly effective, the order of effectiveness being as follows: potassium bicarbonate, sodium chloride, borax, sodium bicarbonate, disodium hydrogen phosphate (hydrate), sodium sulphate (hydrate), sodium carbonate (anhydrous), sodium bisulphate, sodium sulphate (anhydrous), ammonium chloride, ammonium carbonate, zinc carbonate, magnesium hydroxide, ferric oxide, aluminium hydroxide, calcium hydroxide. The tests were comparative and no attempt was made to relate the quantity of diluent required to its thermal or chemical properties.

Table 5.8 Suppression of wheat flour and cork dust flames by diluents [28]

Diluent dust	*Particle size per cent passing 240 B.S. mesh (66 micron)*	*Percentage diluent required to suppress flame propagation*	
		Wheat flour 72 per cent passing 240 B.S. mesh	*Cork dust 34 per cent passing 240 B.S. mesh*
Derbyshire limestone	68	68	74
Magnesian limestone	72	67	72
Calcium carbonate	100	46	63
Sodium chloride	Not measured	18	42
Sodium bicarbonate	75	8.5	43

The effects of alkali metal halides on the lower explosibility limits of several dusts, including aluminium, were compared using a glass flask as explosion vessel and an electric spark as ignition source [7]. The lower limits for aluminium and anthracene, using sodium chloride and ferric oxide as diluents, are shown in Fig. 5.21; the values are expressed as mass of dust in the flask (volume approx. 1 litre). With aluminium the two diluents were equally effective, but with anthracene the sodium chloride was considerably more effective than the ferric oxide. Further experiments with anthracene dust showed that for both potassium and for sodium salts the most effective diluent was the fluoride, followed by the other halides in the order: iodide, bromide, chloride. The potassium halide was more effective than the corresponding sodium halide. Another interesting result was that potassium and sodium nitrates had a marked effect in increasing the lower explosibility limit of anthracene dust; although the salts are strong oxidizing agents there was no evidence that they promoted the combustion of the anthracene. Another oxidizing agent, potassium chlorate, also raised the lower explosibility limit of dextrine dust, but it appears to have little effect in explosions involving aluminium or picric acid. Finally, the organic halogen compound hexachlorethane mixed with anthracene, had a comparable effect to that of sodium chloride except that at percentage additions between 0 and 4 the lower explosibility limit of the mixture was decreased (possibly due to increased dispersibility arising from the presence of the hexachlorethane) before increasing sharply again.

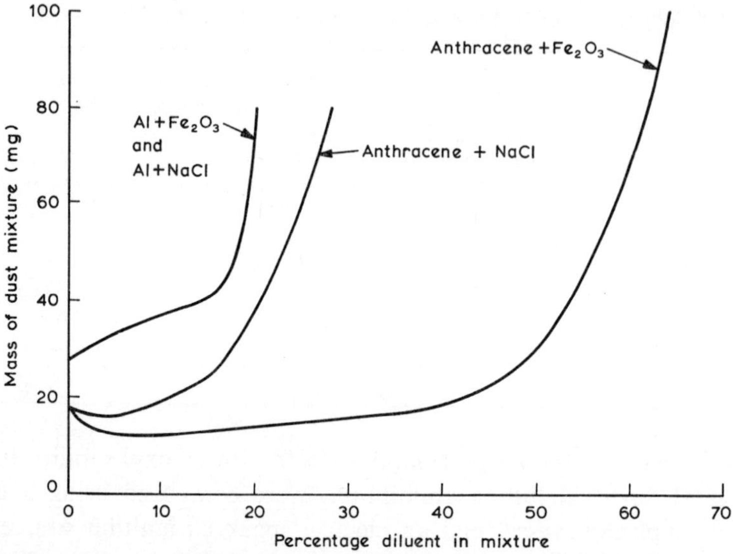

Fig. 5.21 Lower explosibility limits of aluminium and anthracene mixed with ferric oxide and sodium chloride [7].

For many dusts the proportion of diluent required to prevent explosion has been measured in the routine test described in Chapter 3. In most cases the diluent was Fullers earth, but some values were given for other diluents [77]. In particular, a series of tests using corn starch enabled the inerting ability of various powders to be compared. The tests were made with electric spark ignition and the total weight of corn starch with diluent in the apparatus was held constant at 1 g, giving a dust concentration of about 1 g/l. The diluent powders, in decreasing order of effectiveness were sodium carbonate, potassium carbonate, sodium bicarbonate, potassium chloride, barium carbonate, potassium bicarbonate, potassium iodide, potassium tartrate, Fullers earth, sodium chloride, calcium carbonate (limestone), ammonium chloride, silicon dioxide (sand). The alkali metal salts were again the more effective diluents although sodium chloride was relatively less so than reported elsewhere [28] and was indeed less effective than Fullers earth. Also barium carbonate was comparable in effectiveness to the alkali metal salts generally, in marked difference to its behaviour when mixed with phenol formaldehyde resin (Fig. 5.19 and 5.20). The explosibility of sulphur dust has been investigated with the Furnace apparatus at 820°C [78]. Superphosphate and gypsum were more effective in preventing explosion than limestone of the same particle size, and the maximum safe proportion of sulphur in the mixture was about 23 per cent.

One explanation of the apparent contradictions in assessing the relative effectiveness of diluents in suppressing dust explosions is that the diluents could affect the dispersability of the dust mixture. In general, if the dispersability were reduced flame propagation would be made more difficult and hence the diluent would appear to be more efficient than it actually was. Other factors such as the particle sizes and moisture contents of the dust and the diluent influence the dispersability. Until these aspects have been examined it is unlikely that a quantitative explanation of the effects of the diluents could be obtained. However, there is strong evidence that the alkali metal halides can be highly effective in suppressing explosion, and it is reasonable to infer that chemical reaction takes place when these salts are present. What this reaction or reactions are, and with which dusts they occur, cannot at the moment be predicted.

The use of diluent dusts to reduce the explosibility of coal, and to provide safe working conditions in mines, has resulted in numerous inuvestigations of the effect of diluents on coal dust explosions both in large and small scale experiments. The percentage of diluent required to prevent explosion has been used as a measure of the explosibility of the coal (Fig. 5.6 and 5.7) and for this purpose the diluent dusts have been those customarily used in mines. Typical materials were Fullers earth, limestone, shale and ash. In British mines the material used nowadays is limestone and it is recommended that with coals of volatile content above 25 per cent the amount of limestone to be mixed with the coal in the mines should be as high as 80 per cent of the mixture in certain locations [79]. The amounts of diluent dust found experimentally to be necessary in full scale experiments for the prevention of explosions have already been given (Equations 5.1, 5.5, 5.6 and 5.7).

The relationship of full scale to laboratory scale determinations was investigated with the Furnace apparatus, using Fullers earth as the diluent dust [48]. A Furnace temperature of 720°C gave the best correlation between the percentage of diluent required to prevent explosion in large scale apparatus and the Furnace apparatus. Various British and American coals were then compared at this temperature and some differences in the diluent dust requirement were found, even when particle sizes of the coals had been standardized. It was concluded that the results of tests of American coals were in general agreement among themselves, as were those of the British coals, but there was no common agreement between the two groups of coals. Systematic differences were also found in large scale experiments [48, 49] (Fig. 5.7). The ash content of the coal was included in the calculated percentage of diluent.

The explosibilities of various Indian coals have been measured in the Furnace apparatus [80, 81] with similar results when Fullers earth or an alternative diluent was used. The ash in the coals appeared to have an effect similar to that of the added diluent.

The lower and upper explosibility limits of a coal containing 36 per cent volatiles have been measured in a large vertical tube apparatus, 25 cm in diameter [63]. Particular attention was paid to the proportion of diluent (limestone) necessary to prevent propagation of flame in the coal/stone mixture at any concentration of suspension, as in Fig. 5.22 The percentage of limestone necessary was between 60 and 65 per cent and was in agreement with the value of 65 per cent that had been obtained with the same coal and stone dusts [91], in a mine gallery of 4 ft diameter. Further interpretation of these results is given in Chapter 6.

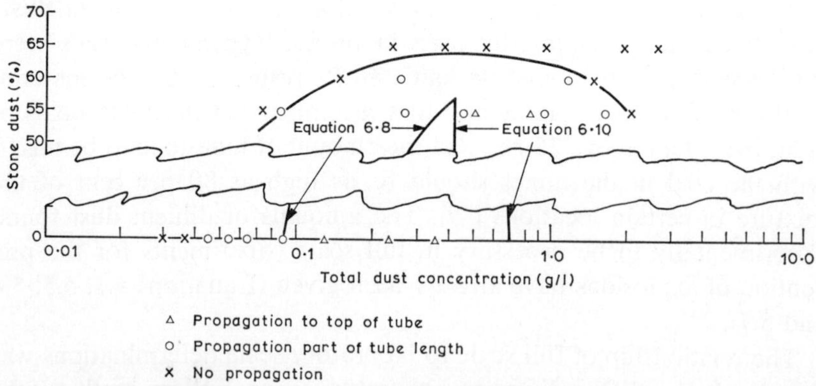

Fig. 5.22 Explosibility limits of coal dust/stone dust mixtures (top of tube open, bottom closed, tube diameter: 25 cm, tube length: 5.2 m).

Addition of diluent dust affected the flame speeds in large-scale coal dust explosions [82]. In an analysis of results obtained previously in other countries a large number of experiments were surveyed and those were selected in which the following conditions were fulfilled:

(1) The dust deposit was either pure coal or a homogeneous mixture of coal dust and diluent dust.
(2) The flame propagated throughout the dusted zone of the experimental gallery.
(3) The flame speed did not decrease as the flame approached the end of the dusted zone.
(4) There were no stone dust barriers.

Two hundred and one experiments were considered from those in the United States (as in Table 5.4 [49]), for which the experimental conditions varied widely:

(1) Volatile contents of coal: 14–41 per cent
(2) Total diluent content of the dust mixture: 8–78 per cent
(3) Coal dust concentration: 0.14 to 1.7 g/l

The flame speed was measured at about 300 feet from the ignition source and the frequencies with which the various flame speeds occurred were calculated. The distribution was such that a mode occurred around a flame speed of 175 m/s and the maximum recorded was about 1350 m/s.

A similar review was made of experiments in Poland [50] (as in Table 5.4) in which the experimental conditions were as follows:

(1) Volatile content: 14–69 per cent
(2) Total diluent content of the dust mixture: 11–85 per cent
(3) Coal dust concentration: 0.08 to 2.0 g/l

The flame speed was measured over the last 50 metres of the 200 metre long roadway. The frequency distribution of the flame speeds gave a mode at about 250 m/s and in these experiments no flame speeds greater than 1000 m/s were recorded.

A common feature was that widely differing flame speeds were produced in repeat tests with the same experimental conditions, and also in the majority of the explosions flame speeds did not exceed about 400 m/s. A correlation appeared to exist between the flame speed and the diluent content, in the direction of increasing diluent reducing flame speed, and further tests were carried out to confirm this correlation using a 4 ft diameter mine gallery (Table 5.4, Ref 48). With limestone as the diluent (57 per cent through 56 micron sieve), and with ignition sources of three different strengths, it was clearly established that the addition of the diluent reduced the flame speed, except in the case of the weakest ignition source where there was no significant effect. With the two stronger ignition sources addition of only 5 per cent of diluent reduced the flame speed by between 20 and 60 per cent of the value for the pure coal dust, and with an addition of 40 per cent of diluent reductions in flame speed of between 50 and 90 per cent were observed. Final conclusions were that the presence of limestone diluent in proportions less than those required to suppress explosions would act as an adequate safeguard against detonations in coal mines, and also that general stone dusting should be made in mines even in the presence of stone dust barriers.

Limestone dust has a tendency to cake when wetted and water-proofing agents such as wool-wax acid, are used to coat the limestone particles in commercial products [83]. The water-proofed limestone is less efficient in suppressing coal dust explosions; the reason for this behaviour has not yet been proved, and to retain the original efficiency the water-proofed limestone needed further modification. Gallery tests showed that a limestone coated with 1 per cent by weight of wool-wax acid and one half per cent of hexachloroethane was at least as effective in suppressing coal dust explosions as limestone dust that had not been water-proofed and was used under dry conditions.

There have been numerous attempts to find diluents having a superior performance in suppressing coal dust explosions but which would also be efficient under mining conditions, would not be affected by moisture, would not have adverse effects on the men and machinery, and would be economically competitive. A number of materials have been tested under explosion conditions and a survey emphasized that alkali halides and some materials subject to thermal decomposition were particularly effective [84]. The decomposable materials were either carbonates or contained water of crystallization; one exception was sodium bicarbonate which although easily decomposed was no more efficient than the usual stone dust. For the hydrated salts gypsum ($CaSO_4$ $2H_2O$), Epsom salts ($MgSO_4$ $7H_2O$), and borax (Na_2 $B_4O_7 10$ H_2O) the increased efficiency in suppressing explosions due to the heat required to drive off the water of crystallization was shown by correlating the heat of decomposition with the amount of diluent required to suppress explosion. In the calculations the specific heat of the anhydrous diluents was taken to be that of limestone, measured at ordinary temperatures, The correlation was good for mixture of these salts with two coal dusts, of volatile contents 22 and 36 per cent.

The marked effectiveness of alkali metal halides in suppressing dust explosions has already been noted with dusts other than coal. Several investigations with coal have been reported in which sodium and potassium chlorides were used; respective proportions of 23 and 15 per cent were reported [84] for suppressing explosions in a coal of volatile content 36 per cent, as compared with a proportion of 65 per cent required when limestone was the diluent, and 18 per cent sodium chloride for a similar coal in a large scale tube apparatus [63]. Full scale explosions in the United States in a mine gallery [49] (Table 5.4), also with a coal of volatile content about 36 per cent, showed that on mixing sodium chloride with the limestone diluent a proportion of 25 per cent of the alkali halide in the limestone reduced the diluent requirement from about 75 per cent to 63 per cent in the combined diluent/coal dust mixtures. In spite of the marked effectiveness, particularly of sodium chloride, its use in mines has not found widespread acceptance because of its unsatisfactory behaviour under working conditions. It is deliquescent and rapidly takes up water from the atmosphere so that it could not be relied upon to disperse with the coal dust during an explosion. It also promotes corrosion of metals.

The action of alkali metal halides and other salts on the propagation of coal dust flames [86] has been studied using an explosion chamber of volume 110 litres and a full scale explosion duct, length 200 m, diameter 1.8 m. In the smaller vessel, which was closed, the maximum explosion pressure was measured as a function of percentage of diluent in the mixture with coal dust; the coal concentration was held constant at 0.6 g/l. Explosion pressures were reduced to about 10 per cent of the value for coal alone by the addition of 30–50 per cent of potassium chloride, sodium chloride, and ammonium carbonate. Between 50 and 80 per cent of diluent was required for sodium bicarbonate, calcium carbonate, barium sulphate, potassium bicarbonate, kaolin, potassium carbonate, potassium sulphate. The alkali halides and ammonium carbonate were thus considerably more effective than other salts of potassium and sodium, including the bicarbonates, and of salts of other metals. Tests with a solution of potassium chloride in water showed that it was less effective that the dry material but more effective than water alone. In the large scale duct, a 50 per cent mixture of calcium carbonate with the coal caused the flame to extinguish about 10 m from the open end of the duct, whereas sodium bicarbonate in the same proportion allowed the flame to accelerate towards the end of the duct, and was thus less effective.

Although the alkali metal halides are particularly effective in preventing flame propagation in coal dust, as compared with other salts and stone dusts, when tested either on large scale or in small scale apparatus with ordinary wall temperatures, different behaviour was observed with a Furnace apparatus [87]. In a range of 31 diluent materials, the proportion of diluent required to be mixed with the coal to prevent explosion ranged from 67 to 91 per cent; the value for stone dust was 75 per cent and for dry sodium chloride was 76 per cent. The combined water in the diluent appeared to be very important in that it reduced the percentage required and the specific heat of the diluent was also suggested as being a factor. A quantitative relationship between percentage of diluent and specific heat and water content was not obtained.

Water

Attention is given here to water present in the dust, either by absorption or by mixing, and to the application of water to dust explosion flames. Some discussion has already been given of the effects of water vapour or steam and of water of crystallization in connection with the effects of inert gases and diluent dusts respectively on explosions.

As would be expected, moisture tends to reduce the explosibility of dust, and the effect is shown by increases in the minimum explosible concentration and minimum ignition energy (Fig. 5.23) and in minimum ignition temperature when these properties are measured in the usual small scale apparatus [77]. With spark ignition the percentages of water required to prevent ignition of dusts varied from 16 per cent for coal (volatile content 37 per cent), 19 per cent for corn starch, and 35 per cent for paper dust. The reduction in explosibility due to water appears to arise from two causes; firstly the heat absorbed by the water in evaporating during combustion of the dust, and secondly the reduction in dispersability caused by the water which tends to prevent the formation of a uniform dust suspension and thereby hinders flame propagation. Observations on the reduction of dispersability have been reported for coal [44] with which a dust finer than 75 microns could no longer be dispersed by blowing on it with the mouth when the moisture content was raised to about 27 per cent. Dust in this condition seemed to be wet on being pressed in the hand and retained its shape afterwards.

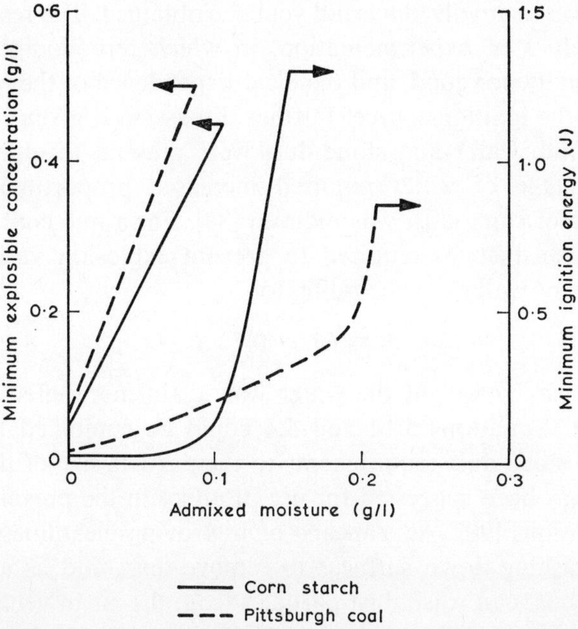

Fig. 5.23 Minimum explosible concentration and ignition energy of moist dusts [77].

The addition of water to a coal dust/stone dust mixture can lead to a complicated dispersability situation because the dispersability of the coal and the stone dust, may be modified to different extents by the water. In addition, wetted limestone dusts tend to cake after drying again [88], although the caking has only a minor effect on its ability to suppress coal dust explosions. The maximum amount of water retained by fine coal varies with the volatile content; for volatile contents up to about 40 per cent the maximum retained moisture is about 36 per cent, but for higher volatile contents the retained moisture rises sharply to about 54 per cent [88].

In large scale experiments the percentage of water required to quench coal dust explosions increased markedly with the energy of the ignition source (Fig. 5.6) and rose to about 48 per cent water in the mixture for fine coal dust, in a favourable position for dispersion, accompanied by a strong ignition source. The corresponding amount of diluent dust required to prevent explosion under these conditions was 88 per cent. In terms of weight it is clear that water is more effective than stone dust, in this case by a factor of about 7, but a detailed understanding of the processes involved has not yet been obtained. The reasons include the difficulties of experimentation, in which reproducibility between tests is usually not good, and the clear dependence of the results on the energy of the ignition source. Further discussion is given in Chapter 6.

When both water and stone dust were present, together with coal, the percentage of water required increased proportionately as the percentage of stone dust was reduced [89]. For a fine coal the percentage of stone dust (S) required to prevent explosion varied with the percentage of water (w) according to

$$w = 61 - 0.74\,S \quad . \tag{5.14}$$

On a basis by weight the water was again more effective than the stone dust. Equations 5.14 and 5.4 could be combined to relate the water and stone dust requirements to the particle size of the coal dust.

Water has been suggested for use in mines in the prevention of coal dust explosions [90]. As a means of improving cleanliness, it may be used in washing down surfaces to remove dust, and as a continually produced mist can wash dust particles from the air in ventilation roadways. Should an explosion be initiated, either in coal dust or in coal/diluent dust mixtures, water sprays or jets may be used to extinguish the flame. Tests in a 4 ft diameter gallery, 330 ft long, showed that both a spray and a jet of water was capable of quenching explosions [91], particularly if the water was projected from a nozzle along the gallery towards the approaching explosion, in such a way that the jet was broken up only by the explosion itself and not through splashing either on an obstacle or through self-impingement. In application of this technique to practice it would of course be necessary to install some form of explosion detection so that the water jet or spray would be established before the arrival of the flame. Alternatively, the water may be installed in mine roadways in the form of barriers or other vessels capable of easy discharge and by these means explosions have been quenched under a range of conditions [92].

Water has not been widely used in the quenching of explosions in dusts in other industries than coal mining, although it is used in the safe handling of a number of very explosible dusts. In the grinding of magnesium, for instance, wet scrubbers are used in the removal of dust formed during manufacturing processes. The deposit obtained from the scrubber is in the form of a wet sludge which is safer to handle and dispose of than a dry magnesium powder (Chapter 10).

Flammable gas

Flame propagation in explosible dust/flammable gas/air mixtures has been investigated in detail for coal dust and methane. The lower explosibility limit of coal dust/methane/air mixtures in vertical and horizontal tubes, 15 cm diameter and 1.8 m long, were clearly defined and the following empirical relations were established [93].

For downward propagation (mixtures containing 0 to 53 mg/1 coal dust)

$$M = 35.1 - 0.24\,c - 0.0056\,c^2 \quad , \tag{5.15}$$

and for horizontal propagation (mixtures containing 0 to 69 mg/1 coal dust)

$$M = 37.0 - 0.49\,c \quad , \tag{5.16}$$

where c and M are concentrations of coal and methane in mg/1. The volatile content of the coal was 37 per cent

The ignition source in these experiments was a hot gas jet produced by the explosion of a methane/oxygen mixture delivered through a narrow cylindrical channel into the explosion tube. This ignition source was relatively powerful and gave smaller lower explosibility limits and greater flame speeds than ignition with a capacitance spark. For example, with downward flame propagation and hot gas ignition a methane concentration of 17.5 mg/l together with coal at 35 mg/l gave flame speeds of about 80 cm/s. With spark ignition a methane concentration of 37 mg/l was required with the same coal dust concentration, and the flame speed was only about 20 cm/s. Ignition by methane/oxygen/nitrogen mixtures [39] or by heated nitrogen [94], in explosion vessels respectively 33 cm long 9 cm diameter, and 80 cm long 15 cm diameter, gave flame propagation which may not have extended beyond the influence of the ignition source. With the combustible gas ignition source, the lower explosibility limits of the methane and coal dust were usually less than those obtained in routine limit measurements in air, and some evidence was obtained of a 'synergistic' effect between coal and methane in certain concentration ranges. For example, with a coal of volatile content 30 per cent, ignition was obtained with the coal alone at a concentration of 24 mg/l, and ignition of methane alone at a concentration of 30 mg/l. When both the coal and methane were present, ignition could be obtained at a total concentration of only 20 mg/l. This behaviour is in contrast to that represented by equations 5.15 and 5.16 and may have arisen from the 'overdriving' of the explosion because of the proximity of the ignition source.

The quenching distances of methane/coal dust/air mixtures have been measured using a tapered rectangular tube [95]. The tube was vertical, the widest cross-section being at the top, and was filled with the mixture to be investigated. After ignition by a methane flame at the open top of the tube, the distance down which flame propagated through the mixture was observed and the width of the tube at the point of extinction was taken as the quenching distance. The quenching distances were larger for mixtures of methane and coal of high volatile content than for the same total concentration composed of methane only. A coal of volatile content 17 per cent gave results similar to those for the methane present alone and the quenching process appeared to be similar to that for a gas mixture containing a relatively small amount of inert dust. Relatively large coal dust particles did not contribute their total weight at the quenching limit, probably because such particles did not remain in the flame front long enough to burn out completely, and it was concluded that the equivalence between coal and methane could not be explained on a purely thermal basis.

The maximum explosion pressures in methane/coal dust/air mixtures have been measured in a closed vessel [68] of 0.33 ft^3 (9.5 1) volume using a coal of volatile content 37 per cent. The methane concentration was varied over the range 0 to 5 per cent by volume in air, and was thus always below the lower explosibility limit. The coal dust concentration was varied over the range 0.1 to 2 g/l and covered the value at which the maximum explosion pressure was obtained in air (about 0.5 g/l). With low dust concentrations (0.1 to 0.4 g/l) small amounts of methane (not more than 3 per cent) increased the explosion pressure, but with higher dust concentrations the explosion pressure was reduced. Direct comparison of the equivalence of coal and methane in generating explosion pressure must take into account the wide range of coal dust concentrations used compared with the narrower range of methane concentrations, all of which were well below the concentration of methane alone giving the maximum pressure (about 10 per cent).

The presence of methane counteracted the quenching effect of diluent dusts mixed with the coal. The explosibility of methane/coal dust/ diluent dust/air mixtures may be gauged from the proportion of diluent dust required to prevent explosion, and the addition of only 5 per cent methane to the air markedly affected the proportion of diluent required [24]. With a coal of volatile content 45 per cent mixed with clay dust in a small cylindrical explosion vessel (diameter 15.6 cm, height 24 cm) and an electric spark ignition source the proportion of diluent required to suppress explosion in a coal was 48 per cent in the absence of methane. With 5 per cent methane in the mixture 85 per cent diluent was needed to prevent explosion.

The effect of the presence of methane was even more marked when a larger explosion vessel was used, of volume 115 litres, and with a pyrotechnic mixture as ignition source [96]. With a coal of volatile content 25 per cent the proportion of diluent dust required to prevent any appreciable explosion pressure development in the vessel, in the absence of methane, was about 80 per cent. In the presence of between 3 and 12 per cent methane in the air, explosion pressures of between 3 and 5 kg/cm^2 were obtained with 80 per cent diluent and a coal dust concentration of 0.2 g/l. In the absence of methane and diluent the maximum explosion pressure was 6.3 kg/cm^2 for the same coal dust concentration. In these experiments, higher methane concentrations were used than in the other investigations and the results may be regarded as showing that a coal/diluent dust mixture containing 80 per cent diluent and 0.2 g/l of coal, were insufficient to quench near-stoichiometric methane/air flames.

Observation by streak photography of spark ignited coal/and aluminium/methane/air mixtures showed that the flame speed increased as the flame front moved away from the ignition point [97]. The photographs also indicated that within the first 5–7 cm of flame expansion, the burning dust particles which were ignited at the flame front did not move with the flame, but remained behind in a stationary position in the central combustion products. With coal flames of constant dust concentration, the calculated burning velocity increased from 12 to 28 cm/s as the methane concentration was increased from 3 to 8 per cent. Both with dust and with dust/methane mixtures the flame speed increased as the flame front moved away from the ignition source. For example, the calculated burning velocity of an aluminium/air flame, at a concentration of about 0.2 g/l, increased from 23 to 55 cm/s as the distance from the spark source increased from 1 to 7 cm. The presence of methane increased the flame speed of aluminium, as would be expected by analogy with coal dust.

Although most of the available information is for methane/dust/air flames, other combustible vapours and gases together with explosible dusts are encountered in industry. For example polystyrene granules containing pentane are manufactured, and catalytic metals mixed with flammable gases are not unusual. The pattern of explosion behaviour of these mixtures is probably broadly similar to that for methane/coal, after allowing for differences in burning velocity which may be substantially higher, but detailed investigations have not been reported.

Turbulence

Relatively little investigation of the effect of turbulence on the propagation of dust explosions has been made, although turbulent conditions are widely encountered both in laboratory apparatus and in industrial plant. The turbulence may arise either from movement of the dust/air suspension along a pipe, from the fall of dust particles under gravity through air, or by the passage of flame through a dust suspension.

Some indication of the effect of turbulence on the maximum explosion pressure and rate of rise in a coal dust/air suspension were obtained using a 0.33 ft^3 (9.5l) explosion vessel within which the dust was dispersed by a blast of air [69]. The initial air pressure within the vessel was reduced below atmospheric, so that the air introduced during dispersion returned the initial pressure to ambient. The turbulence parameters were not measured directly but the degree of turbulence was measured in arbitrary units as the difference between the original pressure in the explosion vessel and that in the reservoir holding the dispersing air. As the turbulence was increased the maximum explosion pressure and rate of rise both passed through a maximum (Fig. 5.24), the rate of pressure rise being particularly dependent on the turbulence. The variation of rate of pressure rise was of particular interest because of its relationship to explosion venting requirements (Chapter 8). It was considered doubtful that the turbulence was a linear function of the pressure differential of the dispersing blast and further attempts were made to evaluate this effect by delaying the time of ignition, and allowing the disturbance to decay. The measurements were made with 9.5 per cent methane present in the air and the rate of pressure rise decreased by a factor of about 2 when the delay between dispersion and the ignition was increased from 0.3 to 0.5 seconds, but decreased by only a further 20 per cent even after a delay of 50 seconds.

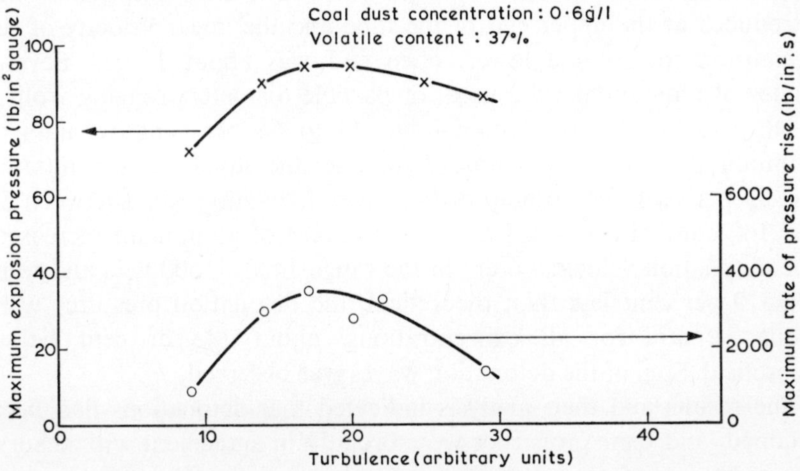

Fig. 5.24 Effect of turbulence on explosion pressures of coal dust [69].

The influence of turbulence on the burning rate of fuel suspensions in furnaces has been reviewed [98], and although the information is not strictly relevant to dust explosion some of the conclusions are likely to be applicable. Turbulence is known to increase the burning rate of flames in furnaces, and with pulverized fuel and heavy fuel oils the greatest part of the combustion time of individual particles is taken up in burning the carbon residue remaining after the evolution and burning of the volatiles. Calculations showed that the increased burning rate could not be accounted for by improved heat and mass transfer arising from more rapid movement of the particles relative to the gas. A better explanation for the increase in burning rate caused by turbulence was that rapid mixing, arising from eddy diffusion, resulted in increased heat transfer rates between pockets of gas already burning and those as yet unignited. In dust explosion flames, no stabilizing burner would usually be present, but turbulence could be caused by normal pipe flow, the presence of obstacles, or by the expansion resulting from the presence of the flame. Mixing by turbulent diffusion is therefore feasible.

Detonation

Aluminium/oxygen mixtures have been shown to detonate in a long tube, using as ignition sources squibs, detonator caps, or silver wire exploded electrically [99] The tube was mounted vertically, was 2.8 m long and 26 mm diameter in most tests. The dust suspension was introduced at the upper end of the tube and the linear velocity of the suspension toward the lower, open end was about 1 m/s. Several grades of aluminium were used, of particle diameters ranging from 5 to 40 microns. Mixtures containing 48 to 64 per cent by mass of aluminium in the oxygen were detonable; the stoichiometric mixture was 53 per cent. The run-up distances to detonation were between 50 and 160 cm and decreased as the percentage of aluminium increased. The detonation velocities were in the range 1400—1600 m/s and were about 9 per cent less than theoretical; the detonation pressures were about 31 atm for all concentrations, about 14 per cent below theoretical. Spin of the detonation waves was observed.

The results and their analysis indicated that detonations had been obtained, and their properties were broadly in agreement with theory. The existence of true detonations in dust/air suspensions is not at present generally accepted, although phenomena akin to detonation have been observed in large scale explosions in coal mines in which the ignition source generated strong shock waves.

SYMBOLS

a	Constant
A	,,
b	,,
B	,,
c	Minimum explosible concentration of dust (mass/unit volume)
c_1	Mean specific heat of volatiles from combustible dust.
c_2	Mean specific heat of diluent dust.
C	Constant.
F	Specific surface of dust.
F'	Fineness factor.
F_c	,, ,, (coal).

F_1 Fineness factor (diluent dust).
F_0 Constant.
F'_0 ”
H Heat of combustion per unit mass of dust.
k Explosion number.
K Constant.
m ”
M Methane concentration at lower explosibility limit (mass per unit volume).
n Constant.
p_a Atmospheric pressure.
p_i Initial pressure (gauge).
P Maximum explosion pressure (gauge).
P_0 ” ” ” ” with dust suspension initially at p_a.
Q Factor representing increase in volatiles production.
r Expansion ratio on combustion.
S Proportion of diluent dust to prevent dust explosion (per cent).
t Time.
T_0 Ambient temperature.
T_1 Flame temperature at explosibility limits.
V Volatile content of dust (per cent).
V_0 Constant.
w Proportion of water to prevent dust explosion (per cent).
W Maximum weight loss on heating.
x Dust concentration (mass per unit volume).
x_1 ” ” ” ” ” ” at lower explosibility limit.
x_2 ” ” ” ” ” ” at stoichiometric.
y Diluent dust concentration (mass per unit volume).
Z Explosibility coefficient.

References

[1] Dawes, J. G. (1952), Safety in Mines Research Establishment, Research Report, 36.
[2] Dawes, J. G. (1952), Safety in Mines Research Establishment, Research Report, 49.
[3] Gerrard, J. H. (1963), *Brit. J. Appl. Phys.*, **14**, 186–92.
[4] Singer, J. M. Greninger, N. B. and Grumer, J. (1969), US Bureau of Mines RI 7252.

[5] Brown, K. C. and James, G. J. (1962), Safety in Mines Research Establishment, Research Report, 201.

[6] Boyle, A. R. and Llewellyn, F. J. (1950), *J. Soc. Chem, Ind.*, **69**, 173–81.

[7] Matla, W. P. M. (1936), *Rec. Trav. Chim.*, **55**, 173–91.

[8] Roxburgh, J. M. (1961), *Canad. J. Chem. Engin.*, **39**, 46–8.

[9] Brown, K. C. and Woodhead, D. W. (1953), Safety in Mines Research Establishment, Research Report, 86.

[10] Wheeler, R. V. (1936), *Trans. Farad. Soc.*, **32**, 1244–9.

[11] Hartmann, I. (1957), US Bureau of Mines, Inf. Circ., 7785.

[12] Cybulski, W. (1952), 7th International Conference of Directors of Safety in Mines Research. Buxton, England, Paper 36.

[13] Brown K. C. (1951), Safety in Mines Research Establishment. Research Report, 22.

[14] Brown, K. C. and Curzon G. E. (1960), Safety in Mines Research Establishment, Research Report, 194.

[15] Brown, K. C. and Curzon G. E. (1963), Safety in Mines Research Establishment, Research Report, 212.

[16] Essenhigh, R. H. and Woodhead, D. W. (1958), *Comb. and Flame*, **2**, 365–82.

[17] Box, W. D. (1959), *Bull. Amer. Soc. Test. Mat.*, 34–7.

[18] Powell, F. (1962), *Comb. and Flame*, **6**, 75–6.

[19] Maguire, B. A., Slack, C. and Williams, A. J. (1962), *Comb. and Flame*, **6**, 287–94.

[20] Palmer, K. N. and Tonkin, P. S. (1967), 3rd Symposium on Chemical Process Hazards with Special Reference to Plant Design. Inst. Chem. Engin., London, 66–75.

[21] Craven, A. D. and Foster, M. G. (1967), *Comb. and Flame*, **11**, 408–14.

[22] Beer, J. M., Thring, M. W. and Essenhigh, R. H. (1959), *Comb. and Flame*, **3**, 557–60.

[23] Howard, J. B. and Essenhigh, R. H. (1965), *Comb. and Flame*, **9**, 337–9.

[24] Ishihama, W. (1961), 11th International Conference of Directors of Safety in Mines Research, Warsaw, Poland, Paper 47.

[25] Helwig, N. (1967), *Mitteil. der VGB*, **106**, 44–7.

[26] Line, L. E., Clark, W. J. and Rahman, J. C. (1957), 6th International Symposium on Combustion, 779–86. Reinhold. New York.

[27] Wheeler, R. V. (1935), Home Office Report on Experiments into the Means of Preventing the Spread of Explosions of Carbonaceous Dust. HMSO, London.

[28] Bowes, P. C., Burgoyne, J. H. and Rasbash, D. J. (1948), *J. Soc. Chem. Ind.*, **67**, 125–30.

[29] Cassel, H. M., Liebman, I. and Mock, W. K. (1957), 6th International Symposium on Combustion. 602–5. Reinhold, New York.

[30] Cassel, H. M. and Liebman, I. (1959), *Comb. and Flame*, **3**, 467–75.

[31] Burgoyne, J. H. and Long, V. D. (1958), Conf. on Science in the use of coal, 28, Inst. Fuel, London.

[32] Markstein, G. H. (1963), 9th International Symposium on Combustion, 137–47. Academic Press, New York.

[33] Marshall, W. F., Palmer, H. B. and Seery, D. J. (1964), *J. Inst. Fuel*, **37**, 342–9.

[34] Cassel, H. M. (1964), US Bureau of Mines RI 6551.

[35] Mason, W. E. and Wilson, J. G. (1967), *Comb. and Flame*, **11**, 195–200.

[36] Nagy, J. and Mitchell, D. W. (1963), US Bureau of Mines RI 6344.
[37] Hartmann, I., Jacobson, M. and Williams R. P. (1954), US Bureau of Mines RI 5052.
[38] Line, L. E., Rhodes, H. S. and Gilmer, T. E. (1959), *J. Phys. Chem.*, **63**, 290–4.
[39] Singer, J. M., Cook, E. B. and Grumer, J. (1967), US Bureau of Mines RI 6931.
[40] Morse, A. R. (1958), National Research Council, Canada. Radio and Electrical Engineering Division. Report 4968. Ottawa.
[41] Jacobson, M., Cooper, A. R. and Nagy, J. (1964), US Bureau of Mines RI 6516.
[42] Zehr, J. (1962), *Staub*, **22**, 494–505.
[43] Rae, D. (1968), Safety in Mines Research Establishment, Research Report, 255.
[44] Cybulski, W. B. (1959), Prace Glownego Inst. Gorn., 231, 3–23.
[45] Raftery, M. M. (1968), Ministry of Technology and Fire Offices Committee Joint Fire Research Organization. Fire Research Technical Paper 21. HMSO, London.
[46] Nagy, J., Dorsett, H. G. and Cooper, A. R. (1965), US Bureau of Mines RI 6597.
[47] Taffanel, J. (1921), Annales des Mines. Eleventh Series.
[48] Mason, T. N. and Wheeler, R. V. (1931), Safety in Mines Research Board Paper 64.
[49] Godbert, A. L. and Greenwald, H. P. (1935), US Bureau of Mines. Bulletin 389.
[50] Cybulski, W. B. (1962), *Arch. Gorn.*, **VII**, 2, 147–81.
[51] Petrukhin, P. M. and Netseplyaev, M. I. (1965), *Bezopas. Truda Prom.*, **9**, 11, 16–17.
[52] Finney, C. S. and Spicer, T. S. (1964), *Pyro.*, **1**, 231–49.
[53] British Standard No. 1016, Part 3, (1957).
[54] Loison, R. and Chauvin, R. (1964), *Chimie et Industrie. France*, **91**, 3, 269–75.
[55] Sharkey, A. G., Shultz, J. L. and Friedel, R. A. (1964), *Nature*, London, **202**, 988–9.
[56] Field, M. A., Gill, D. W., Morgan, B. B. and Hawksley, P. G. W. (1967), Combustion of Pulverized Coal. British Coal Utilization Research Assoc. Leatherhead.
[57] Martin, S. (1965), Tenth Symposium (International) on combustion. Combustion Institute, Pittsburgh, Pennsylvania, 877–96.
[58] Spiers, H. M. (1961), (Ed.). Technical Data on Fuel. Brit. Nat. Committee World Power Conf. London.
[59] Hartmann, I., Cooper, A. R. and Jacobson, M. (1950), US Bureau of Mines RI 4725.
[60] Essenhigh, R. H. (1961), *Comb. and Flame*, **5**, 300–1.
[61] Helwig, N. (1966), *Staub.*, **26**, 65–9.
[62] Hattori, H. (1957), 6th Symposium (International) on Combustion. Reinhold, New York, 590–5.
[63] Palmer, K. N. and Tonkin, P. S. (1971), *Comb. and Flame*, **17**, 159–70.
[64] Cybulski, W. (1954), Prace Glow. Inst. Gorn. Ser. A. Komunikat 147. Stalinogrod.

[65] Cybulski, W. (1954), 8th International Conference of Directors of Safety in Mines Research. Dortmund Derne. Germany. Paper 41.
[66] Jones, E. and White, A. G. (1948), Dust in industry. Society Chemical Industry. London, 129–39.
[67] Palmer, K. N. and Tonkin, P. S. (1967), Fire Research Note 667.
[68] Zehr, J. (1959), *Staub.*, **19**, 204–14.
[69] Nagy, J. and Portman, W. M. (1961), US Bureau of Mines RI 5815.
[70] Shagalova, S. A. and Reznik, V. A. (1965), *Teploener*, **12**, 7, 63–5.
[71] Helwig, N. (1967), *Staub.*, **27**, 349–53.
[72] Artingstall, G. and Corlett, T. C. (1965), Safety in Mines Research Establishment, Research Report, 238.
[73] Long, G. (1961), *Ind. Engin. Chem.*, **53**, 823–5.
[74] Nagy, J., Cooper, A. R. and Stupar, J. M. (1964), US Bureau of Mines RI 6561.
[75] Zehr, J. (1958), *Staub.*, **18**, 77–80.
[76] Zehr, J. (1962), *Berufsgen.*, 4, 150–3.
[77] Nagy, J., Dorsett, H. G. and Jacobson, M. (1964), US Bureau of Mines RI 6543.
[78] Rothbaum, H. P. and Groom, P. S. (1961), *New Zealand J. Sci.*, **4**, 476–88.
[79] Coal Industry National Consultative Council, Safety and Health Committee. Final Report of the Working Party on Coal-dust Explosions. National Coal Board, London (1967).
[80] Paterson, J. H. (1945), *Trans. Min. Geol. Metall. Inst. India,* **40**, 2, 24–54.
[81] Sen, M., Das Gupta, S. K. and Lahiri, A. (1957), *Trans. Min. Geol. Metall. Inst. India,* **54**, 43–62.
[82] Eisner, H. and Shaw, S. K. (1961), Safety in Mines Research Establishment, Research Report, 206.
[83] Eisner, H., Maguire, B. A. and Shaw, S. K. (1963), Safety in Mines Research Establishment, Research Report, 215.
[84] Essenhigh, R. H. (1961)), Coll. Engin., 534–9, (1962) 23–8, 65–72, 103–4.
[85] Greenwald, H. P., Howarth, H. C. and Hartmann, I. (1940), US Bureau of Mines RI 3529.
[86] Helwig, N. (1968), *Staub.*, **28**, 236–40.
[87] Pieters, H. A. J. and Hovers, J. W. J. (1948), *Fuel*, **27**, 109–17.
[88] Mitchell, D. W. and Nagy, J. (1962), US Bureau of Mines IC 8111.
[89] Cybulski, W. (1968), *Arch. Gorn.*, **13**, 1, 3–19.
[90] Kuzetchenko, G. (1968), *Bezopas. Truda Prom.*, **12**, 14–16.
[91] Shaw, S. K. and Woodhead, D. W. (1963), Safety in Mines Research Establishment, Research Report, 221.
[92] Steffenhagen, A. and Meerbach, H. (1966), *Re. Industr. miner.*, **48**, 973–83.
[93] Singer, J. M., Bruszak, A. E. and Grumer, J. (1968), US Bureau of Mines RI 7103.
[94] Singer, J. M. (1963), 9th Symposium (International) on Combustion, Academic Press, London, 407–14.
[95] Singer, J. M., Bruszak, A. E. and Grumer, J. (1966), US Bureau of Mines RI 6761.
[96] Helwig, N. (1967), *Gluckauf—Forschung*, **28**, H2, 81–6.
[97] Singer, J. M. and Liebman, I. (1968), *Comb. and Flame*, **12**, 506–9.
[98] How, M. E. (1966), *J. Inst. Fuel*, **39**, 150–8.
[99] Strauss, W. A. (1968), *AIAA J.*, **6**, 9, 1753–6.

6 *Mechanism of Dust Flame Propagation*

Under this heading are considered theories that have been developed to explain the mechanism of flame propagation in dust explosions, and also the extent to which they have been confirmed by experiment. A sound theoretical explanation of the flame processes is of fundamental importance and, if it could be developed, would have important consequences for industrial problems. As a result, the methods of protection against dust explosions could then be put on a more accurate basis, and information already available in relation to gas explosions might with more confidence also be used for dust explosions.

The following general points can be made concerning theories of dust flame propagation:

(1) As yet no unified theory exists covering the entire flame process from pre-flame reaction through to the final state of the combustion products.

(2) Theories are available for parts of the combustion process, e.g. pre-flame reactions, predictions of burning velocity, times of burning of individual particles.

(3) Identities of the important flame processes and particularly the rate controlling processes are still in doubt.

(4) The combustion of single particles is a simpler problem, but the theory cannot be extrapolated to the general case of the burning cloud.

(5) The theories which have been developed are often for specific dusts and may not be applicable generally.

As it is not yet possible to present a unified account, various aspects of the flame processes will be discussed in turn. These will include pre-reaction and ignition zone, flame propagation through dust clouds, burning time of clouds and single particles, and evidence from other combustion systems such as liquid droplets.

In the ignition and the flame propagation processes there is a stage at which some form of pyrolysis or decomposition by heat, must take place (including vaporization) for the fuel to become involved in the combustion reaction. It may be advantageous to regard pyrolysis as a separate stage from ignition and combustion [1] so that attention is concentrated on the details of the mechanisms involved. Insufficient is known for a clear picture to be drawn, even to the extent to which pyrolysis of a particle could precede or succeed its ignition. The former situation would arise if heat were transferred to the dust ahead of the flame front, the latter if the dust moved into or near to the flame front before rising appreciably in temperature. The pyrolysis could occur throughout the dust particle, or could be governed by the rate of conduction of heat inwards from the particle surface. Emission of volatiles could be the first stage of pyrolysis, leading to or resulting from ignition of the particle, or it could remove oxygen from the vicinity of the particle surface and quench any surface combustion that had ben initiated. Different dusts could have different pyrolysis behaviour. Furthermore, pyrolysis may be only partial, so that a solid residue remains, and it may be affected by variation in the heating rate as in Equations 5.2 and 5.3. The whole problem has been reviewed [1], but further research is needed before the mechanisms of pyrolysis can be established.

Pre-heating and ignition zone processes

In considering flame propagation through a dust cloud it is logical to start by considering a cold region of the dust cloud as a flame approaches. Alternatively, for mathematical convenience, the flame may be regarded as stationary and the cold dust supension move toward it. The velocity of the gas entering the flame will be higher than that of the cold gas [2], due to thermal expansion, and the burning velocity of a plane flame would be defined in terms of movement relative to the cold suspension.

Transfer of heat of the flame to the cold suspension can be by conduction, convection, or radiation, or combined processes. One starting point has been the Mallard and Le Chatelier theory for flame propagation in gas mixtures, although it did not give good agreement with experiment, and modifying the theory for dust suspensions. As developed for gas flames the theory considered conduction to be the mechanism of heat transfer ahead of the flame, and used a simplifying assumption that an ignition temperature of the gas existed below which no reaction occurred, above which the reaction was self-sustaining. The temperature profile is shown in Fig. 6.1. By equating heat flows at the plane T_i the following relation was obtained

$$S_u = \mu(T_f - T_i)/\rho c_p l(T_i - T_0) \quad , \tag{6.1}$$

where the symbols are defined at the end of the Chapter.

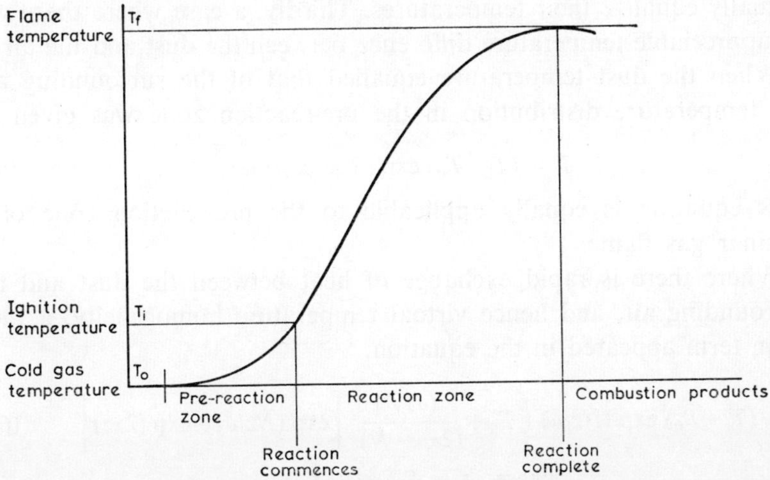

Fig. 6.1 Schematic temperature profile according to Mallard and Le Chatelier.

For dust flames radiation may be an important, but not controlling, factor in the heat transfer and the Equation 6.1 was therefore extended [3] to include a term accounting for radiation, but assuming an emissivity of unity.

$$S_u = \frac{\mu(T_f - T_i)/l + lX\sigma\alpha G(T_f{}^4 - T_0{}^4)/\rho_d r_d}{(c_p\rho + c_0 X)(T_i - T_0)} \quad . \tag{6.2}$$

For aluminium dusts calculated values of S_u from Equation 6.2 were in qualitative agreement with experiment. However, the equation can at best be only an approximation because of the known deficiencies in the thermal conduction model for gas flames, and because it is assumed that the dust particles and the gas surrounding them are at the same temperature.

By broadening the theoretical treatment, it has been possible to build up a semi-quantitative picture of the pre-reaction zone in a coal dust flame [4]. The assumption that no chemical reaction took place in the pre-reaction zone was retained, but three sets of thermal conditions in the zone were considered. Firstly, the dust particles were heated by radiation and the air by conduction from the flame front, and these respective rates of heating to be such that the rise in temperature of the dust was equal to that of the surrounding air. Secondly, the dust was heated by radiation and the air by conduction, but there was sufficiently rapid heat transfer between dust and air to virtually equalize their temperatures. Thirdly, a case where there was an appreciable temperature difference between the dust and the air.

When the dust temperature equalled that of the surrounding air, the temperature distribution in the pre-reaction zone was given by

$$T = (T_i - T_0) \exp (S_u \rho c_p z/\mu) + T_0 \quad . \tag{6.3}$$

This equation is equally applicable to the pre-reaction zone of a laminar gas flame.

Where there is rapid exchange of heat between the dust and the surrounding air, and hence virtual temperature homogeneity, a radiation term appeared in the equation,

$$T = (T_i - T_0) \exp (Nz/\mu) + T_0 + \frac{(\sigma T_f^4)}{(2\alpha\mu - N)} \left[\exp (Nz/\mu) - \exp (2\alpha z) \right] \tag{6.4}$$

$$\text{where } N = S_u(\rho c_p + X_0 c_0) \quad .$$

For an appreciable difference betwen the temperature of the dust particles and that of the surrounding air, the calculations became very complex and were not solved.

Analysis of Equation 6.4, using realistic values for flame properties showed that, for a coal dust concentration of 0.5 g/l, S_u would be greater than 87 cm/s and, for a concentration of 0.05 g/l, S_u would be greater than 120 cm/s. Both values are higher than would be expected from experiment, the latter particularly so because the dust concentration would be near that of the lower explosibility limit.

A theoretical treatment has been given [5] of the ignition of a cloud of particles in a gas, including the induction period between the initial temperature rise and the ignition of the particles. The ignition was assumed to be a heterogeneous reaction on the surface of the particles, and was more akin to the generation of a cloud of incandescent particles than to the initiation of a continuous flame. Other assumptions were that the temperature of a particle was uniform, that heat transfer to the gas was by conduction, and that initially the particles and the gas were at the same temperature. Distinction was drawn between the ignition of a single particle, by which the gas temperature would not be changed, and ignition of a dust cloud by which the gas temperature would rise.

A thermal radiation theory has also been developed for plane flames, on a burner [2]. Because uni-directional flow was assumed there could be no mixing of hot products of combustion with the unburnt mixture; in dust explosion flames this re-circulation may occur due to turbulence in the flame front. Consideration of the plane flame model may throw some light on explosion propagation. Heat transfer ahead of the flame was assumed to be by radiation; the heat was absorbed firstly by the dust which then lost heat by conduction to the ambient air. The theory considered the behaviour of the dust until it reached the ignition temperature, which was assumed to be that at which the dust decomposed. The ignition temperature was assumed to be constant and independent of ambient conditions. The model was developed in terms of a coal dust flame. In the analysis the following alternatives were considered:

(1) Infinite heat transfer rate from dust to air
(2) Heat transfer from dust to air assumed to be zero, with no heating of the air
(3) Heat capacity of the air assumed to be infinite
(4) Dust and gas temperatures not equal, both varying
(5) The distance between the burner and the flame front was assumed to be either infinite or not.

One solution of special interest was that for condition 1 and the infinite system of condition 5, as they yielded a value for the burning velocity

$$S_u = I/(T_i - T_0)(X_0 c_0 + \rho_g c_g) \qquad . \qquad (6.5)$$

The value of S_u, given by Equation 6.5, was considerably greater than 100 cm/s for the stoichoimetric concentration and decreased exponentially at higher dust concentrations. If the ignition time were less than infinite (alternative system of condition 5) the burning velocities progressively decreased to as little as 40 cm/s for an ignition distance of 5 cm, and the maximum velocity was obtained for a dust concentration of about four times stoichiometric.

The analysis involved a number of assumptions, the important ones being that both volatile and non-volatile contents of the dust were considered for the stoichiometric calculation so that a coal dust particle could burn completely in the flame front, there was no relative motion between dust and air, and the ignition temperature was fixed and independent of conditions. Nevertheless it is significant that a heat transfer mechanism consisting of radiation, accompanied by a possible cooling of particles by conduction from the dust, could yield a semi-quantitative account of behaviour of a burner flame.

Ignition of dust clouds by radiation alone has been reported, Chapter 3, where photographic flash bulbs could ignite clouds of metal dust. Successful experiments have also been reported in which a burning dust cloud in air, physically separated from a second cloud in air, has been able to cause ignition [6]. Flames of metal dusts, zirconium and titanium, were able to ignite clouds of these and other metals (aluminium, magnesium, iron) as well as coal dust. Flames of coal dust, which would have lower temperatures, were unable to cause any ignitions.

Results have been given for an experimental arrangement similar to that considered theoretically in the derivation of Equation 6.5, in which a plane coal dust/air flame was stabilized inside a chamber and downstream of a water cooled tube bank [7, 8, 9]. Samples of the coal were taken from various parts of the flame, and analysed for loss in volatiles, probably by the usual means which involve heating for some minutes in a furnace and measuring weight loss. Ahead of the flame front less than one tenth of the volatile content was evolved, and loss of fixed carbon was negligible; on entering the flame, losses were recorded in both the volatile content and the fixed carbon. Ignition appeared to start on the solid particles, rather than in volatiles, and for particle diameters of less than 15 microns the combustion was apparently heterogeneous with no flamelets of volatiles round the particles. However, if allowance is made for the increase in total volatiles evolved during rapid temperature rises, see Chapter 5, the apparent loss of fixed carbon could be explained by a Q value greater than unity [10]. It would not then be necessary to postulate heterogeneous combustion before that of the volatiles. Further direct observations of coal flames are required to settle the point.

The contribution of radiation in the heat transfer processes has been considered in relation to results obtained for the burning of coal dust flames on a burner, using air enriched with oxygen to 28 per cent [11]. The burning velocity increased with dust concentration up to 0.25 g/l (Fig. 5.14) and then remained constant. The stoichiometric composition based on the volatiles, determined by standard proximate analysis methods, was about 0.4 g/l. A theoretical analysis of the results was given, using as a starting point the Mallard-Le Chatelier equation with an extra term for thermal radiation.

$$S_u\rho[c_p(T_i-T_0)+h] = [\mu b\bar{U}(T_f-T_i)/XVS_u]+\epsilon_1\epsilon_0 T_r^4 G \quad . \tag{6.6}$$

Although equation 6.6 contains a radiation term, subsequent estimates showed that in fact the radiation, though appreciable, was probably not nearly so significant in small coal dust flames as it may have been in flames of aluminium or graphite dust in oxygen. By taking typical experimental values for the coal flame the heat fluxes due to conduction and to radiation were calculated, and the contribution of radiation was computed to be about 17 per cent of that of conduction. Further consideration of the processes in the coal flame led to the conclusion that the propagation rate was controlled by the rate of combustion of volatiles, which in turn depended upon the rate of release of volatiles from the coal particles.

A theoretical appraisal of present knowledge of propagation of flame through coal dust clouds also concluded that combustion of the volatiles was the basis of the process [12]. Using the assumption that the flame propagated in combustible gases, it was pointed out that preheated dust ahead of the flame, could undergo distillation. The situation was considered in which each particle was surrounded by gas at a temperature below the ignition temperature, but above that at which distillation would take place. Inside each particle the temperature was assumed to drop rapidly to that of the cold dust so that a proportion of the mass would not be preheated. Ignition was suggested to occur when the volatiles around a particle came into contact with burning gas from a neighbouring particle. Analysis was not developed further in this direction, and experimental confirmation was not attempted.

The ignition of a dust cloud in a heated enclosure has been analysed using thermal ignition theory, based on the Frank Kamenetskii theory for ignition in a homogeneous mixture [13]. The ignition processes were in three dimensions, as compared with the uni-dimensional approach for burner flames. The assumptions included heat transfer through the volume by conduction, equal temperatures for the dust and the air, uniform size and distribution of the particles, and that only the surface atoms or molecules of the dust particles participated in the chemical reaction. The critical condition for ignition was given by

$$2 = a^2 \frac{H^1}{\mu} \frac{E}{RT_i^{\,2}} M_0 \frac{3XM_d}{\rho_d r_d} f \exp(-E/RT_i) \quad . \tag{6.7}$$

According to Equation 6.7 the minimum ignition temperature, T_i, should decrease as the dimensions of the dust cloud were increased. Using typical values, a drop of 70°C was calculated for a tenfold increase in dimensions [13]. Also T_i should increase with particle size but should fall with increasing dust concentration. These predictions were tested experimentally by measuring the minimum ignition temperature of magnesium dust, falling through a vertical furnace tube about 2.5 cm in diameter; the results are shown in Fig. 6.2. Agreement between theory and experiment was not good, but there were indications that high dust concentrations the position might improve [13].

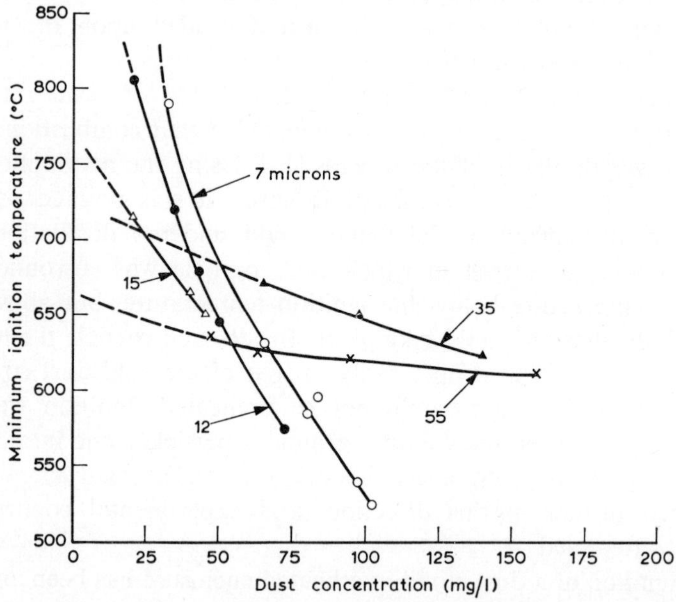

Fig. 6.2 Minimum ignition temperatures of magnesium dust clouds [13].

In similar experiments with magnesium and magnesium/aluminium alloy, and with aluminium and graphite flames on a burner, ignition was not achieved uniformly throughout the dust cloud but originated in a distribution of flamelets [14]. The concentration of flamelets increased with that of the dust, until eventually they seemed to merge. The flamelet formation was attributed to clusters in the dust clouds arising from random motions of the particles. A theoretical analysis was not given.

Ignition of a dust suspension in the Furnace apparatus, Chapter 3, was studied both theoretically and experimentally using corn starch dust [15]. The theoretical model assumed that ignition occurred within the dust suspension, inside the Furnace, and the ignition criteria developed by Semenov were applied to the system. The critical conditions for ignition were that the rate of heat generation through the suspension should equal the rate of heat loss to the walls of the Furnace, and that the differentials with respect to temperature should also be equal. The resulting equations were found to broadly correlate with values obtained with corn starch, after typical values for physical and chemical properties of the dust were assumed.

Further investigations with other dusts in the apparatus showed that the mechanism of ignition was more complex than assumed as the basis of the theory. Thus an inert chemical dust having a melting point of 482°C was dispersed through the Furnace at 500°C and 700°C and in neither test did the chemical melt. Fullers earth, with 25 per cent moisture, was then dispersed through the Furnace at 500°C, and no loss in moisture was detected. Further trials with coal dust, volatile content 35 per cent and ignition temperature 610°C, in which it was dispersed through the Furnace in an atmosphere of carbon dioxide to prevent ignition showed that only slight loss in volatiles occurred even with Furnace temperatures of 700°C and 850°C. The mass of the dust was not heated in these tests, and probably only those particles striking a wall were heated when dispersed through the Furnace. The uncertainties in the operation of the Furnace apparatus have been discussed in Chapter 4.

Evidence on the ignition of dust suspensions in heated enclosures was obtained by observing the impact of particles on the heated wall of a 600 mm globe [16]. With organic dusts, particles which contacted the wall gave immediate flashing and after a short delay, if conditions were favourable, a flame then propagated throughout the globe. With these dusts ignition of the particles themselves did not occur directly, but followed combustion of volatile products generated by the dust as a result of sudden heating. With inorganic dusts, the particles ignited immediately they struck the walls of the globe and flame propagated immediately throughout the volume. The relationship between the volatile content and minimum ignition temperature of several coals showed that the ignition temperature increased considerably as the volatile content decreased, using the heated globe apparatus. As a further demonstration of the relationship between volatile content and ignition temperature, some samples of the coal were partially devolatilized at temperatures of 300°C, 500°C and 1000°C. The residues obtained from heating at these temperatures were pulverized and sieved in order to obtain particles of the same size as the original coal. The minimum ignition temperature of the samples, determined in the heated globe, increased by up to 100°C as the volatile content was diminished.

The different characteristics of dust explosion flames and those in pulverized fuel furnaces may be exemplified by the case of anthracite [17]. The flow patterns of fuel and air in a pulverized fuel furnace are such that hot gases can recirculate into the pre-ignition zone and become an important part of the mechanism of stabilization of the flame on the burner. Ignition delays, the time intervals between particles entering the mixing zone and igniting, were calculated to be an order of magnitude less than the actual delays. The difference was ascribed to the entrainment of particles into the mixing zone being gradual, and to a minimum number being required for a flame to form. Another conclusion was that the ignition delay was of greater significance than the flame speed, because a stationary flame front may not be formed.

Further evidence of the behaviour of dusts under conditions of rapid heating was provided by the ignition of clouds of particles in oxygen in a shock tube [18]. In such experiments ignition of particles did not occur as a result of propagation of flame from one particle to the next, but as a result of heat transfer from the heated oxygen to the particles. Ignition of various coals was reported to occur after very small amounts of volatiles were lost from the particles, and the ignition may have been catalysed by the particle surface. A model was suggested in which the particle cloud was heated by the oxygen to a temperature between 700° and 800°K, depending on the volatile content of the coal, until ignition of the evolved volatiles occurred, near the particle surface. Once ignition occurred, the heat released caused further heating of the particles and acceleration of the loss of volatiles. A theoretical analysis was given which agreed with the trend of the experimental results, and the ignition of volatiles played an important part in the proposed mechanism.

A further means of causing ignition ahead of a flame front, which has not yet received theoretical analysis, is fragmentation or explosion of burning particles leading to burning solid or vapour being projected ahead of the flame [6]. With burning metal dusts, fragments which were probably molten were reported to be projected at high velocity; some of these fragments in turn subsequently disintegrated further. Although the process was not proved to be capable of causing flame propagation it is a possibility if burning fragments of particles can be projected into the unburnt dust cloud.

Flame processes in dust clouds

Although it is convenient to consider the flame processes separate from the pre-ignition and ignition zone processes, they are clearly linked and, inded, it is unlikely that the processes involved in the establishment of the flame would be unconnected with those required for the continued existence of the flame. Consequently in considering flame processes the difficulties are again encountered in selecting the important heat transfer mechanisms, in formulating the contribution of volatile and non-volatile constituents of the dust, and the possibility of variation in mechanism between dusts.

A mechanism of flame propagation in which combustion of volatiles was the basic process was suggested for explosions of phenolformaldehyde resin in a large scale tube apparatus [19], which have already been given some consideration in Chapters 4 and 5. The resin was mixed with various proportions of inert dust, magnesium oxide, and the explosibility limits of the mixture were measured. As the proportion of magnesium oxide was increased the limits narrowed until eventually a peak value was obtained which indicated the minimum percentage of magnesium oxide necessary to prevent explosion in any concentration of the mixture (Fig. 5.18). By assuming that the magnesium oxide acted as a chemically inert heat sink, and that the resin vaporised completely, Equations 5.11 to 5.13 were derived. Two processes for the action of the magnesium oxide were considered. The first was that addition of the inert material caused the flame speed to decrease until it was ultimately less than the velocity of fall of the dust particles. With upward propagation the flame would then be carried downwards into combustion products and would be extinguished. The experimental results did not support this postulate. The minimum values of flame speeds measured relative to the explosion tube were of order 100 cm/s, and thus appreciably greater than zero. In addition visual and photographic observations of the flames did not show a pronounced retardation as would be expected if they were swept downwards by falling dust.

The second possibility was that the presence of the magnesium oxide in the flame acted as a thermal sink and that when a certain proportion of the heat generated by the flame was abstracted propagation would cease. If the flame temperature were reduced by cooling to a value less than that at the lower explosibility limit, the flame was assumed to be quenched. Attempts had been made earlier to apply similar assumptions to the quenching of coal dust explosions by stone dust [20], with limited sucess. However the combustion of coal particles is complicated by the presence of both volatile and non-volatile constituents, and will be considered again below.

The combustion of the resin dust appeared to be principally in the vapour phase and the following overall process was envisaged. The phenolformaldehyde resin and magnesium oxide dust particles entered the front of the flame zone together. The flame zone consisted of the luminous burning region and the free oxygen with which it was in close proximity; gas sampling showed that even with resin concentrations of five times stoichiometric (5×0.12 g/l) only about 50 per cent of the oxygen had been consumed immediately behind the flame front. The resin particles vaporized and burned leaving the diluent dust in suspension in the flame zone acting as a thermal sink. On completion of combustion the volume originally occupied by a resin particle surrounded by cold air would be filled with hot products. For a stoichiometric mixture the diameter of this volume of products is of the same order as the original distance between the resin particles. Hence the transfer of heat to neighbouring particles, which is necessary for flame propagation, would occur.

It was not immediately apparent whether the diluent dust (magnesium oxide) would disperse as the resin vapours burned and expanded, or would remain at the same concentration as before entering a flame zone. An assessment of the probable behaviour was made by comparing the acceleration to be expected from the drag force on the particles due to expanding gas with the acceleration required if the particles were to disperse as the gas expanded. It was found that the acceleration due to drag was small compared with that required to disperse the diluent dust; the conclusion was drawn that the diluent dust particles were not dispersed as the resin burned. This conclusion may only be applicable near the front of the flame zone where combustion is relatively rapid because of the high oxygen concentration, but it is at this position that the heat balance would be critical in deciding whether further propagation could be sustained. Away from the front of the flame zone different considerations could apply but these would not be expected to govern the thermal conditions at the front.

If the resin vapours expanded on combustion, but the diluent dust particles were not dispersed, the relative concentrations of resin and diluent would change, and this is the reason for the introduction of the expansion ratio on combustion, r, in Equations 5.11 to 5.13.

Independent supporting evidence for the above process has been reported for cork dust flames [21]. From photographs of the flames, in which the movement of individual incandescent particles was measured, it was concluded that particles in the 'nose' of the flame were nearly stationary whereas those more deeply within the flame were rapidly accelerated. There was evidence that particles were decelerated as they entered the flame and the thickness of the zone within which the particles were moving slowly appeared to be about 2 cm in a tube of diameter 7.6 cm. In this zone the particles were incandescent but their relative separations had not been increased by the expansion of the gas.

The application of Equations 5.11 to 5.13 to the phenolformaldehyde resin/magnesium oxide system gave reasonable agreement (Fig. 5.18) but if the proposed mechanism of flame propagation had general applicability it should also hold for other fuels and other diluent dusts. On retaining phenolformaldehyde resin as fuel, but varying the diluent, the observed and calculated peak values given in Figs. 5.19 and 5.20 were obtained [22]. With some exceptions agreement was again fair, a marked exception being the behaviour of sodium iodide which, being a halide, gave anomalous behaviour.

On changing the fuel to polyethylene, whilst retaining magnesium oxide as the diluent, the percentage of magnesium oxide required to prevent explosions (the peak value) increased to between 90 and 95 per cent (Fig. 6.3). Here the dust concentration has been given on a kinetic basis, allowing for the velocity of fall of the dust particles against which the flame is propagating. Under these conditions the quantity of dust entering the flame front was higher than would have been the case if the dust were stationary, by a factor involving the velocities of fall and of flame propagation. With polyethylene the stoichiometric concentration in air (x_2) was 0.088 g/l. The lower explosibility limit determined experimentally, and corrected to a kinetic basis was only

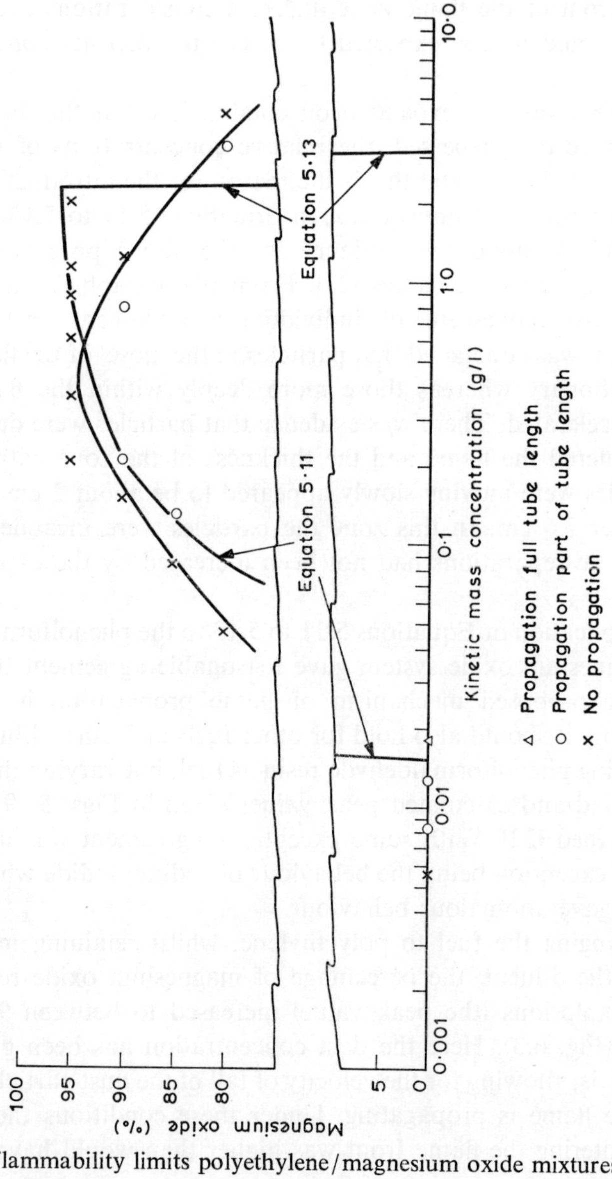

Fig. 6.3 Flammability limits polyethylene/magnesium oxide mixtures [22].

0.015 g/l. On inserting the appropriate values in equation 5.12 the theoretical lines shown in Fig. 6.3 were obtained, and these again are in good agreement with experiment as regards the peak value. The lower explosibility limit of polyethylene was so low that the calculated flame temperature (T_1) was only 790°K. The calculations were based on the assumption that the heat release in the flame was homogeneous. Visual observation of the flame indicated that the temperature was in fact much higher, and a likely explanation is that the flame was heterogeneous and consisted of hot gaseous regions interspersed with cold air.

With coal dust as the combustible some modifications of the above treatment would be necessary. Not all of each particle is likely to vaporize on entering the flame, and some carbonaceous residue would be expected. The solid residue would be surrounded by vaporized fuel initially, which would be burning with a diffusion flame, hence direct access of oxygen to the residual solid would be hindered whilst there was still volatile material burning. After the volatiles had been burned, oxygen could then react with the solid surface. However it is likely that the propagation of the flame is governed principally by conditions at the front of the flame, where combustion of the volatiles would still be in progress. The solid material formed on the devolatilization of the coal will thus, at least initially, act as a heat sink rather than a source of fuel; if coal/stone dust suspension were entering the flame two solid heat sinks would then be present [23]. In this simplified process the possible fragmentation of the carbonaceous residue and combustion within its pores have not been considered. For coal and diluent dust the variables in the heat balance Equations 5.11 to 5.13 would be modified to become:

for concentrations leaner than stoichiometric

$$(y_1c_2+y_2c_3)(T_1-T_0) = \frac{H(x-x_1)}{r} \quad , \tag{6.8}$$

at the stoichiometric concentration

$$(y_1c_2+y_2c_3)(T_1-T_0) = \frac{H(x_2-x_1)}{r} \quad , \tag{6.9}$$

for concentration richer than stoichiometric

$$(y_1c_2+y_2c_3)(T_1-T_0)+\frac{c_1(x-x_2)(T_1-T_0)}{r} = \frac{H(x_2-x_1)}{r} \quad . \tag{6.10}$$

The total dust concentration was given by $(x + y_1 + y_2)$. For Equations 6.8 to 6.10 to be applied, information was required on the amount of volatiles produced from the coal under flame conditions and on the lower flammability limit of the volatiles in the absence of non-volatile residue.

Under conditions of rapid heating, of order 10^4 or 10^5 deg C/s, the amount of volatiles produced from coal differs from the volatile content measured by standard procedures; for swelling coals the volatiles evolved may be increased by up to 70 per cent with rapid heating (Equations 5.2 and 5.3). These equations should be used in the calculations of concentrations of volatiles and non-volatiles residues from the coal.

For the purposes of calculation it is also necessary to know the lower explosibility limit of the volatiles from coal, in the absence of non-volatile component, and with negligible air movement ahead of the flame. The lower limit of the volatiles cannot be measured directly from the experiments but the results obtained from experiments with a large scale tube [23] have been used as a basis for an estimate. To obtain conditions of negligible air movement, the results for ignition near the open lower end of the tube have been taken (Table 5.6 line 2). The lower limit of the coal dust was 0.09 g/l, the coal having a volatile content of 36 per cent. Because of the continued downward movement of burning particles after ignition, the dust concentration as measured was taken as the relevant value; the flame was envisaged to propagate from particle to particle, and as the relative separation of particles would remain constant because of the residual solid during devolatilization, the separation would not be affected by the downward movement in the dust cloud. A dust which volatilized completely, with no solid residue, would change the separation between original particles because the vapour would fall under gravity more slowly, if at all, than would the original particles.

The fraction of volatiles produced by the coal at rapid rates of heating would be $0.36 \times 1.7 = 0.62$, from Equation 5.2.

Hence the concentration of volatiles in the observed lower limit flame was $0.09 \times 0.62 = 0.056$ g/l. This value is not the true lower limit concentration of the volatiles because of the heat sink effect of the non-volatile fraction of the coal. An estimate of this effect was made by calculating, for the lower limit flame, the equivalent sink effect of surplus air to that of the non-volatile fraction.

The load of non-volatiles is $0.034 \times 5 = 0.17$ g; the concentration of non-volatiles is the difference betwen the observed lower limit concentration and the calculated volatiles concentration. The expansion ratio of 5 was applied because the volatiles expanded on heating whereas the non-volatiles were assumed not to change their concentration.

The heat sink effect arising from 0.17 grams of non-volatiles is equivalent to that caused by $0.17 \times 0.23 \times 22.4/(0.27 \times 28.8)$ litres of air at atmospheric temperatures and pressure. The specific heat of air was taken as 0.27 and the mean specific heat of the non-volatiles was derived as 0.23. The latter value was obtained from a heat content of 261 cal/g for a temperature rise of 1160°C [24]. The heat content was obtained by extrapolation from an observed temperature range up to 756°C of a char from a high volatile bituminous coal previously carbonized at 927°C.

The additional heat sink effect due to the non-volatiles was therefore equivalent to that of 0.11 litres of air. Thus the lower limit of the volatiles, in the absence of non-volatiles, was estimated to be $0.56/1.11 = 0.050$ g/l.

With a lower explosibility limit of 0.050 g/l, and taking the heat of combustion of coal and coal products as 8300 cal/g, the mean flame temperature at the lower explosibility limit was calculated to be 1510°K. Hence $r = 1\,510/300 = 5.0$.

In applying Equations 6.8 – 6.10 to the results obtained with the large scale tube, the following values were taken for the properties of the coal:

$x_1 = 0.050$ g/l
$x_2 = 0.11$ g/l
$y_2 = 0.38x/0.62 = 0.61x$ g/l
$c_1 = 0.35$, the value for phenolformaldehyde resin vapour [19] being used, as no direct value for coal volatiles was available. An approximate value for c_1 was adequate for Equation 6.10.
$c_2 = 0.29$ for stone dust
$c_3 = 0.23$
$T_0 = 300$°K
$H = 8300$ cal/g

The calculated lines obtained by inserting the above values in Equations 6.8 – 6.10 are shown in Fig. 5.22; the horizontal axis was calculated as $(x + y_1 + y_2)$ and the verticle axis as $100\, y_1/(x + y_1 + y_2)$. The observed explosibility limits were wider than those calculated, particularly with the upper limits. The minimum proportion of stone dust required to prevent propagation at any concentration of the coal dust/stone mixtures, calculated from Equation 6.9, was 57 per cent at a total dust concentration of 0.40 g/l. The former value may be compared with 60–65 per cent obtained in the experiments, and with 65 per cent obtained with the stone dust in a full scale mine gallery [25]. The peak value (57 per cent) is a more severe check on the validity of theory than are the explosibility limits which are subject to chance variations in concentration of the suspension. The limits would also be affected by the fact that explosions were initiated at a closed end of the tube, thus altering the dust concentration during the experiment because of movement of suspension ahead of the flame caused by expansion due to the combustion. No correction for this expansion effect was made to the calculated lines. Considerable quantities of both dust and air could pass through the flame front unburnt, particularly at the higher concentrations, and non-uniformity in the flame would tend to increase the upper explosibility limits, the experimental values of which were much higher than those calculated. Further experiments in which the diluent dust was sodium chloride, instead of stone dust, yielded a peak value of about 15 per cent sodium chloride in the mixture, whereas the calculated value based on a specific heat of 0.25 would be 60 per cent. Clearly the sodium chloride was far more effective in preventing explosion than would be expected on the basis of its thermal properties alone, (as has been remarked earlier).

The minimum proportion of stone dust required to prevent explosion in coal has been calculated from Equation 6.9, and this treatment was extended to enable the stone dust requirement for coals of other volatile contents to be estimated. The extrapolation necessarily involves further assumptions, namely that the stoichiometry and the heat of combustion of the coals and their volatiles remained constant as did the mean specific heat of the non-volatile residue from the coals. Lines calculated from equation 6.9 for both swelling and non-swelling coals are shown in Fig. 6.4, together with the line representing gallery tests with coal dust/Fullers earth mixtures [26] (see also reference 48 in Table 5.4). The diluent dust was assumed to have a mean specific heat of 0.29, i.e. that of limestone, and the lines show the proportion of diluent calculated to be required to prevent propagation of explosions in coals of various volatile contents. The volatile contents were assumed to have been determined by the standard method [27].

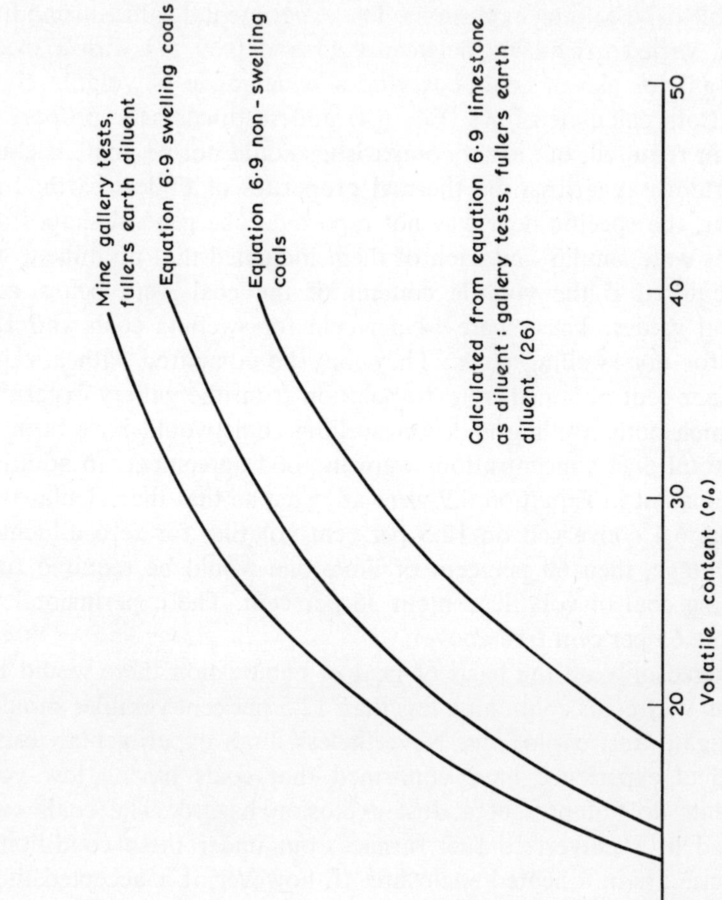

Fig. 6.4 Variation of minimum percentage of diluent to prevent explosions with volatile content of coals.

The calculated total dust concentrations varied slightly, from 0.40 g/l for high volatile content coals to 0.47 g/l for low volatile coal just capable of sustaining explosions. The experimental values, using Fullers earth, varied irregularly between 0.25 and 0.60 g/l with a mean of 0.47 g/l for eleven coals covering a wide range of volatile contents [26]. Both calculated lines (Fig. 6.4) underestimated the proportion of diluent required, but exact comparisons could not be made because of uncertainty regarding the thermal properties of Fullers earth. In particular, the specific heat was not reported. The general shapes of the curves were similar and each of them indicated that no diluent would be required if the volatile content of the coal were below certain critical values. These were 14 per cent for swelling coals and 18 per cent for non-swelling coals. They may be compared with a value of 12.5 per cent obtained by extrapolation from the gallery experiments, in which both swelling and non-swelling coals would have been used. The total dust concentrations were in good agreement. In addition, if the constant in Equation 6.9 were adjusted so that the calculated lines in Fig. 6.4 converged on 12.5 per cent volatiles for zero diluent dust percentage, then 60 per cent of limestone would be required for the swelling coal of volatile content 36 per cent. The experimental value was 60–65 per cent (see above).

Judged only on the basis of heat of combustion there would be no reason why coals containing less than 12.5 per cent volatiles should not propagate dust explosions. Nevertheless, both experimental tests and practical experience have confirmed that coals having low volatile contents do not present a dust explosion hazard. The coals can be burned in a pulverized fuel furnace, but under these conditions the particles are in a heated enclosure. If, however, it is accepted that the non-volatile residue formed when the coal dust particle enters a flame can act as a heat sink, then a physical reason for the observed behaviour in gallery and furnace is apparent. Although the agreement between the calculated and observed values was only moderate, the general trends were correct and would justify further measurements of the thermal properties of the dusts under explosion conditions.

The percentage of stone dust required to prevent explosion in a given coal in a mine gallery depends upon the energy of the ignition source (Fig. 5.6). In the application of Equations 6.8 to 6.10 the value of the expansion ratio r has been taken as 5.0. This value is based on the assumption that the pressure throughout the period of the explosion remained at atmospheric, so that the vapours generated from the coal could expand freely in the flame front. If conditions were such that expansion of the combustion products was hindered so that the pressure behind the flame front was higher than that in the unburnt dust cloud, a coal particle entering the flame front would be exposed to increased pressure and the value of r would be reduced. Such conditions would be most likely to arise in practice when a rapidly propagating flame was initiated by a powerful igniting source. In the extreme case, where no expansion was permitted behind the flame front, r would be unity instead of 5.0. Then, using Equation 6.9, the percentage of limestone diluent required to suppress explosion would be increased from 57 to 89 per cent for a coal of volatile content 36 per cent. For this situation to exist the pressure would need to increase sharply fivefold at the flame front. The flame speeds relative to the wall of the enclosure would need to be at least sonic (330 m/s) and an ignition source would be required sufficiently powerful to bring about these conditions. For comparison, a value of 81 per cent diluent dust can be obtained from Fig. 5.7, for a system with a powerful ignition source. A further comparison can be made with results from a gallery 200 m long, and with a cross-sectioned area of 5 m^2, for coal dust explosions suppressed with stone dust and water [28]. Five types of ignition source were used and with the most powerful source pressure rises of several atmospheres were recorded near the flame front as well as flame speeds of the required range. The proportion of stone dust required for suppression of explosion was 88 per cent. The value calculated from Equation 6.9, with $r = 1$, for a swelling coal of 41 per cent volatile content would be 90 per cent. The agreement between calculated and observed values was good but account should be taken of uncertainty regarding explosion pressure, the direct influence of the ignition source throughout the length of the gallery, and the difficulty of control of large-scale experiments. Equation 6.9 appears to give insight into the effect of the power of the ignition source, which at present can only be considered on an *ad hoc* basis.

Additional evidence on flame processes in dust clouds was provided from studies of the lower explosibility limits of coal/methane/air mixtures [29] where the conclusion was drawn that the coal dust suspended in the air ahead of the flame began to devolatilize because of heat from the flame front. Combustion of the volatiles was then initiated and the heat transfer to the dust particles caused further emission of volatiles. It was suggested that combustion in the gas phase was fast, whereas that at the surface of the non-volatile residue was slow.

Hence the flame front was propagated mainly by gas phase combustion and the solid residue contributed relatively little to the mechanism of flame propagation.

Equation 6.9 can be slightly modified to permit speculation as to the role of water in suppressing coal dust explosions [23]. Consideration must be restricted to the case when water is in intimate contact with the coal dust before the explosion, and would not apply to suppression by water sprays or other means of dispersing water during an explosion. Because of the large reduction in density of water on conversion to steam, the calculation of the cooling effect has been based on heating of the water to 100°C and its subsequent vaporization at this temperature. The cooling effect of the steam would be relatively small. The water required to suppress an explosion in a coal dust of volatile content 41 per cent was calculated as 47 per cent. The experimental value was 48 per cent [28]. The agreement was good, but it must be borne in mind firstly that coal containing 48 per cent of water is difficult to disperse, which would reduce the water requirement. Secondly, because of the difficulty of dispersal, a powerful ignition source was used which could produce pressure effects tending to increase the water requirement. The extent to which these effects cancelled could not be

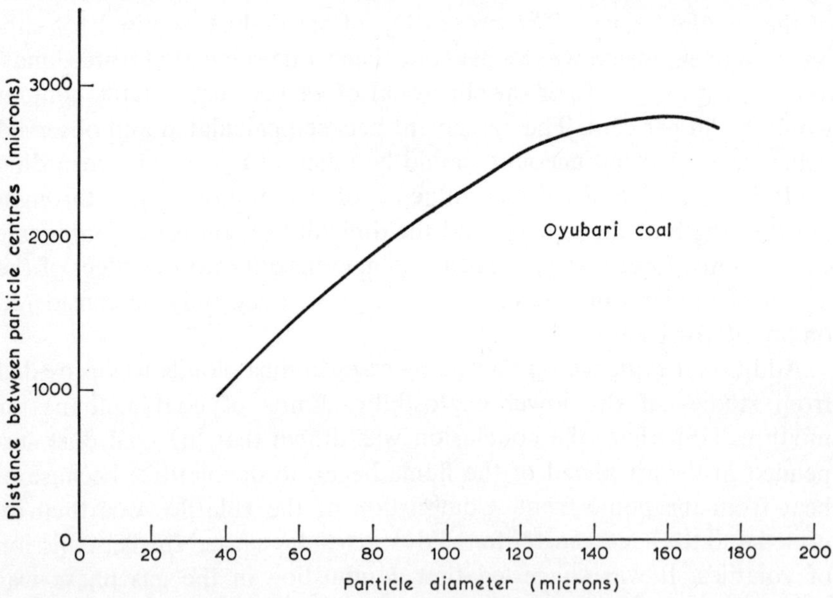

Fig. 6.5 Particle separation at lower explosibility limits [30].

estimated. Use of Equation 6.9 infers that the water acted as a heat sink only, and this is reasonable. Further experimentation is needed to provide more definite evidence.

For flame to propagate from one dust particle to the next in a suspension, the separation of the particles is likely to be an important factor. The separation between the centres of neighbouring particles, as a function of particle diameter, at the lower explosibility limit of a coal dust (volatile content about 37 per cent) is shown in Fig. 6.5; for the purposes of the calculation the dust dispersion was assumed to be uniform and the particles disposed as in a cubic lattice [30]. For particles of diameter up to 100 microns, the separation was proportional to the diameter; the separation reached a maximum for particle diameter about 160 microns and then started to decrease. Thus for particles below 100 microns diameter the separation was about 22 diameters, but at 160 microns was only 18 diameters and for coarser particles the separation rapidly decreased. With the particles of large diameter, only a part of the coal was contributing to the flame propagation. Further analysis of this aspect of flame propagation could be of value, because if it is assumed that the hot gas envelope around a burning particle ignites the neighbouring particle then a method for calculating the lower explosibility limits of dusts from a knowledge of their stoichiometry might be developed. If the diameter of the envelope of hot gas round a completely burned particle were twice the separation between particles, then it would, on the basis of a cubic lattice, just touch six neighbouring dust particles. At this dust concentration flame propagation between particles should be easy. Alternatively, if the diameter of the envelope equals the separation between particles, then because of expected random movements, the envelope is likely to touch at least one of the neighbouring particles so that flame propagation could again proceed. These two separations could represent limiting cases within which actual dusts are likely to fall, so that the measured lower explosibility limit should be between those calculated by this method.

The burning of plastics in bulk may be mentioned because it could assist in understanding flame propagation in plastics dust clouds. The burning of rods of polymethyl methacrylate, polyoxymethylene, and polyethylene indicated that the plastics did not react chemically with the gas around them but vaporized in heat from the surrounding diffusion flames and the pyrolysis products, then oxidized subsequently in the flames [31]. However, with polytetrafluoroethylene in pure oxygen there was evidence that the burning polymer reacted directly with the gas. Evidence has also been put forward that in the burning of polyethylene there was considerable oxidation in the molten liquid phase, as well as in the gas phase [32], but the relative importance of the two processes has yet to be decided. At least, however, the evidence points to oxidation in the vapour or liquid phases with these polymers rather than in the solid phase.

The combustion of pulverized coal in furnaces may be a process allied to a controlled dust explosion, and a comprehensive account of the present state of knowledge on the subject is available [33]. Combustion proceeds in two stages, firstly the volatiles which is followed by that of the fixed carbon. The rate controlling processes in the combustion of the volatiles are the devolatilization, and the diffusion of oxygen to the volatiles. The chemical reaction rates of the volatiles, once mixed with oxygen, appear to be rapid and not rate controlling. As regards combustion of the fixed carbon, the transport of oxygen to the solid surface and its reaction rate at the surface both need to be considered.

The chemical properties of dust flames have received little attention, reflecting the uncertainty as to the rate determining steps in flame propagation, in particular whether these steps are physical or chemical. The reaction kinetics of a magnesium/oxygen diffusion flame, in which the magnesium was carried in a stream of argon and burned in a surrounding oxygen atmosphere at pressures of a few mm mercury [34], using a flame stabilized on a burner, enabled the equilibrium constants for the reaction of magnesium with oxygen to form oxide to be calculated. Although there is insufficient information to enable the full reaction mechanism to be proved some steps in the reaction were proposed. A magnesium/oxygen flame generates a condensed combustion product and the reaction of magnesium with oxygen could take place heterogeneously on the surface of oxide particles. Although the reaction system is superficially simple, with only two elements involved, in fact the kinetics may be complicated.

Another investigation in which the chemical composition of the dust was considered involved the measurement of maximum explosion pressures of dispersions of three sugars in air. The sugars were dextrose, sucrose, and raffinose, and a range of particle specific surfaces was studied for each sugar [35]. The explosions were produced in a

Flour mill explosion. Collapse of brickwork after secondary explosion. (*HMSO*)

Penthouse structures above open hoists damaged by secondary explosions involving oilseed meal. (*Liverpool Fire Brigade*)

3 Distortion of building and window damage caused by esparto grass explosion. (*Fife Fire Brigade*)

4

Casing of bucket elevator destroyed by sugar explosion. (*Liverpool Fire Brigade*)

Multi - storey brick building demolished by provender explosion.

5

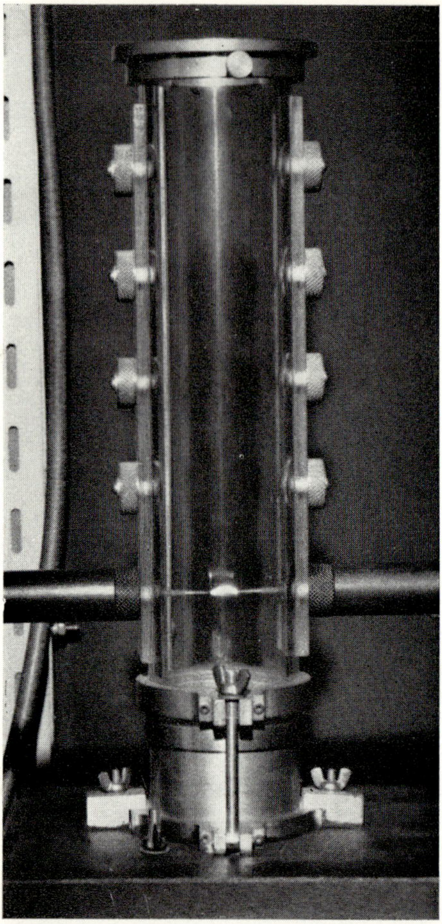

6

6 Dust explosibility testing: Vertical Tube. (*HMSO*)

7 Dust explosibility testing: Horizontal Tube. (*HMSO*)

7

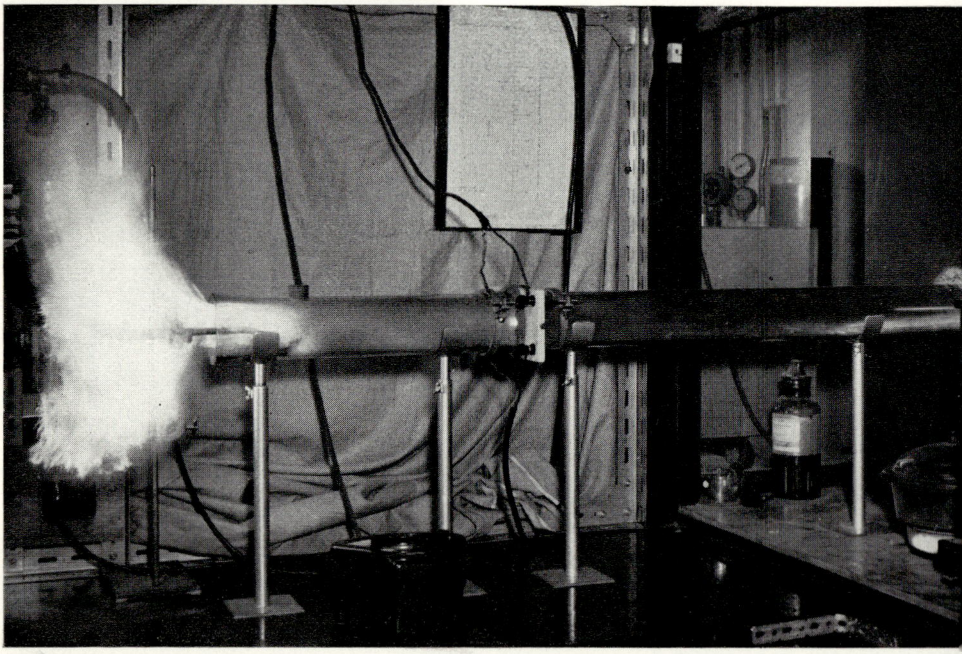

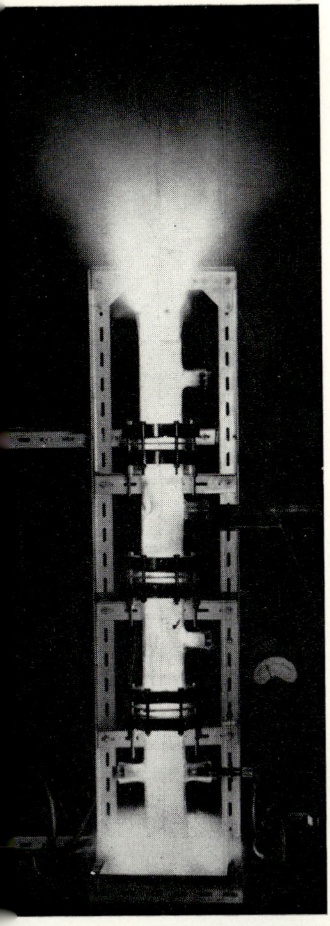

8

8 Dust explosibility testing: Inflammator. (*HMSO*)

9 Dust explosibility testing: Furnace Apparatus. (*HMSO*)

10 Dust explosibility testing: Explosion pressure measurement. (*HMSO*)

9

10

11

12

11 Damage to storage section of adipic acid plant, following explosion in dryer initiated by cutting equipment.

12 Cork dust explosion in vented cyclone plant. (*HMSO*)

13 Stainless steel cover (26 gauge) held by clamping ring on scroll outlet of cyclone. Cut-away base portion of vent ducting also shown. (*Rosin Engineering Co Ltd*)

14 Lightweight tops of silos displaced by oilseed meal explosion. (*Liverpool Fire Brigade*)

15 Two spring-loaded explosion reliefs, in cylindrical shroudings, on coal pulverizing mill. (*Central Electricity Generating Board and International Combustion Ltd*)

16

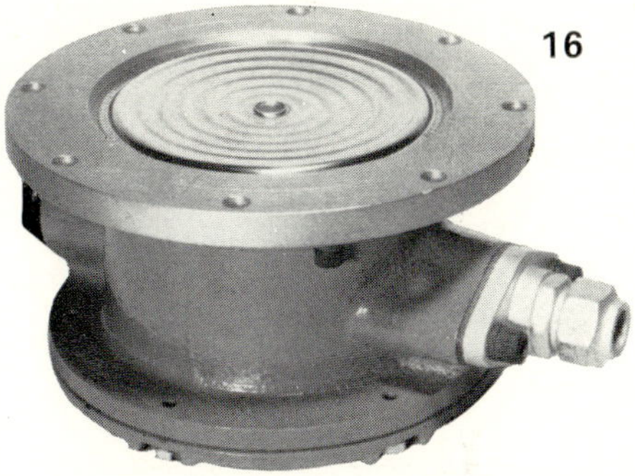

17

16 Automatic explosion suppression: explosion detector. (*Graviner Ltd*)

17 Automatic explosion suppression: cutaway view of 500 cm^3 hemispherical suppressor unit. Cylindrical housing for detonator mounted centrally. (*Graviner Ltd*)

18 Automatic explosion suppression: high rate discharge bottles mounted on ducting. (*Graviner Ltd*)

18

19

19 Wood sawdust ignition experiment using 3 ft × 3 ft × 3 ft specimen in air oven. (*HMSO*)

20 Soya flour explosion. Damage to quay by underground conveyor tunnel which was partly vented by manholes. (*Glasgow Fire Brigade*)

20

21

22

21 Soya flour explosion. Top of bucket elevator fallen on to silos. (*Glasgow Fire Brigade*)

22 Distortion of return leg of bucket elevator by provender dust explosion. (*Liverpool Fire Brigade*)

23

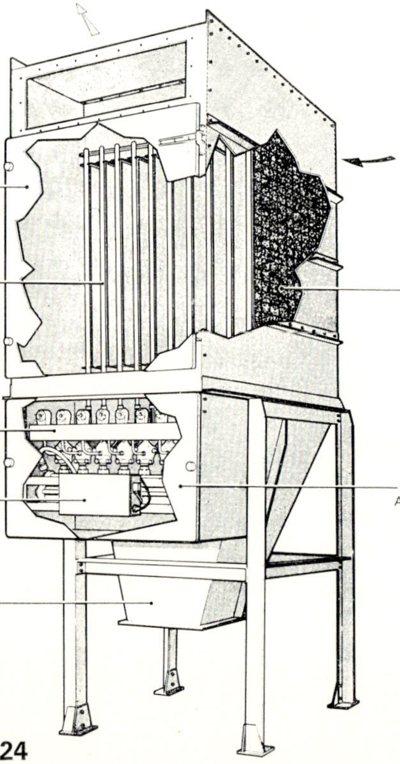

24

23 Unit dust collector (*left*) serving sawing machine for plastics tubing (*right*). (*Dust Control Equipment Ltd*)

24 Cutaway drawing of filter unit with reverse air cleaning. (*Dust Control Equipment Ltd*)

25 Wet collector for fettling and grinding of aluminium. (*J B Thorne & Co Ltd*)

26 Collection of plastics dust by wet scrubber. Waste container underneath for removal of sludge. (*J B Thorne & Co Ltd*)

27 Windows damaged by sugar explosion in framed building. (*Liverpool Fire Brigade*)

28 Explosion in cork grinding plant, with pressure relieved by lightweight cladding. Building framework and glazing intact.

29 Damage to lightweight wall cladding and windows by flour explosion. Steel building frame intact. (*Belfast Fire Brigade*)

cylindrical pressure vessel, 14 in. diameter and 23 in. high, ignited by a high voltage spark. The rates of pressure rise (dp/dt) at any pressure p were shown to follow an equation of form

$$dp/dt = Ap - Bp^2 \quad . \tag{6.11}$$

In addition the maximum explosion pressure, at a given dust concentration, decreased as the number of carbon atoms in the sugar molecule increased from 6 to 18.

Burning times of single particles and clouds

The measurement of the time required for dust particles to burn away and the relation of this measurement to the physical and chemical properties of the particle, are clearly of fundamental importance in flame propagation. The practical problem with dust explosion involves the combustion of a large number of particles which interact with each other, in the process leading to complications in both the experiments and in theoretical analysis. Two approaches have therefore been taken, one being consideration of a burning cloud, the behaviour of individual particles not being taken into account. This approach inevitably means that any resulting analysis is unlikely to explain the behaviour of the dust in fundamental terms. Alternatively the combustion of single particles has been considered, and this has been the favoured approach. The advantages are that the system is much simpler and therefore presumably more likely to yield successfully to analysis, but difficulties are at once met when attempting to proceed from consideration of the single particle to the cloud and this step has not yet been successfully and comprehensively managed.

The burning of individual particles, and particularly the measurement of the time for complete combustion, has been studied by two main techniques. These are the captive particle, in which the dust is attached to a relatively fine support, and a free falling method in which the particle drops under gravity or is carried in a gas stream either through a furnace or a flame. The burning of single particles has been studied both with solids and with liquids (i.e. single droplets) and there are basic similarities in some of the results. In considering solids most of the information relates to coal and allied materials, which have frequently burned in two stages. The first stage is the emission and burning of volatiles, followed by a second stage of the combustion of solid residue. Full consideration of the situation cannot be given here but the field will be briefly surveyed and key review references quoted (Table 6.1.).

Table 6.1 Reviews of single particle combustion

Reference	*Fuels considered*	*Type of review*
36	Mainly coal	General description.
37	Coal and allied materials, various liquids.	Mathematical analysis.
33	Coal	Analysis of combustion of volatiles and of carbon.
38	Coal	Comparison of diffusional and chemical rate control.
39	Liquids	General survey, with mathematical analysis.
40	Coal	Comparison of experimental results and analyses.
41	Liquids	Wide review, including droplet arrays and simplified sprays.

In the burning of the volatiles, the first stage of combustion, the theoretical model is of a sphere of volatiles surrounded by a concentric spherical flame zone from which the combustion products flow outwards into the surrounding atmosphere. The rate controlling process is the diffusion of oxygen to the flame, against the counterflow of combustion products, and the chemical reactions in the flames are assumed to be fast and not rate-controlling. Using this model the burning time (i.e. the total time required for combustion) is given by

$$t = Kd^2 \quad . \qquad (6.12)$$

For the volatiles from coal, and for many liquid fuels, the value of K is of order 10^{-6} s/microns^2 i.e. a particle of diameter 1000 microns would have a burning time of order 1 s [37]. The basic theory may be modified by considering the effect of chemical reaction on the burning time [33, 38] and also the effect of ambient temperature [39]. Furthermore if soot is formed during the combustion of the volatiles, and is itself then consumed, the burning time may be substantially extended [33] unless temperatures are above 2000°K.

For captive spheres of thermoplastic polymers the value of K in Equation 6.12 is in many cases only a quarter to a half the value for oil drops [42]. Their burning rates are thus two to four times greater than those of oil drops. This unexpected behaviour was attributed to the much closer location of the flame zone to the particle surface observed for these materials than for oil drops.

During the combustion of coal, the burning of the volatiles is a well defined first stage, and in the burning of liquids it may be the only stage. Other dusts have not received detailed consideration but those which devolatilize completely, yielding no solid residue, would in effect undergo only the first stage of combustion. Dusts which do produce solid residues, could then yield a second stage as does coal. A further complication with dusts other than coal is that some can give solid products of combustion, whereas those of coal are mainly gaseous. Solid products of combustion would be formed at the zone where fuel vapour and oxygen meet and can interfere with the diffusion of oxygen to the fuel, and the permeability of the oxide to the passage of oxygen becomes important. If the fuel is burning in pure oxygen, instead of air, the situation could rise where all the combustion products are solid or molten (as with metals such as magnesium, zirconium) and where the combustion products would not move counter current to the oxygen flow. The rate determining processes under these conditions could be the resistance to diffusion of the oxygen, combined with the flow of heat required to produce further vapour for combustion. The combustion of metals under these conditions has assumed a practical importance in rocket propulsion and rapid advances in theory and in experiment are being made.

With coal, the second stage of combustion may be regarded as involving the fixed carbon residue after devolatilization. The process is basically that of reaction of a solid surface and several steps are involved any of which may be rate controlling. The steps are transport of oxygen to the surface against a counterflow of combustion products, chemical reaction at the particle surface, some of which may occur in pores within the particle, and movement of combustion products away from the particle surface [33]. The burning time is governed by the slowest of the steps. The reaction rate on the particle surface may be governed either by the chemical process involved, or by the kinetics of adsorption [37].

Experimental evidence indicates that the burning time for second stage combustion follows Equation 6.12, except that K is now of order 10^{-5} s/microns^2 so that the burning times are about ten times as long as for the volatiles [37]. The theoretical analysis of second stage combustion is more complex than that of first stage combustion, and simplified solutions are not generally applicable. A detailed consideration of the present position is available [33].

Where the burning time is controlled by diffusion, the transport of oxygen to the fuel may be dependent upon the velocity of the gas relative to the fuel particle [38]. If the gas is turbulent it can accelerate the removal of combustion products and the arrival of oxygen so that combustion is faster. The calculated decrease in the burning time is up to 90 per cent, although it depends to some extent on the air/fuel ratio and the particle diameter. Scavenging is however, not likely to be complete, and so smaller reductions in burning time would in practice be expected.

The combustion of volatiles and of coke residue may be sufficiently separated for them to be regarded as completely independent processes, and as being the result of a double ignition [43].

The relation between burning time and particle diameter is superficially straightforward, but when the heating process is examined in detail it is found to be complex. When single coal particles were heated at rates of temperature rise of 1800°C per second in a hot stage microscope some coals expanded forming a cellular structure [44]. Such particles burned away faster than those which showed only slight expansion, with slower heating rates less expansion was obtained. Expansion was also relatively greater for finer particles, and combustion of the coal took place not only on the surface of the particle but within the interior.

Similar behaviour was found for captive coal particles heated by radiation [45]. With particles initially shielded from the radiation, and then suddenly exposed, the flash of the volatiles burning was observed almost immediately and this was followed by the second stage combustion, usually after some delay. During the second stage, the diameter of the particles decreased by only about one third, indicating that they were hollow and burning at the surface only. The burning time varied with initial diameter according to Equation 6.12.

The burning time of coal particles is also a function of the ambient pressure and oxygen concentration [46]. An increase in the oxygen concentration reduces the burning time considerably more than when air is compressed to give the same total oxygen pressure. For example with charcoal, increasing the oxygen concentration from 21 to 98 per cent reduced the burning time to about a quarter, and the temperature of the particles rose considerably. Increase in ambient pressure of air had little effect on the burning time with fine particles, but the effect of pressure did increase with particle size. The burning times of both anthracite and charcoal were similar to each other.

Decreasing the oxygen concentration from that in air, by the addition of nitrogen, showed that the structural changes in coal rapidly heated by being dropped through a furnace depended upon the oxygen level [47, 48]. Particles treated in nitrogen possessed a more open structure and were more swollen than those heated in air. Swelling was also influenced by the duration and temperature of treatment.

The combustion processes of metals such as aluminium and magnesium in oxygen have received attention [49]. The principal combustion product, the oxide of the metal, is solid or molten, and the flame temperature is probably governed by its boiling point. At these high temperatures the importance of radiation in heat transfer must be considered, and a fairly comprehensive mathematical treatment has been put forward. A study has also been reported of the combustion of boron particles injected into gas streams at 1800–2900°K, with mole fractions of oxygen in the range 0.08–0.37 [50].

The burning times of individual particles present in suspension during dust explosions have not been studied, and the nearest approach for free burning flames is the combustion of pulverized fuel in a furnace [33].

Particle burning times in flames on burners have been measured experimentally and analysed theoretically. One such analysis considered the behaviour of a single particle and then integrated for the cloud [13], assuming that the behaviour of a cloud was the sum of that of its individual components. Interactions between dust particles were not considered. For the stoichiometric composition the calculated burning time was infinite, because reaction rates would continuously decrease as the concentrations of dust and oxygen were reduced. At the other limiting condition of a particle in infinite excess of air the calculated burning time was that of a single particle. The analysis took no account of explosibility limits. The predicted burning times of magnesium and aluminium particles, of diameter 5 microns were fractions of milliseconds for flames leaner than stoichiometric. The burning times increased to several milliseconds for particles of 50 microns diameter.

The results from an investigation of coal dust/oxygen flames on a burner [51] showed that the burning time of the particles varied with initial diameter according to

$$t = Ad + B \quad . \tag{6.13}$$

The volatile content of the coal was 38 per cent, and the value of A and B varied with the coal/oxygen ratio. Because of the linear relationship between burning time and particle diameter the rate controlling process was believed to be chemical reaction, rather than transport of oxygen, at least for particles smaller than 30 microns.

Evidence from burning droplets

Flame propagation can occur through suspensions of liquids in gases, even with liquids of low volatility, and it is useful to summarize the available evidence on mechanism of propagation, for comparison with dust flames. Usually the combustion of liquid droplets is taken to be principally in the vapour phase, although with some liquids a coke residue may be left after the volatiles have been removed, and combustion of the solid phase may then occur. Only the combustion of vapour from the liquid has usually been considered. In comparing flames in dusts with those in droplets the analogy is likely to be closest for those dusts which would be likely to volatilize completely in the flame. However there may be a substantial number of such dusts.

The experimental evidence on the combustion of sprays of droplets has ben summarized [39] and two modes of combustion have been identified by direct observation and by photography. The two modes have been termed homogeneous and heterogeneous combustion. Homogeneous burning has been produced when the droplet diameter was small, 10 microns or less, using liquids such as tetralin and oils. With this mode of combustion the droplets vaporized completely before igniting, and appeared to form a continuous flame front. The extent to which the vapour mixed with air before ignition has not yet been investigated, and it is often tacitly assumed from the appearance of the flame that pre-mixing is good. However it is likely that the vapours are not well mixed with air and that combustion is essentially that of an array of small pockets of vapour, surrounded by air, burning with diffusion flames which have linked together. The flames are observed often to be blue, and this has favoured the assumption of premixing, but the evidence is not conclusive because very small gas diffusion flames can also be blue. Additional evidence that could support the hypothesis of premixing is that the minimum explosible concentration of tetralin droplets of 10 micron diameter was in the range 0.045–0.050 g/l, and were close to the values obtained with many premixed hydrocarbon/air flames [52]. If complete premixing did occur, the flames would be homogenous and all the vapour and gas would be heated uniformly as it passed through the flame front. At the other extreme, if no mixing of the vapour occurred, and the flame was an array of diffusion flamelets, some air might be able to pass through the flame front unheated, particularly at low droplet concentrations.

The heterogeneous mode of combustion was that in which evaporation of the droplets was not complete before they entered the flame front. Each droplet then becomes surrounded by its own diffusion flame, and complete premixing cannot occur. Heterogeneous flames are brush-like and thick, and are often yellow. In such flames the rate of evaporation of the droplets, and the transport of oxygen to the vapour, are linked as the controlling process, and a theoretical analysis giving reasonable agreement with the experiment is available [39]. The continuity equations assume that all vapour and liquid entering the flame becomes heated uniformly; a similar theoretical approach has been proposed, introducing what is termed relay transfer [53]. It is assumed that as the droplets are ignited, diffusion flames form around them which heat by conduction the air ahead of the flame. The droplets in the heated air, after a time lag, then ignite and so the flame propagates. The assumption is again made that all the gas and droplets entering the flame front are heated. A mechanism which does not seem to have been proposed is that individual particles ignite and the flames surrounding them impinge on neighbouring droplets causing ignition and propagation of the flame. With this mechanism not necessarily all the air surrounding the droplets would be heated as it entered the flame front.

The minimum explosible concentrations for heterogeneous droplet sprays can be much lower than for homogeneous flames. With droplets of diameter 100 microns, flame propagation was obtained at concentrations as low as 0.01 g/l [52, 53]. However after allowing for the velocity of fall of the droplets, against an upward propagating flame, leading to higher concentrations of droplets in the flame front the concentrations reverted to the values in the region 0.04 to 0.05 g/l.

For large droplets the term minimum explosible concentration may have little meaning [39]. A stationary flame has been reported in a one dimensional stream of droplets and the stability of the flame depended on the separation between the droplets. In such a situation the droplet concentration is not calculable on a volume basis but the mode of combustion can be regarded as heterogeneous.

Further study of the combustion of a mono-sized droplet stream, at diameters 10–500 microns, showed that the droplets were surrounded by a blue flame above which was a yellow flame region [54]. The size of the yellow flame increased with droplet diameter and as the spacing between the droplets was reduced. The experimental method enabled the burning rates of the droplets to be measured and led to the conclusion that the combustion was a non-steady state process because the average temperature of the burning droplets remained below the boiling point for much of their lifetime.

The extent to which vaporization occurred ahead of the flame, in a suspension of fuel oil (No. 2 heating oil), was estimated by diluting the suspension with methane and ethylene [56]. The mean droplet diameters were between 100 and 250 microns, and were relatively large. The mean flame velocity, in a vertical tube, varied with the concentration of flammable gas and, with methane, was affected by the presence of suspensions of the smaller diameter droplets. Some pre-vaporization of the oil was believed to be the cause. No significant effect was detected with ethylene. The presence of the oil caused the flames to become much thicker, with luminous zones extending up to 30 cm behind the flame front, indicating vaporization of the oil during the combustion process.

Turbulence affects both homogeneous and heterogeneous combustion [39, 57], and is of particular importance in oil fired combustion. However the conditions are remote from those in dust explosion flames, and will not be considered further.

Summary of present position

A fundamental difficulty is that the mechanism whereby flame propagates through a dust suspension, in an explosion, has not been firmly established. There is a clear priority for investigation of this problem, either in small scale laboratory work or in large scale field tests. The possible existence of combustion mechanisms similar to the homogeneous and heterogeneous modes fairly clearly established for liquid droplet systems should have priority in investigations. The experimental investigations need not be complex, and convincing evidence might be obtained by optical means. Observations would be necessary on a range of dusts, particularly those which volatilize completely in a flame, e.g. some plastics, those which could be expected to give both volatiles and a solid residue, e.g. coal, and those forming non-gaseous combustion products, e.g. some metals. If some definite conclusions could be drawn on such a basis, then the theoretical analysis could be made on firm foundations. At this stage it should perhaps be pointed out that in the past, more investigations have been made with coal dust, than with any other single material, but there are indications that the behaviour of coal is more complex than that of some other dusts, such as some plastics, and investigations with a simpler fuel may be advisable.

Once the mechanism of flame propagation has been established with reasonable certainty, even if for only one dust, the number of possible rate determining physical processes is reduced and this simplifies theoretical analysis. Such analysis is more likely to be successful if a wide range of variables does not have to be considered, and could lead, at least in general terms, to equations representing the steady state of flame propagation, from which flame speeds and thicknesses would be predictable. The relative roles and conduction in flame propagation should then become clarified and this would be of considerable assistance in considering the preheating and ignition zone processes. A semi-quantitative treatment of these proceseses already exists, but they are based on the Mallard and Le Chatelier equation which is known to be unsatisfactory even for gas mixtures. A theoretical analysis of the steady flame propagation would be a basis on which to consider the burning time characteristics of particles in explosion flames. The present situation regarding this aspect is very unsatisfactory and greatly needs improvement. It is of both theoretical and practical interest because it could lead to calculation of flame thickness and the rate of burning could be related to the rate of pressure rise for the dust and therefore to explosion venting requirements in industrial plant.

Fundamental information on the flame processes, as outlined above, would act as a framework into which the large amount of empirical information on factors affecting flame propagation, given in the preceding Chapter, could be assessed and fitted. Thus, effects such as those of particle diameter, dust concentration, ambient temperature and pressure, and the presence of solid, liquid or gaseous diluents should be dealt with by the analysis. Already, there are numerous quantitative empirical relationships, indicating that it should be possible to provide a more rigorous theoretical background. Other factors, such as the influence of the ignition source and of turbulence on dust explosions have been investigated qualitatively, but general relationships have not been obtained. This may be either because the experiments have not been sufficiently controlled and selective, or, that the effects are complex and not amenable to *ad hoc* treatment. A theoretical framework would give some pointers towards a solution, even if the analysis was too complex for a complete solution.

The most likely direction in which a major advance into understanding the mechanism of dust flame propagation could therefore lie is the direct examination of flame processes, rather than obtaining further evidence on the factors affecting the flames by *ad hoc* techniques. Such investigations could bring rewards both in the understanding of fundamental processes and also in providing a more detailed picture of conditions in practical situations.

SYMBOLS

a	Radius of cylindrical vessel.
A	Constant.
b	,,
B	,,
c_o	Mean specific heat of dust.
c_1	,, ,, ,, ,, volatiles from combustible dust.
c_2	,, ,, ,, ,, diluent dust.
c_3	,, ,, ,, ,, non-volatile residue from coal.
c_g	,, ,, ,, ,, gas.
c_p	,, ,, ,, ,, mixture over range T_0 to T_i.
d	Initial diameter of dust particle.
E	Activation energy.
f	Frequency factor.
G	Geometrical view factor.
h	Heat absorbed during devolatilization before ignition.
H	Heat of combustion per unit mass of volatiles.
H^1	Heat of reaction per mole of dust.
I	Radiation intensity at flame front.
K	Constant.
l	Thickness of reaction zone.

SYMBOLS

M_d	Number of moles per monolayer of dust particle.
M_o	Molar oxygen concentration.
N	Dependent variable.
p	Explosion pressure variable.
r	Expansion ratio on combustion.
r_d	Radius of dust particle.
R	Gas consant.
S_u	Burning velocity.
t	Time.
T	Temperature variable.
T_o	Ambient temperature.
T_1	Flame temperature at explosibility limits.
T_f	Final flame temperature.
T_i	Ignition temperature.
T_r	Radiation temperature of inner cone.
$\overline{U}$	Mean rate of evolution of volatiles.
V	Volatile content of dust (per cent).
x	Volatiles concentration (mass per unit volume).
x_1	,, ,, ,, ,, ,, ,, at lower explosibility limit.
x_2	,, ,, ,, ,, ,, ,, at stoichiometric.
X	Dust ,, ,, ,, ,, ,,
X_o	Dust concentration (mass per unit volume) at temperature T_0.
y_1	Diluent dust concentration (mass per unit volume).
y_2	Concentration of non-volatile residue from fuel (mass per unit volume).
z	Distance variable through flame.
α	Correction factor for radiation from products.
ε_0	Absorptivity of unburnt cloud.
ε_1	Mean emissivity of inner flame cone.
μ	Thermal conductivity of mixture at temperature T_i.
ρ	Density of mixture at temperature T_0.
ρ_d	Density of dust particle.
ρ_g	Density of gas at temperature T_0.
σ	Stefan's constant.

References

[1] Essenhigh, R. H. (1967), 11th International Symposium on Combustion. Comb. Inst. Pittsburgh, Penn., 291–305.

[2] Essenhigh, R. H. and Csaba, J. (1963), 9th International Symposium on Combustion, Academic Press, London, 111–24.

[3] Cassel, H. M., Das Gupta, A. K. and Guruswamy, S. (1949), 3rd International Symposium on Combustion, Williams & Wilkins Co., Baltimore, 185–90.

[4] Long, V. D. and Murray, W. L. (1958), Safety in Mines Research Establishment, Research Report, 145.

[5] Klyachko, L. A. (1968), *Pyro.*, **6**, 19–38.

[6] Leuschke, G. (1965), *Staub.*, **25**, 180–6.

[7] Howard, J. B. and Essenhigh, R. H. (1965), *Comb. and Flame*, **9**, 337–9.

[8] Howard, J. B. and Essenhigh, R. H. (1966), *Comb. and Flame*, **10**, 92–3.

[9] Howard, J. B. and Essenhigh, R. H. (1967), *Ind. Engin. Chem.*, (Process Des. Dev.), **6**, 74–84.

[10] Kimber, G. M. and Gray, M. D. (1967), *Comb. and Flame*, **11**, 360–2.

[11] Marshall, W. F., Palmer, H. B. and Seery, D. J. (1964), *J. Inst. Fuel,* **37**, 342–9.

[12] Schneider, G. (1958), *Ann. Min.*, Paris, 77–92.

[13] Cassel, H. M. (1964), US Bureau of Mines, RI 6551.

[14] Cassel, H. M. and Liebman, I. (1959), *Comb. and Flame*, **3**, 467–75.

[15] Nagy, J. and Surincik, D. J. (1966), US Bureau of Mines, RI 6811.

[16] Costa, D. and Cerniani, A. (1952), Pub. Fac. Sci e Ing. Univ. Trieste Ser. B., 106, 463–72.

[17] Cutress, J. O., Peirce, Y. J. and Tucker, A. C. N. (1964), *Comb and Flame*, **8**, 289–98.

[18] Nettleton, M. A. and Stirling, R. (1967), *Proc. Roy. Soc. A.*, **300**, 62–77.

[19] Palmer, K. N. and Tonkin, P. S. (1967), 3rd Symposium on Chemical Process Hazard with Special Reference to Plant Design, Inst. Chem. Engrs., London, 66–75.

[20] Essenhigh, R. H. (1961), Coll. Engin., 534–9, (1962), 23–8, 65–72, 103–4.

[21] Essenhigh, R. H. and Woodhead, D. W. (1958), *Comb. and Flame*, **2**, 365–82.

[22] Palmer, K. N. and Tonkin, P. S. (1967), Fire Research Note No. 667.

[23] Palmer, K. N. and Tonkin, P. S. (1971), *Comb. and Flame*, **17**, 159–70.

[24] Gomez, M., Gayle, J. N. and Taylor, A. R. (1965), US Bureau of Mines RI 6607.

[25] Shaw, S. K. and Woodhead, D. W. (1963), Safety in Mines Research Establishment, Research Report, 221.

[26] Mason, T. N. and Wheeler, R. V. (1931), Safety in Mines Research Board Paper 64, HMSO, London.

[27] British Standard 1016:Part 3:(1957).

[28] Cybulski, W. (1965), *Selected Translations on Explosives*, Scientific Publications Foreign Co-operation Center of the Central Institute for Scientific, Technical and Economic Information. P.177. Warsaw.

[29] Araki, S. (1959), 10th International Conference of Directors of Safety in Mines Research, Pittsburgh, Penn., Paper 29.

[30] Ishihama, W. (1961), 11th International Conference of Directors of Safety in Mines Research, Warsaw, Poland. Paper 47 October.

[31] Fenimore, C. P. and Jones, G. W. (1966), *Comb. and Flame*, **10**, 295–301.
[32] Burge, S. J. and Tipper, C. F. H. (1967), *Chemy. Ind.*, 9, 362–3.
[33] Field, M. A., Gill, D. W., Morgan, B. B. and Hawksley, P. G. W. (1967), Combustion of Pulverized Coal, British Coal Utilization Research Assoc., Leatherhead.
[34] Markstein, G. H. (1963), 9th International Symposium on Combustion, Academic Press, London, 137–47.
[35] Meek, R. L. and Dallavalle, J. M. (1954), *Ind. Engin. Chem.*, **46**, 763–6.
[36] Brown, K. C. and James G. J. (1962), Safety in Mines Research Establishment, Research Report, 201.
[37] Essenhigh, R. H. and Fells, I. (1960), Disc. Farad. Soc., London, 30, 208–21.
[38] Essenhigh, R. H. (1961), *J. Inst. Fuel*, **34**, 239–44.
[39] Williams, A. (1968), *Oxid. and Comb. Reviews*, **3**, 1, 1–45.
[40] Effenberger, H. (1963), *Energietech.*, **13**, 85–9, 125–9, 162–7, 208–12.
[41] Hedley, A. B., Nuruzzaman, A. S. M. and Martin, G. F. (1971), *J. Inst. Fuel*, 38–54.
[42] Essenhigh, R. H. and Dreier, W. L. (1969), *Fuel*, **48**, 330–42.
[43] Zembrzuski, M. (1967), *Energietech.*, **17**, 205–6.
[44] Shibaoka, M. (1969), *J. Inst. Fuel*, **42**, 59–66.
[45] Brookes, F. R. (1969), *Fuel*, **48**, 139–49.
[46] Katsnelson, B. D. and Marone, I. Y. (1964), Thermal Engin. Jan., 12–7.
[47] Lightman, P. and Street, P. J. (1968), *Fuel*, **47**, 7–28.
[48] Street, P. J., Weight, R. P. and Lightman, P. (1969), *Fuel*, **48**, 343–65.
[49] Brzustowski, T. A. and Glassman, I. (1964), *Heterogeneous Combustion*, Academic Press, New York, 75–115.
[50] Macek, A. and Semple, J. McKenzie, (1969), *Comb. Sci. and Tech.*, **1**, 181–91.
[51] Chamberlain, C. T. and Gray, W. A. (1967), *Nature*, London, **216**, 1245.
[52] Burgoyne, J. H. and Cohen, L. (1954), *Proc. Roy. Soc. A.*, **225**, 375–92.
[53] Mizutani, Y. and Ogasawara, M. (1965), *Int. J. Heat Mass Transfer*, **8**, 921–35.
[54] Nuruzzaman, A. S. M., Hedley, A. B. and Beer, J. M. (1970), *J. Inst. Fuel*, 301–10.
[55] Nuruzzman, A. S. M. and Beer, J. M. (1971), *Comb. Sci. and Tech.*, **3**, 17–24.
[56] Wood, B. J., Rosser, W. A. and Wise, H. (1969), *Comb. and Flame*, **13**, 663–6.
[57] Cekalin, E. K. (1962), 8th International Symposium on Combustion, Williams & Wilkins Co., Baltimore, 1125–9.

7 *Sources of Ignition and their Avoidance*

In the safe design of industrial plant, where explosible dusts are handled, sources of ignition must be excluded as far as possible. In practice their presence cannot be completely eliminated, because they may arise from a breakdown or may even be introduced maliciously, and so the exclusion of sources of ignition cannot be relied upon as the sole method of protection of the plant. The likelihood of a source of ignition being present can however, be much reduced by good design and management and should be regarded as an essential requirement for plant safety.

Although there are a vast number of different sources of ignition having various temperatures, energies, durations etc., they can be conveniently grouped as follows:

Flames
Glowing combustion and smouldering
Hot surfaces
Welding and cutting
Friction and impact
Electric sparks
Spontaneous heating.

Consideration will also be given to the means whereby ignition from these sources may be prevented. The extent to which the various sources have been studied in the laboratory, as regards their mode of action and their properties measured, varies considerably. Hence, the treatment here will be principally in terms of the practical implications, rather than the technical properties of the sources.

Flames

The flames produced from the burning of gases, liquids or solids are potent sources of ignition for dust suspensions. This applies whether the flames are diffusion, that is unmixed fuel burning in air, or premixed, in which the air and the fuel are mixed before entering the flame. Because flames are such effective ignition sources, little investigation has been made into the minimum size and duration of flames needed to ignite dusts, because these minima are likely to be so small as to have little practical application. Ignition of a dust suspension can be made by a permanently established flame, as on a gas burner or torch, or from a pilot light, or from transient flames such as could be produced by a blow-back, ignition of a cloud of vapour, or a nap fire propagating over fine fibres.

A gas flame is used in one of the standard tests applied to dust suspensions, in the Federal German Republic (see Chapter 3). Visual observation is made of the dust flame produced, in the flame formation test, to give an approximate overall assessment of the explosibility of the dust. The existence of this test illustrates the effectiveness of a small gas flame in igniting dust suspensions.

Flames may also be generated if a dust layer becomes ignited and is then dispersed. In particular, a dust layer can be ignited to give glowing or smouldering combustion, see below, and the smouldering can be converted to flaming by the action of an air current [1] or if the dust is dislodged and falls through the air. If a dust layer ignites and smoulders, its adhesion to the wall of the plant may be reduced and the burning material is then dislodged by either a slight external stimulus or even by its own weight. Although the flame produced may only be transient, it could ignite a dust suspension.

The reasons for the great effectiveness of flames as ignition sources have not been fully explored, but they probably derive from a combination of several factors. These include the relatively high temperature, usually in the range 1000 to 2000°C, the appreciable duration in time, and the magnitude of the volume which may be heated practically simultaneously. In contrast, a hot surface may persist for a longer duration but the temperature would often be lower, and an electric spark would have a higher temperature than the flame but its duration would be considerably less. In addition, with the hot surface and the electric spark, the volume of suspension heated at a given instant would be limited. As a general conclusion, if a dust suspension can be ignited, a flame is likely to cause ignition and should be regarded as extremely hazardous.

Because of the effectiveness of flames for igniting dust suspensions, the main general safeguard is to prevent their coming together. This may mean excluding flames from within and from the vicinity of the dust handling plant. The flames may be involved in other processes, or may be introduced by personnel, e.g. matches, cigarette lighters. The degree of supervision required is a matter for management, and the creation of 'no smoking' areas may be necessary, on entry to which all smoking materials must be given up. Control would also apply to machines and equipment which could cause a flame in the area, such as engines and hot cutting gear.

The presence of a flame and an explosible dust suspension may be an inherent part of the process. For example in the burning of wood waste in a furnace the wood dust must be dispersed in the presence of a flame inside the furnace. Combustion under these conditions is designed to be well controlled, but care must be taken to ensure that all the wood is burnt inside the furnace and none passes through to the flue system, and that burning dust cannot blow back from the furnace.

Where a dust suspension is being deliberately ignited by a flame, as in a furnace, the dust feed must be strictly controlled and monitored and great care must be taken to ensure that no unburnt dust accumulates, or is stored, in the vicinity of the furnace. Further considerations of the hazards is given in Chapter 10. A room containing a furnace or boiler should be kept free of dust, apart from that which is in process of passing to the furnace.

When flames are required for process heating it is preferable to use indirect heating via a heat exchanger rather than direct firing. Even so, careful maintenance of the exchanger is necessary to keep the surfaces free of dust and because repeated heating and cooling cycles may lead to distortion of the exchanger and provide communication between the two circuits within it.

To summarize, flames should be prevented from coming into contact with explosible dust suspensions because otherwise ignition is likely. With furnaces and other installations where the dust suspension is deliberately ignited the combustion must be well controlled and ignition of extraneous dust avoided.

Glowing and smouldering combustion

Ignition of a dust suspension can be caused by glowing or smouldering materials. Combustion by these processes is characterized by the absence of flame, and visible signs of the burning are not always apparent. Glowing combustion may be the burning away of a mass of ignited particles, or may be visible processes associated with smouldering, which is the propagation of flameless combustion over or through a layer or deposit of material.

Glowing or smouldering can arise in a variety of ways, some of which will be considered in relation to dust fires, Chapter 9, and which include direct heating of the material, ignition of the material by flame followed by extinction of the flame, contact with another smouldering material, or friction or impact. The subsequent ignition of a dust suspension can be caused by:

(1) the dust suspension coming into contact with glowing or smouldering material deposited on a surface.
(2) the smouldering being converted to flaming by the action of an air flow, the flame then igniting the dust suspension.
(3) the glowing or smouldering material itself being dispersed into a dust cloud, producing a flame.

Because of the ease with which glowing or smouldering can be converted into flame [1] they should be regarded as ignition sources of comparable effectiveness to flames. Steps can be taken to minimize the likelihood of glowing or smouldering being initiated by the exclusion of other ignition sources, prevention of deposition on hot surfaces, elimination of hot particles etc. Also, unwanted deposits of dust should not be allowed to accumulate within the plant, or outside the plant in the workroom. In the food industry, for example, the prevention of accumulation is often necessary on hygienic grounds, but in other industries frequent removal of dust deposits is also necessary for fire and explosion prevention.

Hot surfaces

Under this heading are considered surfaces which are heated by energy sources external to the dust suspension or deposit. The practical situations would arise from:

(1) presence of the dust suspension within a heated enclosure, e.g. a furnace or dryer
(2) deposition of dust on a surface within a heated enclosure, e.g. dust on internal wall of dryer
(3) dust dispersed around, or deposited on a heated surface with cooling to atmosphere, e.g. dust on overheating bearing, or on an electric lamp

Tests are available for measurement of minimum ignition temperatures of dust suspensions in heated enclosures using small furnaces (Chapter 3) and published values are available for many dusts (Appendix 2). These values can be determined precisely in the test apparatus, and although the interpretation is sometimes not straightforward (see Chapter 4) a reasonable estimate can be given of hazardous temperatures for this type of risk. As regards fires originating in dust deposits in hot enclosures or on hot surfaces, theoretical analysis of the thermal ignition properties is available, Chapter 9, but values for very few materials have been published. *Ad hoc* tests with dust deposits in furnaces are made in certain countries, and minimum ignition temperatures for a number of dusts have been so measured, Chapter 3.

Ignition and explosion can be avoided by ensuring that the hot surfaces are at temperatures well below the minimum for ignition.

For surfaces whose temperatures can be controlled, such as heaters, dryers, etc., guidance on appropriate safety margins may be obtained from the Federal German VDE Regulation No. 0165 (Chapter 4). For dust layers 0.5 cm thick the maximum permissible surface temperature is 75°C below that required for the ignition of the dust. The ignition temperature of the dust layer falls as the layer thickness is increased, and the maximum permissible surface temperature must be reduced if thicker dust layers than 0.5 cm are encountered. In situations where the surface temperature cannot be controlled, as with an overheated bearing, accumulations of dust should be minimized. The use of temperature sensors to detect overheating is advisable.

Lighting fittings and lamps, particularly of the filament type, can be expected to heat appreciably whilst running. Preferably such fittings should be mounted external to the room, bin, etc. containing the dust, separated by a dust-tight armoured glass window. Light can thus penetrate to the enclosure but dust is prevented from coming into contact with the heated lamp. Access to the lighting fittings for servicing should be from the outside of the enclosure, preventing the entry of dust during maintenance operations.

Inspection lamps attached to wander leads are hazardous, and should not be provided; recommended alternative maesures are either permanent external lighting or the provision of battery operated hand lamps. Portable lamps attached to wander leads are hazardous because the heated lamp may come into direct contact with dust whilst inspections are being made, and secondly because the lamp may inadvertently be left switched on in the enclosure and subsequently buried in dust.

Hot surfaces can originate in electrical equipment designed for the purpose, such as heaters, and also in equipment such as motors which heat during normal working. Wherever possible these items should be excluded from contact with dust.

Welding and cutting

Similar hazards to those with welding and cutting arise from operations such as soldering, burning, and use of power tools. In all cases a localized heating of plant is produced which is known to have caused dust explosions [2, 3], Plates 1 and 11.

Specialist publications are available covering all aspects of safety in welding and cutting [4, 5], and only the principal aspects concerned with explosion and fire will be summarized here. These are:—

1. No welding, cutting or other repair work involving the application of heat shall be started until the plant has been shut-down and all practicable steps taken to remove explosible or flammable dust from the interior and exterior of the plant. Failure to observe this requirement may lead to a breach of the Factories Act, in the United Kingdom. To encourage compliance with this requirement the introduction of a 'permit to work' system (Chapter 12) is useful, because if properly carried out the system should provide a positive check on the state of the plant before repairs commence. The system is particularly advisable when the repairs are being carried out by outside contractors, who may be working at times when the degree of supervision of the factory is reduced e.g. at weekends, and where confusion may arise as to the relative responsibility of factory management and contractors.
2. Any combustible materials present in the vicinity of the plant should be removed beyond the reach of flying sparks from the welding or, if this is not possible, should be adequately protected by covering with non-combustible sheeting (ordinary tarpaulins must not be used). Any nearby wooden structures should also be protected similarly and wooden floors should be covered or wetted before repairs commence. Precautionary measures should extend to consideration of openings or cracks in walls and floors, and flammable material in adjoining rooms or on the floor below should be removed to a safe position [5].

3. Suitable fire fighting equipment shall be to hand, and the operatives informed of its presence. If a fire hazard is still deemed to be present, even though the requirements of (1) and (2) above have been fulfilled, it is good practice to employ a watcher to look for incipient fires. The site should be kept under continuous observations for at least a half-hour after work has ceased [5], because small fires can easily smoulder for this period before bursting into flame.
4. Particularly where welding is carried out in a confined space, exhaust ventilation may be necessary to keep the atmosphere fresh. This should be provided before work starts, and should be installed by a competent person [5]. On no account should oxygen be released into the air in a confined space to freshen it, because combustibility of materials is greatly increased by oxygenating the atmosphere. Similarly, because of the danger of accumulation from leaks over long periods gas equipment should not be left unused in confined spaces.
5. No operator should be required to enter any confined space, large vessel, etc. until it has been certified as safe for entry and to work in by a responsible person.
6. With electric welding, a live electrode should not be laid down and left unattended near any container in which there is or has been any flammable substance, because of the danger of ignition following localized heating.

In addition to the above requirements there may well be other steps which must be taken, because of local conditions. If in doubt, the advice of the appropriate responsible authority should be asked. Stringent precautions are needed with welding, cutting etc. because of the known explosion and fire risks entailed.

Friction sparks

Sparks arising from friction or impact are usually considered together, because of the difficulty of differentiating between them in many practical cases. There is strong evidence from industrial experience that such sparks are able to ignite dust suspensions [2], although relatively little laboratory evidence is available to provide a detailed technical background. Consequently information is lacking on the levels of mechanical energy needed to ignite various dusts, the energies of the sparks produced by friction or impact, and the quenching effect on the development of the explosion arising from the presence of the striker close to the spark. Ignition of a grain dust cloud by a large metal striker, impinging on a rusty steel plate, has been detailed [6]. In these experiments the striker fell 26 m (85 ft) before hitting the plate, but even so ignitions were not obtained using a stone striker.

In industrial processes frictional ignitions can arise in five main circumstances:

1. The material being processed itself generates excessive friction in the plant. For instance, a failure to regulate the quantity of material being processed can lead to the overloading of grinding plants, conveyors etc. Because of its resistance to movement the dust may clog the machinery instead of passing smoothly through it, and localized heating can result. Preventative measures include the installation of feed regulators, and the automatic shut-off of power if the design load on the machine is exceeded.
2. The material being processed may contain a foreign object such as tramp metal, stones, etc. Magnetic or pneumatic separators can be installed to remove these objects, Chapter 10, and separators are particularly advisable if the material is to pass to a grinding mill, disintegrator, or other unit in which friction or impact is part of the operation.
3. Friction or impact can be caused by the breakdown of the plant. The danger of an overheated bearing has already been mentioned, and the main preventative measures are good design, with bearings outside the dust conveyor, and good maintenance. The plant should be designed and laid out so that access can be obtained to bearings, without great difficulty. Inaccessible bearings are unlikely to be favoured with regular inspections and may also be vulnerable to the accumulation of dust over long periods. In such cases constant monitoring of the temperature should be considered.

   Fracture of machine components cannot be foreseen, but the result may be a severe hazard particularly if a fracture occurs on rapidly moving parts, such as those in mills, fans, etc. A similar risk can arise from small objects such as screws, rivets, working loose and falling into the process stream, eventually coming into contact with a rapidly moving part. Separators give protection, where they can be installed, but there may be other parts of the plant where positive protection cannot be reasonably obtained.

   Another cause of breakdown of plant arises from choking by the feed stock, as (1) above, where with belt driven plant the increased resistance caused by choked dust can cause the plant to stall, and if the power is not shut off, localized overheating of belts near pulleys is likely. Badly adjusted and loose belts can also give frictional heating. Preventative measures include regular inspection and good mantenance, coupled with automatic switch-off if the plant becomes overloaded.

Friction and impact can also develop if the plant becomes misaligned so that moving parts strike the casing; this may well be accompanied by stress and overheating of the bearings. The precautions are again good design followed by frequent inspection and good maintenance.

4. The use of tools, powered or hand, by operatives can give rise to friction. Power tools such as drills can give intense localized heating, and should only be used under stringent supervision. Prior removal of dust from the vicinity both within the outside the plant is required, accompanied by the shutting down of the plant before operations commence. The use of cartridge tools where these can be permitted, should also be strictly controlled. With powered tools the basic protection is to restrict their use to situations where they can be stringently controlled, involving good supervision. Non-powered hand tools such as shovels, scoops, etc can be a hazard with dusts of low ignition energy, e.g. sulphur. With such dusts the danger of ignitions can be reduced by careful use of the tools, but even so occasional ignitions may be expected. Providing the operation is being carried out in open or semi-open conditions, which will usually be necessary for manual work to be practicable, ignition of the dust would lead to a small fire. Suitable fire fighting equipment should be provided and the operatives informed of the procedures to be adopted (Chapter 9). Under confined conditions, or where the dust suspension is above the minimum explosible concentration, alternative mechanical means of processing should be used.
5. A dust layer, or possibly a suspension, can be ignited by thermite sparks caused by the impact of a striker, e.g. stainless steel, on rusty mild steel carrying an aluminium smear [7]. The use of a 'non-sparking' striker, of copper-beryllium, did not significantly decrease the hazard.

Although the occurrence of friction and impact sparks can sometimes be foreseen and the plant and its operation designed to avoid the risk, there will be cases where due to failure of plant or personnel ignition can occur. Attempting to prevent ignition is likely to be only partly successful, and alternative means of protecting against the consequent fire or explosion must be taken, Chapters 8 and 9.

Electric sparks

Dust explosions and fire hazards due to electric sparks can arise from two principal origins, electric power and electrostatic charging. Electric power, from the mains or batteries, is a powerful source of ignition because the energies available are usually greatly in excess of those required to ignite dusts, unless special precautions have been taken in the design of the equipment.

The sparks produced during the normal working of switches, contact breakers, commutator motors, fuses, etc. can ignite dusts and need to be protected. The principal precautions are firstly to install the sparking equipment in clean air, away from the presence of dusts. Secondly, where the equipment must be used in the presence of dust it should preferably be of a dust tight design, for example in accordance with the appropriate British Standard [8]. The design of dust tight equipment should also be such that if covered with a layer of dust, hazardous surface temperatures do not develop. The ignition temperature of the thickest deposit of dust likely to be encountered is then required; this may be obtained either from the literature or by testing. As a safety factor, the surface temperature of the equipment should not approach within 75 °C of the ignition temperature (see above).

Electrical instrumentation and monitoring equipment can be installed in dusty atmospheres, providing it is suitably enclosed, non sparking, and does not heat. For safety the number of such instruments should be minimized, and they should be installed in positions where they are easily accessible for maintenance and inspection. Permanent wiring should be to the appropriate standards, such as the IEE Regulations, and should as far as possible be run away from the presence of dust.

The production of hazardous sparks from electrostatic charging of dusts has been studied for many years. It is common knowledge that in the processing, transportation, and general handling of dusts electrostatic charges are generated. These charges arise from the separation of surfaces which may either be of dust from a containing surface or of individual particles from each other. In consequence the dust particles become charged; the sign of the charge can be positive or negative in the same dust. Thus, in practical situations, it must be regarded as inevitable that dusts can develop a static electricity charge.

The extent of the charging of dust, whether explosible or not, has been the subject of numerous laboratory investigations [9] but simple predictions for practical situations cannot yet be made. The procedure to be adopted must be to assume that static charges will be generated and to arrange that explosion hazards are minimized. The charge may usefully be regarded as residing in three situations:

(1) In the dust itself.

(2) In the plant in which the dust is contained, and associated equipment.

(3) In nearby operatives and insulated objects, charged by induction.

The charge on the dust can be considered when the dust is in suspension, or when it has been deposited. Although dust particles in suspension are very likely to be charged, there is no definite experimental evidence to prove that a discharge from one particle to another in the suspension can cause ignition and lead to the generation of an explosion throughout the suspension. But this mechanism has not been definitely disproved either, and so must remain a possibility, particularly in large-scale installations [9, 10]. It may be considered unlikely on physical grounds, when account is taken of the low electrical capacity of a fine particle of dust and the high voltage that would be required to produce a spark capable of igniting a neighbouring particle.

A totally different situation may arise when flammable gas or vapour is also present in the dust suspension. There would then be a considerably greater possibility of ignition of the gas or vapour by discharge between two dust particles. This proviso also applies to the remarks in the succeeding paragraphs.

If the dust has settled out of suspension, discharge within the layer is unlikely to cause ignition, but the charge may reside in the dust for many hours, or even days, before leaking away, even if the dust is within a vessel bonded to earth. The relaxation time depends on the electrical resistivity of the dust, which can vary over a wide range. The greater the resistivity, the longer the time required for the charge to leak away. Thus although a charged dust may be put inside an earthed container, the electric charge may leak away only slowly and thereby permit only a slow decay in voltage.

In industry the greatest danger from static electricity arises because of the transfer of charge to isolated metallic conductors that are not bonded to earth. The reason for the hazard is that such metal components can store enough electrical energy to exceed the minimum ignition energy of the dust, Chapters 3 and 4, and because the metal is a good conductor the charge will drain readily from it on sparking to earth. Isolated metal units can be present in plant as a result of inadequate bonding during the construction of a plant, or by the subsequent introduction of insulating sections, e.g. in pipes, or by inadequate maintenance of the bonds e.g. across pipe flanges, or by their introduction during process operations. All of these are obvious hazards and must be prevented.

The reason for the particular hazard of unearthed metal, and other conducting objects is that when discharge to earth does occur the charge flows readily from the whole of the object to the point from which the spark passes. With poor conductors which become charged, any discharge to earth is likely to be only from parts adjacent to the point of discharge and sparks of lower energy, from the same voltage, are thereby produced. The division between good and poor conductors is arbitrary, but if the resistance from a charged point to earth is less than 10^8 ohms the path along which the electricity flows may, from the point of view of static electricity, be regarded as a good conductor. The resistance to earth from a bonded object should not exceed a few ohms, to allow for rapid passage of charge and to provide a safety margin. Where the plant installation is extensive, and particularly where pipe lines or ducts may run for distances of hundreds of feet, the bonding should be to a common earth throughout, as otherwise differences in voltage may persist between one part of the plant and another even though both are nominally earthed. The usual means of bonding is by copper strip or wire which is attached to a special point on the object to be earthed, preferably by a permanent connection. Where a number of units are joined together, as in a pipe line, bonding should be incorporated at each flange joint as it is not sufficient to assume that electrical continuity through nuts and bolts between flanges will be adequate. If the units are connected to each other by special bonding strips then frequent connections to earth are unnecessary provided that sufficient connections are made to ensure that all conductors are earthed.

Certain items of plant such as conveyor belts are often made of non-conducting materials which pass rapidly over conducting rollers. High charges can build up on the conveyors and as a safety measure it is good practice to use anti-static or similarly conducting materials. Although the conveyors themselves may not be able to pass sparks of sufficient energy to ignite dust suspensions, the high voltages generated may be induced onto unearthed conducting objects inadvertently nearby. Sparks from the conductor may then of course ignite the dust.

As the main defence against ignitions from static electricity lies in bonding the plant, it is essential that electrical continuity should be checked both during the erection and frequently during the lifetime of the plant. It should also be checked after any maintenance on the plant, even for operations such as painting which may not have required any dismantling, but which may have affected the resistance to earth.

In recent years a number of hoppers and silos have been constructed from plastics. The question has then arisen as to whether this form of construction increases the hazard from static electricity. Some information is available from tests in which oils were introduced into glass-reinforced plastic tanks and the potentials measured [11]. It was concluded that hazardous discharges from glass reinforced polyester tanks would be improbable although all metal components on the tanks, such as manhole covers, should be bonded to earth. A more hazardous situation may arise from plastics of higher resistivities, such as polyethylene, polypropylene, or polyvinylchloride [12] where dangerous sparks might be drawn off limited areas in large vessels. Suggested solutions to this problem were the incorporation in the container of a bonded liner or inner coating of moderate resistivity [13] or, if the material is of laminated construction, the incorporation of a conductive filler such as carbon black or a wire mesh screen in the laminations [11]. The hazards arising from the indiscriminate use of metal dipsticks will of course still exist.

The ignition of dust clouds by sparks from charged operatives was discussed in Chapter 4. The main situation in which this hazard should be considered is where the dust has been shown by test to have a minimum ignition energy of less than 25 mJ. Actions involving close contact between moving dust and the operatives should be avoided as much as possible. Such actions include shovelling, sieving or classifying and bagging off. If the energy is below 10 mJ then anti-static clothing and footwear for the operative is advisable, and the floor on which the operatives stand should likewise be anti-static [14, 15, 16]. Particular care should be taken where the dust has been dried, or is warm, and is finely divided. In general, higher charges are to be expected on fine dusts because of their greater surface area, and they are also more readily ignited.

The greater of static electricity can be minimized by increasing the relative humidity of the atmosphere at least to within the range 60–75 per cent RH. In so far as this is practicable it is worthwhile particularly for operations in heated buildings where in cold weather the relative humidity may fall to low values. Alternatively, the air in the building may be made conducting by the use of electrical discharges or ionizing radiations, and these will again encourage the rapid dissipation of charge. These techniques are specialized, and may be hazardous if not properly installed, so that expert advice should be sought either from the manufacturer of the equipment or the appropriate safety authority. Minimizing the generation of static electricity should be regarded as useful but must not be relied upon for complete protection. It is essential that all metal components of plant handling dusts should be bonded to earth, if ignition and explosions are to be avoided, and other safety measures do not permit relaxation of these requirements.

Static electricity is particularly expected to be in pneumatic conveying systems, and for these the advice on bonding given above should certainly be followed. It is again the case that relatively little detailed investigation of static electricity generation has been carried out with these systems, but the hazard has been recognized [17, 18]. A vulnerable point of such a system is where the conveyed dust is discharged into a storage bin or hopper, because it is likely to be charged; there is also likely to be an explosible dust suspension somewhere within the hopper while it is being filled, and operatives are likely to be nearby and may inadvertently introduce unearthed objects. Discharges from the surface of polypropylene pellets (particle diameter 3 mm) in a hopper have been measured [19]. The pellets were conveyed to the hopper pneumatically in a closed circuit. The electric discharge to an earthed sphere (20 mm diameter) suspended over the surface was of a brush form.

Spontaneous heating

If a dust can undergo chemical reaction with the surrounding atmosphere, as with the oxygen of the air, or if it is unstable and decomposes, then heat may be produced in the reaction. The extent to which the heat can raise the temperature of the dust depends upon its rate of generation and rate of loss to the surroundings. If, at a given temperature the rate of heat generation is greater than the rate of loss, then the temperature of the dust will rise to a new value. As the rate of a chemical reaction increases with temperature, the generation of heat at the new temperature will be more rapid than previously, but the temperature differential to the surroundings will also be increased, so that the losses will also rise. If, however, the rate of generation still exceeds the rate of loss then the temperature of the dust can rise still further. The process is termed spontaneous heating. If heat generation always exceeds heat loss, the rate of heating will accelerate to a runaway condition and eventually ignition will develop. This process is termed spontaneous ignition. The combined process of spontaneous heating and ignition may be regarded as spontaneous combustion. If the ignited dust is dispersed in air a flame may propagate through it and cause a dust explosion. If the dust is a deposit or layer, an outbreak of fire will be caused which can be either smouldering or flaming combustion, or both.

Spontaneous heating may also be caused by biological activity, particularly when the dust is moist, but this activity can only raise the temperature of the dust to a limited extent. There is a temperature, usually around 80°C, above which biological organisms cannot remain active and at this temperature the evaporation of moisture may also be an important factor restricting further temperature rise. With some materials the rise in temperature caused by biological activity may be high enough for chemical reaction with the oxygen of the air to set in and carry the spontaneous heating to higher temperatures. The stage in the process at which ignition occurs is then governed by chemical rather than biological action.

Although a dust may undergo reaction, giving out heat, several factors may intervene to prevent heating to ignition. The factors may be put into two broad groups, those which influence the rate of generation of heat, and those which influence the rate of loss of heat.

The factors governing the rate of generation of heat are the quantity of reactant present, the access of oxygen for reaction (where oxygen takes part), the heat of reaction and its variation with temperature. The rate of loss of heat will depend on the temperature differential between the dust and ambient, the dimensions of the dust sample, the mass rate of flow of air through the dust, and the quantity of moisture present. In simple terms, for spontaneous heating to be hazardous there must be adequate reactant present in the dust, although this reactant may be only a minor constituent of the material, there must be sufficient access of air at all stages to allow the heating to accelerate, and heat losses to the surroundings must be insufficient to control the process. If any of these conditions is not met then the temperature of the dust may rise above that of its surroundings, but the rate of rise will slow down and eventually the temperature will reach a peak value after which it will cool down again. The time for which the temperature stays at the peak may vary from seconds to days, or even weeks.

The extent to which a particular dust will spontaneously heat under a given set of circumstances cannot at present be predicted solely in terms of the composition of the dust and the temperature of the surroundings. However, if simple tests, usually limited in number, are carried out then, for dust layers or deposits, useful predictions can be made on the basis of the results of these tests, Chapter 9. At present this approach is limited to dusts which contain one or two active components which undergo reaction, and for conditions in which the access of oxygen to sustain the heating is not a controlling factor. It can also be applied to certain materials which evolve heat on decomposition without the oxygen of the atmosphere participating. From tests on cubes of the dust heated in an oven, the mathematical analysis permits predictions to be made of the spontaneous heating behaviour of large heaps, or layers such as may be deposited on the walls of a dryer. For dust suspensions no comparable analysis, with the same degree of accuracy, is available and consequently recourse has to be made to tests such as the Furnace apparatus (Chapter 3). Extrapolation to larger volumes, or different shaped enclosures, has then to be done empirically. This is clearly unsatisfactory, because it is known from the behaviour of dust layers that the heating and ignition processes are favoured by increase in size of the specimen, and vary considerably with the shape. For the case of dispersions the extrapolation from results obtained with the Furnace apparatus may well introduce important errors.

From the practical viewpoint there are three initial situations in which spontaneous heating and ignition can occur. These are:

(1) The dust in a hot environment.
(2) Hot dust in a cold environment.
(3) Cold dust in a cold environment.

Considering firstly dust in a hot environment, the dust may be initially hot or cold and dispersed or in a layer. If heat transfer from the environment to the dust is rapid, or if the dust has been in position a considerable time, or if it is introduced whilst hot, the situation can favour spontaneous heating providing the chemical reactions have not been completed. This applies to either dispersions or layers of dust. To prevent spontaneous heating to a hazardous extent, or spontaneous ignition, the amount of dust present and the temperature of the environment must both be controlled. Usually the appropriate overriding control is that of the environment, so that a limiting maximum temperature is established on the basis of previous tests, outlined above, and a safety margin of 75°C is applied. As the temperature of the environment at which ignition of the dust will occur depends on the amount of dust present, in layers, and probably also in dispersions, the total amount of dust permitted in the environment must also be stipulated. For instance, in the working of a dryer in which deposits of dust may build up, either a maximum thickness must be established at which deposits no longer continue to accumulate, and this thickness governs the operational temperature of the dryer, or if the drying temperature must be at a certain value, then the maximum thicknesses of deposits which are permitted to accumulate must be stipulated and observed. In general the minimum environmental temperature at which a dust layer will ignite is considerably below that at which a dust suspension can ignite, so that where both layers and dispersions can be present the design of the plant should be related to the ignition of the layers. Typical plant in which dusts are in a hot environment include ovens, dryers and furnaces. In the latter instance the function of the plant demands combustion within it so that the main hazard arises should the flame be inadvertently extinguished. Continued flow of dust into the furnace can then build up a dispersion in a hot environment which may subsequently ignite explosively and cause damage. For this plant the relevant ignition temperature should be that of the dust dispersion rather than that of the dust layer (see also Chapter 10).

If a hot dust is deposited in a relatively cold environment, spontaneous heating can still occur. An example would be the removal of dust from an oven or dryer, whilst still hot, and its deposition in a collecting bin in a room at atmospheric temperature. The hazard arises because the dust, when in the hot environment, is in a different geometry less favourable to spontaneous heating. For instance dust in an oven may be in thin layers on shelves and the layer is thin to ensure that drying proceeds rapidly. Alternatively, the dust may be in suspension in a pneumatic dryer, where conditions are such that heat losses from the dust particles into the walls are higher than in dust layers. When transferred to the collecting bin, however, the hot dust is put into a vessel all three dimensions of which are of comparable size. Heat losses from the centre of the dust in the bin are thereby reduced and, if reaction is proceeding, the temperature in the centre of the dust may rise. Under favourable conditions spontaneous ignition may result. There is at present no theoretical analysis which enables the process to be predicted. If it is suspected, or needs to be tested for, a simple method is to install thermocouples or thermometer probes into the dust within the collecting bin and to monitor the temperature over a period of hours, and perhaps days if the linear dimension exceeds 0.5 m. If spontaneous ignition is occurring it will be a relatively slow process and should be detected without difficulty. After spontaneous heating is detected, it can be halted either by removing the dust, to smaller volumes or sometimes by just mixing it well. The manufacturing process should be altered so that hot dust is not put into compact shapes before being allowed to cool to room temperature.

Spontaneous heating and ignition of a cold dust in layers in a cold environment can occur. The situation arises either when highly reactive materials are present, or when large masses are stored for long times. Typical examples of the former are dusts impregnated with reactive vegetable oils, such as linseed, where the oil is dispersed over particles of large surface area. The rate of reaction with oxygen in the air is fast and heating proceeds to ignition. Alternatively certainly reactive metals, such as magnesium, when finely divided and newly produced under inert gas will also react with the oxygen in the air and spontaneously heat. For this reason, in the manufacture of magnesium and similar metals under inert gas, it is necessary to introduce a small amount (less than 1 per cent by volume) of oxygen to oxidize the metal surface as soon as it is formed. The hazards with these highly reactive materials can be avoided either by allowing oxidation to proceed under controlled conditions before storage, or by storing in sealed containers under inert gas, or with the vegetable oil/dust mixture by storing in closed cans with limited air access so that once the oxygen within the can is used up the contents are effectively inerted and will cool to atmosphere.

Dispersion of a cold dust in a cold environment is very unfavourable to ignition and very few instances have been reported. The most detailed has been that of the ignition of the metals zirconium and uranium, and uranium hydride [20] but this is a rare phenomenon. More common is the spontaneous heating and ignition of large deposits of combustible dusts, such as wood sawdust, colliery waste, and coal. Spontaneous heating occurs not because the dusts are highly reactive, but because the deposits are so large that the thermal insulation at the centre of the deposit is extremely high, and even a low rate of generation of heat exceeds the rate of loss.

Direct experimentation on this scale, e.g. with heaps of height many metres, is impracticable under controlled conditions so that recourse is made to small scale oven tests and the available theoretical analysis used to predict the large scale behaviour, Chapter 9. The spontaneous heating process under these conditions is very slow so that if it is suspected, direct evidence can readily be obtained by inserting probes and recording temperatures over long periods.

Although spontaneous heating and ignition have been investigated over many years there has been no centralized correlation of data. For instance information is available on coal [21, 22], starch [23] and calcium stearate [24] as well as a general list which includes some dusts [25] but in most practical cases the information required will need to be obtained by test on the actual materials involved.

References

[1] Palmer, K. N. (1957), *Comb. and Flame*, **1**, 2, 129–53.

[2] Dust explosions in factories. Department of Employment. Health and Safety at Work. No. 22. HMSO, London, (1970).

[3] Dust explosions in flour mills and bulk flour containers. Incorporated National Association of British and Irish Millers Ltd. London, (1967).

[4] Electric arc welding. Department of Employment and Productivity. Safety Health and Welfare New Series No. 38. HMSO, London, (1968).

[5] Health and Safety in Welding and Allied Processes. 2nd Edn. Inst. of Welding. London, (1965).

[6] Morse, A. R. (1958), Nat. Res. Council Canada. Radio and Electrical Engin. Div. NRC No. 4968. Ottawa.

[7] Gibson, N., Lloyd, F. C. and Perry, G. R. (1968), Third Symposium on Chemical Process Hazards with special reference to plant design. Inst. of Chemical Engineers No. 25. London, 26–35.
[8] British Standard No. 3807. (1964).
[9] Freytag, H. H. (Ed) (1965), Handbuch der Raum explosionen. Verlag Chemie. Weinheim.
[10] Heidelberg, E. and Schon, G. (1961), *Die Berufsgen*, (5), 196–9.
[11] Tinson, R. (1967), *J. Inst. Pet.*, **53**, 303–11.
[12] Gibson, N. (1968), *J. Inst. Pet.*, **54**, 19.
[13] Tinson, R. (1968), *J. Inst. Pet.*, **54**, 140.
[14] British Standard No. 2050. (1961).
[15] British Standard No. 3187. (1959).
[16] British Standard No. 3398. (1961).
[17] Abramyan, V. K. and Krapivin, L. E. (1968), *Bezopas. Truda Prom.*, **12**, 43–6.
[18] De Geest, W. (Ed) (1971), Advances in Static Elecricity, **1**, Auxilia SA Brussels, Belgium.
[19] Popov, B. G. and Taubkin, I. S. (1969), Soviet Plast., 10, 66–8.
[20] Hartmann, I., Nagy, J. and Jacobson, M. (1951), US Bureau of Mines RI 4835.
[21] Spontaneous ignition of coal. National Fire Protection Association. Code No. 493. Boston, USA. (1936).
[22] Costa, D. and Cerniani, A. (1952), Pub. Fac. Sci e Ing. Univ. Trieste Ser B., 106, 463–72.
[23] Costa, D. and Costantinides, G. (1949), *Annali. Chim. appl.*, **39**, 3–26.
[24] Schmutzler, G. and Wetzel, H. (1967), *Plaste u. Kautschuk*, **14**, 11, 827–9.
[25] Table of Materials subject to Spontaneous Heating, National Fire Protection Association Code 492, Boston, USA. (1949).

8 *Methods of Dust Explosion Protection*

Although stringent precautions may be taken to avoid the presence of sources of ignition in dust clouds, the protection obtained is not complete. In some instances sources of ignition are present as part of the process, as in a furnace, and protection against initiation of explosion is not obtainable. Further, because of the difficulty of controlling dusts owing to their ability to settle out and become re-dispersed at different concentrations, protection cannot be obtained by trying to ensure that the dust concentration is outside the range of explosibility. Consequently, when explosible dusts can be dispersed in air the possibility of a dust explosion must be considered. The question then arises as to methods of safely dealing with explosions, so that personnel and equipment, neighbouring plant etc., are not damaged, either by the explosion or by a subsequent fire.

The first step should be to consider the extent of the problem. With a given dust, the explosion hazard increases with the volume of dust cloud which becomes involved because of the proportional increase in the energy released. This means that large volumes can present a very serious hazard but conversely, small volumes may give a negligible risk. Accurate generalization cannot be made because the explosion hazard depends on the properties of the dust, the design and construction of the plant or equipment containing it, the hazard of neighbouring plant, the proximity of operatives, and the value of equipment at risk. However, for situations where a minor 'puff' would not present a hazard to neighbouring plant or to personnel, some general guidance can be given, although it must be emphasized that there may be local factors which would not permit any relaxation. In cases of doubt the ruling of the appropriate safety authority should be sought.

Considering dusts of vegetable or organic origin, either natural or synthetic the first step is to ascertain the maximum rate of pressure rise in the routine test apparatus (Chapter 3). When the maximum is below

5000 lb/in^2, and if the *total* volume of the dust handling plant is less than 20 ft^3, constructional precautions of the plant against dust explosion are usually unnecessary. Exceptions would be where dust explosions are expected regularly, or if the equipment were constructed with the intention of initiating explosions in it. This approach means broadly that laboratory scale apparatus and equipment for dust handling would not need precautions, but pilot plant scale equipment may and full industrial scale equipment probably will need protection. For dusts with maximum rates of pressure rise between 5000 and 10 000 lb/in^2, which include a number of plastics, special precautions may not be necessary on equipment of total volume up to 15 ft^3 provided that there is at least 1 ft^2 of area which can readily open in the event of an ignition of dust cloud. For dusts with high maximum rates of rise, above 10 000 lb/in^2, such as the light metals, it is advisable to provide some form of explosion protection, as discussed below.

The reason that many dusts do not need explosion protection, in equipment up to 20 ft^3 total volume, is that sufficient venting area to keep pressures to low values is likely to be obtained by the openings in the equipment required for the entry and delivery of dust, ventilation etc. Provision of extra vents would be of limited value, unless the inlet and delivery areas were automatically closed in the event of an explosion. Otherwise, flame would be released both through the vent and through the working openings, so the total vent area is likely to be more than adequate. Whilst acting as vents, all openings are liable to discharge flame and burning dust, which can cause injury or small local fires, but with small scale equipment the hazardous region is likely to be acceptable. The small scale plant and equipment is, if made of metal, relatively strong compared with much plant of larger scale, and may withstand pressures up to 5 lb/in^2 without serious damage. The risk of flying metal is thus also less than with larger scale plant. It should be emphasized that this approach should be used with caution, bearing in mind local circumstances, but providing it is used sensibly it can be economic and logical.

Although with small volumes the explosion hazard may be minimal, there is, nevertheless, the possibility of an outbreak of fire following a 'puff' which is the likely effect of the explosion. If not tackled promptly these fires could spread and become large, so that suitable hand extinguishing equipment should be available (Chapter 9).

The relevant volume of plant or equipment to be considered is that within which the explosion may propagate, and not necessarily that in which the dust dispersion is initially present. If there are weak partitions etc. within the plant which could break down under pressure expected in the explosion, then a larger volume of plant becomes involved and this should be the appropriate value. Likewise if the plant is connected to an exhaust trunking system, then the volume should be that of the plant together with the trunking and any dust collector.

Unless it is possible to make special exceptions, as outlined above, positive steps must be taken to deal with explosion, when explosible dusts are present which may be dispersed into a cloud in industrial plant. Suitable methods are conveniently grouped under the following six headings, but all these methods are not alternatives. Some of the methods, if used in conjunction with others, are beneficial but singly do not give adequate protection. The methods are:

Minimizing dust cloud formation
Containment
Separation of plant
Venting
Inerting, by gas or dust
Automatic suppression

In many practical situations there may be a choice as to the method of dealing with explosions, and to obtain the optimum solution consideration would involve the design of the plant, its situation, and its economics. General rules cannot be laid down so the procedure in each case would be to consider the various alternatives and make the final decision in the light of the particular circumstances of the installation (Chapter 12).

Minimizing dust cloud formation

Although some dust dispersion may be inevitable, and in any case should be expected, careful design and layout of the plant can often reduce the likelihood of dust clouds forming and minimize their volumes. For instance, in delivering dust into a large hopper the dust should preferably not be allowed to fall freely from a high point to the surface of the deposited material. Such a situation ensures that a large proportion of air space above the deposit within the hopper is filled with a dust cloud, which could be of an explosible concentration. A better arrangement would be to have the dust enter at various openings in the side of the hopper.

Careful design of the plant can also ensure that unnecessarily large volumes of dust cloud cannot form due to individual units being too large for the process and never becoming filled to their design level. Large volumes of suspension in individual plant units may also be reduced by installing several smaller units in place of a single large one. The likely higher costs initially may be recouped in the long term by only a part of the plant being affected by an ignition of the dust. Where the dust can be mixed with a liquid, and handled in the form of a slurry, the hazard of a dust explosion can be eliminated. However, this approach is possible in only limited circumstances.

Where dust dispersions are formed the plant should be designed so that removal of the dust is made as soon as possible. For instance, ventilation air containing dust should be cleaned as soon as is practicable after it is removed from the plant, rather than connected by long lengths to a central cleaning point at a distance. Where dust settles out of suspension in the plant, except in units specially provided for long term storage such as hoppers, the deposited dust should be removed regularly. This requirement also applies to units such as filters, cyclones, etc. designed to remove dust from suspension and which should never be allowed to accumulate more dust than they are designed to handle. Otherwise dust will not be effectively removed from the air and will pass on in suspension and cause an explosion hazard elsewhere. The minimizing of dust cloud formation is strongly linked to the design of the plant handling the dust, the units of which are considered separately later (Chapter 10).

In minimizing dust cloud formation in plant, attention should be paid to eliminating it completely from buildings containing the plant, because secondary dust explosions, in buildings, are particularly destructive (Plates 1, 2 and 5). Accumulation of dust in buildings can be prevented by combining the good design of plant with efficient cleaning procedures.

Containment

Logically, the first step to consider in protecting against the effects of dust explosion within plant is whether or not the explosion can be completely contained, without pressure release. In practice, containment is only likely to be attractive on plant of small dimensions, because the cost of building large units to withstand the pressures is usually prohibitive and not competitive with alternative methods. The maximum explosion pressure which may occur within the plant can be estimated using the small scale test (Chapter 3), and higher pressures would only be likely if strong shockwaves or detonations were generated. These have been reported with very large scale installations such as coal mines or with extremely rapidly burning dust suspensions such as aluminium/oxygen (Chapter 6). For the general case a safety factor must be added to the measured maximum explosion pressure, and a value of 50 per cent is often taken. The resulting figure is the static pressure which the plant must be designed to withstand.

To contain a dust explosion safely, the individual units of the plant must not only be designed to withstand the maximum pressure plus safety factor, but the connections between units must also be of the same strength. Particular attention should be paid to the points at which the dust is put into, and delivered from, the plant to ensure that there is never direct access from the interior to the exterior of the plant even during an explosion. Any discharge of gas from within the plant during the explosion is liable to contain burning dust, and to be at high velocity, and so is of particular hazard. The procedure adopted during start up and closing down operations should also be closely scrutinized, to ensure that containment is not nullified even though the plant is not running at its normal rate. Dust clouds can persist after the plant has been shut down and opening up under these conditons would be hazardous.

In any event, opening up should be done cautiously because if an explosion has occurred there may be residues of dust or explosible partial products of combustion, together with residual burning and toxic gases. The sudden access of air could lead to a second explosion. Finally although the plant may be sufficiently strong to stand the full explosion pressure, burning dust may persist within the plant for a long period and serious heating of the metalwork could lead to a reduction in its strength.

In the maintenance of the plant, particular attention should be given to any items which show signs of abrasion or corrosion. The movement of dust over prolonged periods can cause the walls of the plant to be worn away and to reduce their strength. For instance, it has been estimated that steel piping conveying pulverized coal continuously can be worn away at exposed points, such as bends, by 50 per cent of the thickness within a year. Where such serious abrasion occurs, periodic renewal of the item is necessary.

Because of the initial cost of strong plant, and the restrictions caused by the necessity for containment to be fully effective at all times, the method is not widely used. A commoner modification, is that in which single units in particularly vulnerable places are constructed very strongly, whereas adjacent units within the plant are of lighter construction and given alternative protection. Should explosion then occur, the plant is still fully protected, although not giving complete containment. The economics of this modification may be attractive, particularly where parts of the plant are in locations where alternative methods are difficult.

Separation

In order to restrict the amount of dust cloud undergoing explosion, and hence the total energy released, and also to prevent ignition from propagating throughout an entire plant, separation of the units of the plant is desirable. The principle of separation can also be applied to a group of plants handling explosible dusts, in that each plant may be put in its own building, which is then spaced apart from its neighbours. Separation of plants may also be necessary to ensure that adequate fire protection is obtained (Chapter 10). Where separation of plants is possible, the important factors involve the design of the plants and their neighbours, the hazards to operatives and passers-by, and the processes and materials involved. Detailed consideration must be given separately to each installation and expert guidance should be sought. For separation of units within a plant, generalizations may be given, as indicated below, but special factors may be present on an individual plant for which expert guidance should again be sought.

For maximum protection against dust explosions, the units within the plant should be isolated from each other in the event of an explosion. On the other hand, production requirements may demand that individual units shall be permanently connected and placed close to each other. These requirements must be acknowledged, and steps taken to ensure that even plant units within close proximity of each other are effectively protected against the transmission of explosion between them. Two forms of choke commonly used for this purpose are:

Rotary valves
Worm conveyers

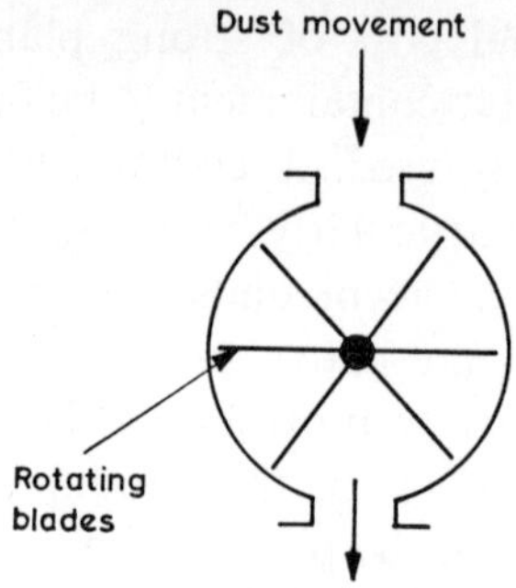

Fig. 8.1 Rotary valve.

The rotary valve, rotary choke, or star valve, consists of a power driven multi-bladed paddle inside a cylindrical housing (see Fig. 8.1). The number of blades varies, but six is the usual minimum. This number prevents a direct path opening betwen the upstream and downstream sides of the valve whilst it is rotating. The drive is usually by electric motor. The blades are accurately machined so that their tips clear the internal surface of the housing by a small margin, usually about 0.2 mm. The speed of rotation is commonly about 50 rpm. With these valves the dust movement must be vertical, the dust falling into the valve under gravity, being conveyed by the rotary action, and then dropping away from the valve under gravity. The valves are designed to be strong so that in the event of an explosion they are not deformed, and act as an obstacle to the passage of flame from one side to the other. Unless appropriate arrangements are made the rotary valve continues to turn during the explosion, but the clearance between the blades and the housing is too small to allow flame to pass between them. It is not a requirement of the design that there should always be a head of dust above the valve, to ensure that all segments between the blades are completely filled as the dust is conveyed, but the presence of such a layer acts as an additional barrier against dust flame propagation.

Systematic investigations of the effectiveness of rotary valves in stopping dust explosions have been few. Tests with rice meal, where the explosions was initiated below the valve so that the flame propagated back against the delivery of the dust [1], showed that the flame did not propagate through the rotary valve to the dust above it. These test conditions were not the most severe, because the flame was travelling against the flow of dust. More recent tests, with a rotary valve mounted at the base of a vented cyclone, and delivering dust into a hopper below, have shown that when a dust explosion was initiated in the cyclone and a pressure of 1 to 2 lb/in^2 developed, explosion was transmitted to the hopper. The dusts in the tests were cork and phenol-formaldehyde resin. The rotary valve was not covered by a layer of dust at the base of the cyclone, because otherwise dust might have been picked up by the vortex within the cyclone and delivered through the central exit duct, and the segments between the blades of the rotary valve were probably not completely filled. The dust explosion could have been transmitted either by a burning suspension being carried around between the rotary valve blades, or by smouldering dusts being carried which inflamed when delivered from the valve into the hopper beneath it. Transmission of the explosion could be prevented by arranging that when the cover on the explosion vent in the cyclone began to move, a trip was operated which shut-off the motor driving the rotary valve. Protection was obtained as long as the interval between the initial movement of the vent cover and the switching off of the motor was not greater than 0.5 seconds. Provided the rotary valves are installed in this way, they are reasonably effective in preventing propagation of flame and are convenient to use in the separation of units of dust handling plant.

Where rotary valves are installed at both the inlet and outlet of a storage unit, as in a hopper, care must be taken to ensure that the rotary valve at the inlet is stopped before the unit becomes overfilled. Otherwise the valve would continue to turn through a stationary layer of dust and there would be a risk of either the dust becoming overheated, or of overloading the drive to the valve. Either of these possibilities could lead to an ignition.

An alternative means of separating units of plant is the worm or screw choke (Fig. 8.2 and 8.3). The choke consists basically of a helix in a cylindrical casing, the dust being fed in at one end and out at the other, and all available space between the inlet and the outlet being filled with dust. Where the axis of the helix is vertical or inclined (Fig. 8.2) one turn of the helix is removed to ensure that even if the dust feed to the inlet ceases, the choke will not completely empty and open up a path along which a flame could propagate. Worm chokes can be used to raise dusts through considerable heights, e.g. from one floor to another in a building, and are preferable on safety grounds to other types of elevator which permit a through path for the passage of flame. When horizontal movement is required one turn of the helix is again removed and, in addition a baffle plate is necessary (Fig. 8.3) to ensure that when the dust feed to the inlet ceases a gap does not open up between the top of the dust layer and the roof of the casing containing the helix.

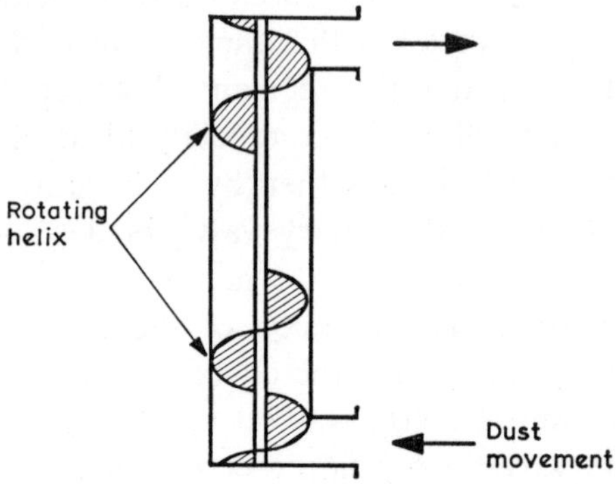

Fig. 8.2 Vertical choke with one turn of the helix removed.

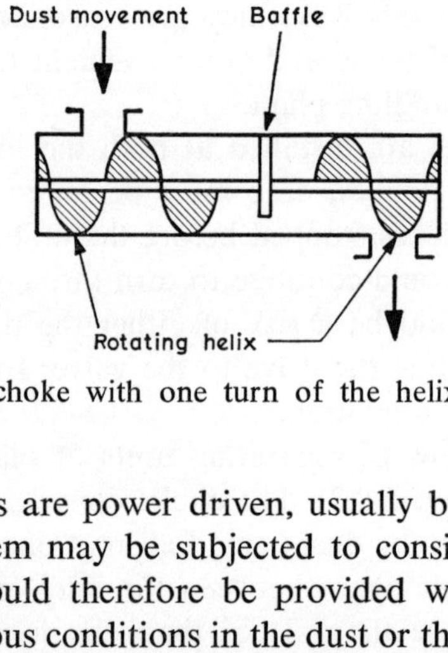

Fig. 8.3 Horizontal choke with one turn of the helix removed and a baffle inserted.

All worm chokes are power driven, usually by electric motor, and the dust within them may be subjected to considerable friction. The driving motors should therefore be provided with overload trips to ensure that hazardous conditions in the dust or the motor, do not arise. Worm chokes may not be suitable for dusts which become less free flowing when heated, e.g. thermo-setting resins, because local overheating of the dust can lead to increased resistance to movement and further overheating. Eventually the choke may seize up, and its contents ignite or the drive motor may overheat.

There have been few tests made on the effectiveness of worm chokes in preventing the transmission of explosion between units of plant.

Some tests have been reported [1] with rice meal, but the incentive for this type of work is less than with rotary valves because it is obvious that if the choke is full of dust flame cannot readily penetrate through it. In addition, the residence time of the dust in the choke is much longer than in rotary valves and therefore the likelihood of a dust flame being carried through is low. Glowing dust could be transported, even though there is no air space when the choke is filled with dust, because sufficient air may be present between particles to enable the smouldering to continue whilst the dust is carried through. Flame may possibly be produced when the smouldering dust is discharged from the choke.

As rotary valves and worm chokes are both rotating machinery, usually constructed of steel, it is important that foreign objects whether of metal or stone cannot enter them. A hazard due to friction or impact sparking could arise and could ignite dust within the valve or choke. If they are near the point of entry of the dust to the plant, so that foreign objects may be carried to them, the installation of separators or screens before the valves or chokes may be necessary (Chapters 7 and 10).

Venting

EMPIRICAL APPROACH

Explosion relief venting involves the provision of apertures on a plant, or in a building, so that if an explosion occurs and pressure starts to rise the rise is limited to a pre-determined value. Venting does not prevent the propagation of the flame through an explosible dust cloud but, if correctly designed, permits an explosion to take place without damaging the structure of the plant or building. After the explosion has been vented, there may be residual fires, against which explosion relief does not protect.

In general, the aim should be to provide venting for the plant, rather than for the building containing it. Whenever possible plant should be built in the open air, or under semi-open conditions, avoiding the possibility of explosion within a building (Chapter 10). If the plant must be installed within a building, which is often necessary, the object should be to keep the dust within the plant, and, if an explosion develops, to relieve it safely to atmosphere outside the building. Spillage of burning dust into the building, or general accumulation of dust within it, should be avoided as far as possible. In this way the hazard to operatives and to other plant is minimized. The venting of buildings may be necessary where hazardous processes involving severely explosible dusts are carried out, but in such cases the plant is best operated by remote control and should be in individual enclosures. The risks to other plant and to operatives are then avoided.

The pre-determined maximum pressure which the relief venting is designed to provide is directly related to the strength of the plant or building. In the general case, unless special provision has been made for strengthening, a maximum safe pressure for plant would be 2 lb/in^2. Where special provision is made for strengthening, higher pressures may be acceptable and in the limiting case of containment, discussed above, the plant is capable of withstanding the full explosion pressure. At the other extreme, plant constructed of thin metal sheeting may be unable to withstand pressures of more than 1 lb/in^2 without damage, and even so large expanses of flat metal, e.g. panelling may need special stiffening or bracing. Many buildings are of similar strength. Provided the correct area of venting is installed, weak constructions can be safely protected against the pressure effects of dust explosions.

Relief venting is generally a cheap method of obtaining protection and is likely to be considered every time a new plant handling explosible dusts is designed. Experience over the years has shown that providing the original design and the subsequent maintenance were adequate, protection can be reliably obtained. There are circumstances where the venting of plant cannot be permitted; these include the presence of toxic dusts which cannot be discharged to the atmosphere and also the awkward siting of the plant. Preferably, vented plant should be in the open air, or with ready access to it, and in such a position that the burning material ejected through the vent does not endanger life, or neighbouring plant, or buildings. If the plant is indoors, the burning material from the vents should be ducted away to the outside air, and the ducting should be short. The ejected material can be of sufficiently large volume to be hazardous to operatives or neighbouring plant and, in addition, can cause a secondary dust explosion if released inside the building. If the plant is awkwardly placed, so that ducting from the vents to atmosphere cannot be provided, venting should be discarded and another method of protection used unless the vent can be situated so that operatives are unlikely to be in the vicinity and dust deposits in the workroom are rigorously excluded.

In the normal working of the plant the vents are normally required to be closed, to prevent emission of dust, but they should open quickly as soon as ignition has occurred and the pressure starts to rise. Vent closures are considered in more detail below. For safety reasons it is important that dust should not escape from the vents during normal working, as even a very small leak over a long period can allow a hazardous amount of dust to come out. Leakages can be minimized by running the plant at slightly negative (i.e. sub-atmospheric) pressure, e.g. 1 in. water gauge. This lowering of the pressure would have a negligible effect on the maximum pressure obtained in the plant should an explosion occur. The area of the vents is of controlling importance in determining the maximum explosion pressure in the plant, but the distribution of the vents is also important. In addition to a regular distribution of vent area throughout the plant or building, vents should also be installed near positions where sources of ignition can be expected. For instance, in an exhaust ventilation system constructed of a long length of ducting through which air is drawn by a fan vents are desirable on both sides of the fan. In such a system the shape of the vent is of less importance; a circular and a square vent of the same area would be equally effective.

The area of vent required for a particular installation has at present to be prescribed empirically. Although numerous investigations have been reported, the data made available are very incomplete and a satisfactory basis for the design of relief venting has not yet been published. The main requirement is to relate the maximum pressure in a vented explosion to the explosibility characteristics of the dust as measured in small scale tests, the area of vent, and the size and shape of the plant. The effect of the vent cover and any ducting leading from the vent to the outside atmosphere are additional variables. Venting has often been characterized on an empirical basis in terms of the vent ratio, that is the area of vent per unit volume of plant. The vent ratio has the dimensions of length^{-1} when the maximum pressure is related to the vent ratio in the design of plant there is doubt about the validity of applying empirical relationships to plant of different scale.

The absence of a well founded method for the calculation of venting requirements has important practical consequences because venting is commonly used for the protection of plant, but the degree of safety obtained cannot be adequately judged at present. The situation is steadily worsening because the scale of industrial processes involving dust is increasing rapidly, but the amount of design information for safety precautions is not growing correspondingly. Extrapolations are therefore becoming greater and causing increasing uncertainty in the effectiveness of the explosion protection measures applied. The situation can only be remedied by a better understanding of the principles involved in explosion relief venting, coupled with confirmatory experiments on a larger scale than that customarily used. A recent review summarized the available data [2].

For plant of relatively light construction, dealing with carbonaceous materials, a vent ratio of 1 ft^2 per 20 ft^3 of plant volume is recommended [3].

The recommended vent ratio may be expressed more generally in terms of the maximum rate of pressure rise as measured in the routine test apparatus (Chapter 3) and values in Table 8.1 may be used as an empirical guide.

Table 8.1 Vent ratios for dusts of different rates of pressure rise in test apparatus.

Maximum rate of pressure rise lb/in²s	*Vent ratio*
< 5000	*1 ft²/20 ft³*
5000—10 000	*1 ft²/15 ft³*
> 10 000	*1 ft²/10 ft³*

Dusts with maximum rates of pressure rise of less than 5000 lb/in² s include the majority of common carbonaceous materials, such as wood sawdust, coal, flour and coffee. There are however numerous exceptions and direct measurements should be made in any cases of doubt.

Some plastics materials, for example phenolformaldehyde resin, have maximum rates of pressure rise in the range 5000–10 000 lb/in² s. With such materials, and dusts other than plastics having similar pressure rise characteristics, the vent ratio is 1 ft^2 per 15 ft^3 of plant. Plastics dusts having maximum rates of pressure rise less than 5000 lb/in² s can be given a vent ratio of 1 ft^2 per 20 ft^3.

Severely explosible metal dusts such as magnesium, and aluminium, giving maximum rates of pressure in excess of 10 000 lb/in² s, and other dusts with the same behaviour, require a vent ratio of 1 ft^2 of vent area per 10 ft^3 of plant volume. If the rate of pressure rise is very high, say above 12 000 lb/in^2 s, relief venting should be used with caution, because of a serious lack of practical data for this region. Additional strengthening of plant may be necessary, or an alternative method of explosion protection used.

The vent ratios in Table 8.1 are intended to give protection against severe dust explosions in volumes up to 1000 ft^3, and are not dependent on the volume of the plant. In practice the vent ratio is customarily reduced for compact enclosures of greater volume, but not for ducting. For example, with a large silo of volume more than 10 000 ft^3, the vent ratio may be reduced from 1 $ft^2/20$ ft^3 to 1 $ft^2/80$ ft^3. The reason for the decrease is that large enclosures are less likely by chance to contain the most explosible concentration throughout the volume, whereas this is more likely with small enclosures and ducting, and the rate of pressure rise in an explosion is likely to be less. Also the free volume within a collector may be reduced by the collected dust. The less severe condition with the large enclosures should be considered with caution, because traditionally such enclosures will be filled by dust falling under gravity from the point of delivery, and the dust cloud would decrease in concentration with distance from the column of falling dust. With modern pneumatic systems a dust cloud is more likely to be maintained continually throughout the free volume of a collector and the conditions for a severe explosion can exist in a larger proportion of the volume. The structural strength of the enclosure is also significant, for if light construction has been used the strength would be low compared with that of a smaller enclosure built with the same materials.

Because of the empirical approach, the recommendations of various authorities differ slightly. The National Fire Protection Association [4] recommend as follows:

1. Enclosures of volume 1000–25 000 ft^3, vent ratio 1 ft^2 per 30–50 ft^3. Attempt must be made to predict the point of origin of explosions with respect to the vent, and the actual volume of suspension which may be involved.
2. Rooms and buildings over 25 000 ft^3 volume, containing hazardous equipment comprising a small fraction of the entire volume.
   (*a*) For heavy reinforced concrete walls, vent ratio 1 ft^2 per 80 ft^3.
   (*b*) For light reinforced concrete, brick or wood, vent ratio 1 ft^2 per 60–80 ft^3.

(c) For lightweight constructions such as prefabricated panels, vent ratio 1 ft^2 per 50–60 ft^3.

3. Large rooms or buildings over 25 000 ft^3 volume, containing hazardous equipment comprising a large part of the entire volume should be vented as generously as possible, vent ratio 1 ft^2 for each 10–50 ft^3.

The above vent ratio recommendations refer to plants and buildings in which all three dimensions are of the same order, e.g. bag filter units, hoppers, silos, rooms. When vents are to be installed in ducting, or other units having one dimension much greater than the other two, the vent ratio in Table 8.1 may be used as a basis, with additional vents where necessary. For example, with dusts having a maximum rate of pressure rise below 5000 lb/in^2 s, in a straight length of ducting a vent equal in area to the cross-section of the ducting should be installed every 20 diameters. If the ends of the duct are not restricted, they can be taken as two of the vents. In many cases ducting will not be straight throughout its length and where bends, elbows, restrictions, orifice plates etc, are present an additional vent of the same area should be installed for each. Likewise, if there is a potential source of ignition in the ducting, such as a fan, vents should be installed on each side of this also. Where convenient, smaller vents may be used, provided that the total area available is not reduced, and it would be beneficial to space them more frequently along the length of the ducting.

The recommended vent ratios are generalizations, and should be interpreted with caution and not followed blindly. The values in Table 8.1 originated from tests using compact enclosures, whose three dimensions were all similar. If they are applied to situations greatly different from which they originated, anomalous results may follow. For instance, consider a cylindrical tank of cross-sectional area 1000 ft^2 and height 50 ft. The volume of the tank is 50 000 ft^3 and the maximum vent ratio physically possible, even with the whole of the top of the tank available for venting would be 1 ft^2 per 50 ft^3. This may be regarded as a generously sized vent and, indeed, may be excessive for dusts of low explosibility, although not generous for rooms and buildings (see above). In such a case the generalization must be modified to give a sensible answer. A more realistic approach, with dusts of maximum rate of pressure rise less than 5000 lb/in^2 s, would be to specify that at least half the top of the tank should be the vent, and design accordingly.

For dusts having greater maximum rates of pressure rise the whole of the top of the tank should be available for venting. If there is doubt about the adequacy of this arrangement, because the dust is very vigorously explosible, rather than attempting to increase the relief area by installing vents in the side of the tank it would be preferable to re-design the plant so that several smaller tanks were used instead. In this way, providing the tanks did not intercommunicate, the amount of energy released in any explosion would be proportionately reduced and the installation would be inherently safer.

It is not usual for vents to be specially manufactured commercially. Frequently some form of vent cover is required, see below, and the design of the vent itself is incorporated in the mounting of the cover. Most vents are simply apertures in the wall of the plant or building which are framed by a robust structure, usually of metal, capable of withstanding the stresses developed when the vent cover opens. It is not usual for the edges of the vent to be of special aerodynamic design, to minimize friction and drag of the expelled combustion products. The empirical rules for vent ratio take no account of aerodynamic shaping. Vents are usually circular, square, or rectangular but other shapes are aceptable provided that the required total area is available. Indeed, a vent for ducting in the form of a long slot, with straight parallel sides, has the technical advantage that some venting is more likely to be close to the ignition source; disadvantages are that the plant may be weakened excessively and that vent covers capable of opening at a low predetermined pressure may be difficult to design.

TECHNICAL BASIS FOR VENTING

The technical literature of the venting of dust explosions is chaotic in that it is inadequate, fragmentary and to some extent contradictory. The situation with dusts compares unfavourably with that for gas explosions where more investigations have been carried out, experimentation is easier and more attractive, and there are underlying theoretical concepts which enable results to be linked together. The absence of a theoretical background for dust explosions may be the basic reason why the total information is inadequate; there is little incentive to carry out work to establish general relationships between venting requirements and the properties of dusts until there are at least some theoretical indications as to important quantities to measure. Because of the general inadequacy, such investigations as have been undertaken tend to have been concerned with certain narrowly defined practical questions needing immediate solutions. This has resulted in available information being fragmentary and often only applicable with confidence to the equipment with which it was obtained. A simple example is that venting experiments have been made with cubical chambers, and with long lengths of ducting, but by different investigators using different dusts and experimental techniques. Tests have also been made on actual items of industrial plant but again, in the absence of a theoretical background, the results are only strictly applicable to the plant tested. The maximum pressures obtained in vented explosions are noticeably dependent upon the power of the igniting source, as well as on the vent ratios and properties of the dust. By varying the ignition source the venting requirements also change, and lead to contradictory conclusions. The whole subject of dust explosion venting is clearly in need of technical ideas and research effort, and it is to be hoped that this need will be recognized. After summarizing the available data on explosion venting, consideration will be given to attempts to provide a theoretical background to the data.

The available information on venting, obtained from experiments, may be grouped under the following four headings depending upon the equipment used.

(1) Compact enclosures, such as cubical boxes
(2) Long ducting
(3) Laboratory scale apparatus
(4) Individual plant units.

For open vents (i.e. vents without covers) it is for compact enclosures that probably the most systematic series of experiments were undertaken using corn starch [5] and cellulose acetate [6]. The tests with corn starch were made in cubical galleries, of volumes 1 to 216 ft^3 (Fig. 8.4) and showed three features plainly. The scatter of the experimental points was small, indicating good reproducibility; the relation between maximum explosion pressure and vent ratio was independent of the volume of the gallery, indicating an absence of scale effect, and finally that the maximum explosion pressure, plotted on a logarithmic scale, varied directly with the vent ratio. The relationship broke down at pressures of 3 lb/in^2 and below, and the observed pressures were higher

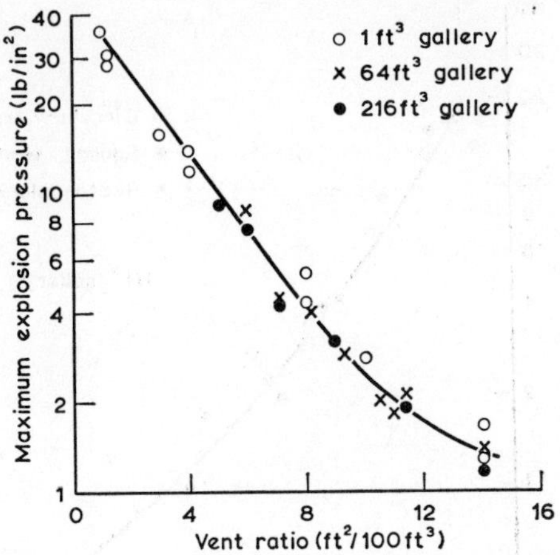

Fig. 8.4 Maximum explosion pressures in vented corn starch explosions [5].

than would be predicted. This is unfortunate because these pressures are of particular practical interest in the safe design of plant and an empirical relationship would have been valuable. The maximum rate of pressure rise of corn starch in the routine test apparatus was reported [5] as 8700 lb/in^2 s. From Table 8.1 the vent ration for plant would be 1 ft^2/15 ft^3 and, from Fig. 8.4, the corresponding explosion pressure would be about 4.5 lb/in^2. This value is high for practical use, but it was measured under experimental conditions deliberately made as severe as possible, and which would be unlikely in practice. However, it indicates that the recommended vent ratios (Table 8.1) are not excessively cautious. Cellulose acetate dust gave a similar relationship between explosion pressure and vent ratio (Fig. 8.5); results are only available for the 1 ft^3 gallery. The same equipment was used for tests on a range of other dusts, under less severe conditions [6].

Corn starch was also exploded in an oblong explosion gallery, volume 4170 ft^3, fitted with vents [7]. Although the quantity of corn starch dispersed was sufficient to consume only half the oxygen in the gallery, it was believed that the quantities tested represented reasonably severe field conditions because optimum conditions would rarely if ever be approached accidently. The intensity of the explosions was measured using a gauge designed to register the same deflection as an average 12 in brickwall 16 ft between rigid supports. As the actual pressures were not measured, direct comparison cannot be made with

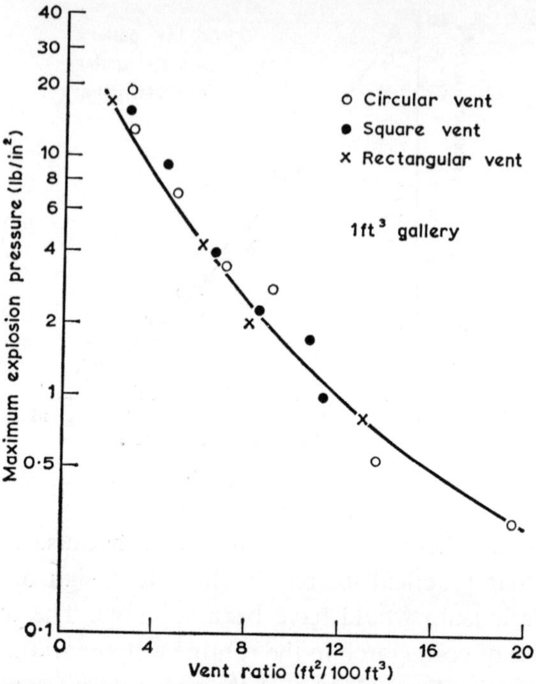

Fig. 8.5 Maximum explosion pressures in vented cellulose acetate explosions [6].

the results in Fig. 8.4. Comparative tests were however made, with a laboratory scale apparatus, as mentioned below.

Explosions in long ducting, using open vents, have been reported following a prolonged series of experiments [8]. The ducting, or gallery, was horizontal, of length 55 ft 6 in, and diameter 4 ft. Uniform dispersion of dust was difficult with this geometry, and the method finally used consisted of suspending controlled amounts of dust, in paper bags, at intervals along the ducting. One or two detonators were placed in each bag and fired electrically. The detonators did not ignite carbonaceous dusts. A check showed that dust dispersed in this way, and allowed to settle, gave a uniform weight of deposit along the ducting length. The ignition source for the explosions was a small cloud of magnesium dust, ignited by gun cotton which in turn had been fired electrically. The results available are for wheat, provender (animal feed), and cork dusts with various areas of venting at different positions of the ducting, Table 8.2. Relatively high pressures were obtained when the dust was ignited remote from a vent, whether or not the full cross

Table 8.2 Explosion pressures in vented long ducting

General arrangement	Percentage of cross-sectional area vented	Maximum explosion pressure lb/in²		
		Wheat dust	*Provender dust*	*Cork dust*
One end of ducting vented, other end closed	* 14%	16	20	24
	* 43%	—	13	13
	* 100%	2.1	8.4	7.4
Both ends of ducting vented	11% * 14%	4.7	1.6	5.7
	11% * 14%	12	2.3	10
	11% * 100%	1.0	1.5	1.5
	100% * 100%	0.4	1.8	1.9
Additional vent in centre of ducting	* 100% (centre 14%)	—	6.4	4.6
	* 100% (centre 43%)	—	3.3	2.4
	* 100% (centre 100%)	—	1.3	2.0
	100% * 100% (centre 100%)	—	—	0.4

* Indicates position of igniting source.

sectional area of the ducting was available for venting. A vent near the ignition source was clearly beneficial in reducing pressures, as would have been expected, and a vent halfway along the ducting also gave benefit. The length of the ducting was about 14 diameters and so with a single vent equal to the cross-section of area at the end of the ducting the vent ratio would be 1 ft²/56 ft³. This is less venting area than suggested in Table 8.1, for carbonaceous dusts, and in fact gave pressures well above the arbitrary figure of 2 lb/in². The recommended vent ratio, 1 ft²/20 ft³ requires at least three vents in the ducting for the necessary vent area. If these are provided, Table 8.2 shows that the necessary reduction in pressures can readily be obtained. When considered together, the data again indicate that the vent ratio in Table 8.1 is not unduly cautious.

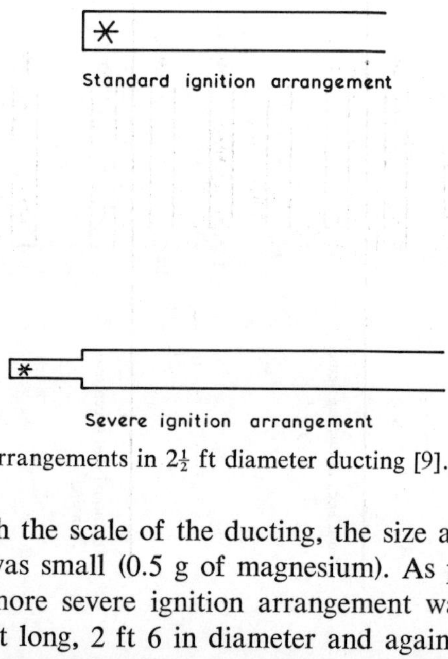

Fig. 8.6 Ignition arrangements in 2½ ft diameter ducting [9].

Compared with the scale of the ducting, the size and power of the ignition source was small (0.5 g of magnesium). As part of a further investigation a more severe ignition arrangement was tried [9]. The ducting was 50 ft long, 2 ft 6 in diameter and again horizontal. The new ignition source consisted of a 10 ft length of 10 in diameter ducting bolted co-axially onto the main ducting, Fig. 8.6, and a cork dust cloud was produced within this additional duct, and fired by a magnesium igniter near the closed end. The flame produced then acted as an ignition source for the dust in the main ducting. Using this severe ignition arrangement the maximum explosion pressure was 16.3 lb/in² instead of 7.9 lb/in² obtained with the magnesium igniter in the main ducting. Although the additional ducting would increase the total volume of the system slightly, the increase could not account for the greatly increased explosion pressure. These results demonstrated that the explosion venting requirements depended on the power of the ignition source.

Surprisingly there is relatively little information available on explosion venting tests using laboratory scale apparatus. The obvious apparatus to use would be that for the routine measurement of maximum explosion pressure (Chapter 3), particularly as apparatus of similar design is used in several countries. The main reported investigation involved the explosion venting of metallic stearates, and of polymethyl methacrylate, with various circular vents in the top of the tube [10]. From the results nomograms were produced which enabled the relationship between vent ratio, maximum explosion pressure, and maximum and average rates of pressure rise to be predicted. In using the nomograms for the design of full scale plant, considerable extrapolation of scale is necessary and, so far, no confirmatory tests of the nomograms on larger scale have been reported. In the absence of any venting, the maximum explosion pressures were less than 60 lb/in^2, even for fine dusts. These maxima are significantly lower than those reported in the routine tests for readily explosible dusts. The nomograms offer a simple approach to vent ratio requirements in plant design but further confirmation using larger scale tests is required.

In another investigation corn starch was tested in a larger laboratory scale apparatus [7], which consisted of a vertical tube 18 in. high and 4 in. internal diameter with the vent at the top. The dust was dispersed upwards from the bottom of the tube, and the ignition source was a heated wire coil near the top. The ignition source was thus close to the vent and as the flame would need to propagate downwards during the explosion the test conditions were not of maximum severity. The vents were small, so that the vent ratios were below 3 $ft^2/100\ ft^3$, and the explosion pressures were consequently high (Fig. 8.7). Direct comparisons cannot be made easily with data in Fig. 8.4, and so consideration of scaling cannot be made. Other experiments on a larger scale [7], see above, using the same corn starch, did not directly give explosion pressures.

The simplest form of individual plant unit tested has been a vertical cylindrical silo, of height 60 ft, and diameter 7.5 ft [1]. Different areas of vent were provided on the top of the silo, and the dust was dispersed

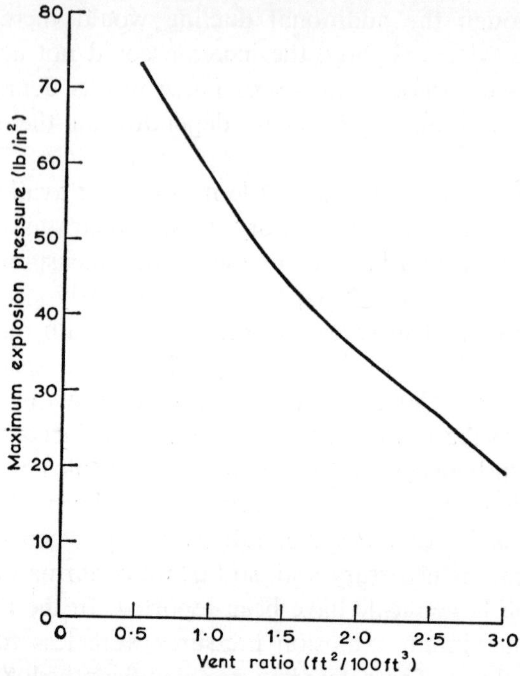

Fig. 8.7 Maximum explosion pressures in vented corn starch explosions. Vessel volume is 0.13 ft^3 [7].

by tipping it off shelves at the top and allowing it to fall downwards. The ignition source was a gas flame across the diameter, about 2 ft from the bottom, and flame could be expected to propagate upwards as soon as the dust concentration reached the lower explosible limit. In these tests the dusts were unlikely to have been uniformly distributed throughout the volume. With rice meal, at a dust concentration of 0.45 g/l, the explosion pressure was barely measurable when the top of the cylinder was fully open, and was about 0.5 lb/in^2 when the top was two thirds open, i.e. two thirds of the area available for venting. When the top was one third open the explosion pressure ranged up to 6 lb/in^2 and with the top one ninth open the pressure was 17 lb/in^2. Further tests with dusts manufactured from ground nuts, sun flower cake, and palm kernel cake gave lower pressures. As the dust concentration at the moment of ignition was probably far from uniform, the marked increase in explosion pressure as the vent area was reduced may have been influenced by the greatly increased turbulence that would be produced in the silo when the vents were small. This turbulence would tend to improve the mixing of the dust, and thereby cause more vigorous explosions, apart from any increase in burning rate due to turbulence.

A plant unit of more complicated shape was a coal pulverizing mill, of throughput 3 tons/hour, with distributed vents [11]. The mill basically consisted of two chambers, a separator unit mounted above a milling section. The height of the mill was about 14 ft, and the diameter 6 ft. The net internal volume was about 183 ft^3 and the openings that were available for experimental explosion vents totalled 16.7 ft^2. If all these openings were used simultaneously as vents, the vent ratio would be about 1 ft^2/11 ft^3. The pressures developed, with the same coal dust, with an internal ignition source in the mill (2 detonators) were substantially less than when the ignition source was external. In practice both types of ignition would be likely. A fire in the mill could lead to an explosion, and ignition would be internal. A blow back from a pulverized fuel furnace down the ducting and back into the mill would represent an external source of ignition. The method of producing ignition from an external source was to connect to the mill a cylindrical vessel (a 10 ft length of 2.5 ft diameter ducting) by means of a 20 ft length of 2 ft diameter pipe. Coal dust was dispersed in the ducting but not the pipe, and ignited by a fusehead. The flame passed along the pipe into the mill, and acted as a large external source of ignition. With a vent ratio of 0.84 ft^2/100 ft^3, based on the area and volume of the mill only, a maximum explosion pressure of 28 lb/in^2 was obtained. With a vent ratio of 3.2 ft^2/100 ft^3 the explosion pressure was only 8 lb/in^2. The recommended vent ratio for coal, Table 8.1, would be 5 ft^2/100 ft^3 to keep explosion pressures down to 2 lb/in^2. Direct comparison cannot be made with the results for the mill because this amount of venting would not be practicable, and in any case was not of practical interest because the mill was sufficiently strong to withstand pressures up to 50 lb/in^2. The complicated geometry of the mill, and the effect of the massive metal components within it on the dust explosion flame, hinder correlation of the results with those from experiments with simpler systems.

In some plant units the distribution of dust is, by design, far from uniform. A typical example is the cyclone separator, in which the dust suspension is concentrated near the walls, by centrifugal action, and the pressures developed in dust explosions may well be different from the usual case where the dust is relatively uniformly distributed. A closed circuit plant for experimental purposes, involving a cyclone [2, 12], is sketched in Fig. 8.8. The cyclone was 6.3 ft in height, 4 ft diameter at the top, and of total volume 43 ft^3, with an outlet at the bottom 7.6 in diameter. The inlet of the cyclone was connected to a

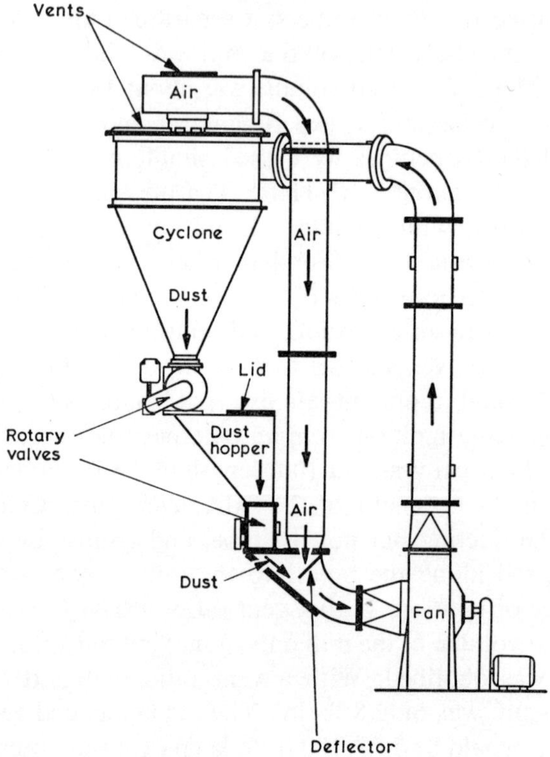

Fig. 8.8 Experimental cyclone plant [12].

fan by ducting of internal diameter 1 ft, a 6 ft length of which was horizontal and the remaining 12 ft was vertical. The bottom of the cyclone was connected by a 6 bladed rotary valve to a hopper which was in turn connected by another rotary valve to the inlet of the fan. The air outlet at the top of the cyclone was connected by 1 ft diameter ducting to the fan inlet. Facilities were provided for 6 sector vents on the top of the cyclone, a circular vent on the return air outlet, and a rectangular vent on the ducting near the inlet of the fan. With this plant a steady dust suspension could be generated and measured and the suspension was highly turbulent. The ignition source was a propane/air flame which was momentarily injected into the ducting 4.5 ft from the cyclone inlet (Plate 12). The explosion pressures of cork dust, with various vent arrangements, are shown in Fig. 8.9. For experimental purposes the vents were covered with brown paper, which burst at a convenient pressure of 0.5 lb/in². An unexpected result, in which explosions in the cyclone differed from those in other compact vessels, was that when the vent was moved from a position near the dust/air inlet to a position remote from the inlet, but kept at the same area on the top of the cyclone, the maximum explosion pressure was increased. For the larger vents the effect was less important but with the smallest vent a marked increase in pressure occurred (Fig. 8.9). An even higher pressure was obtained when the vortex pipe was used for venting.

The vent ratio corresponding to an explosion pressure of 2 lb/in², Fig. 8.9, was about 1 ft²/50 ft³. This was considerably less than the

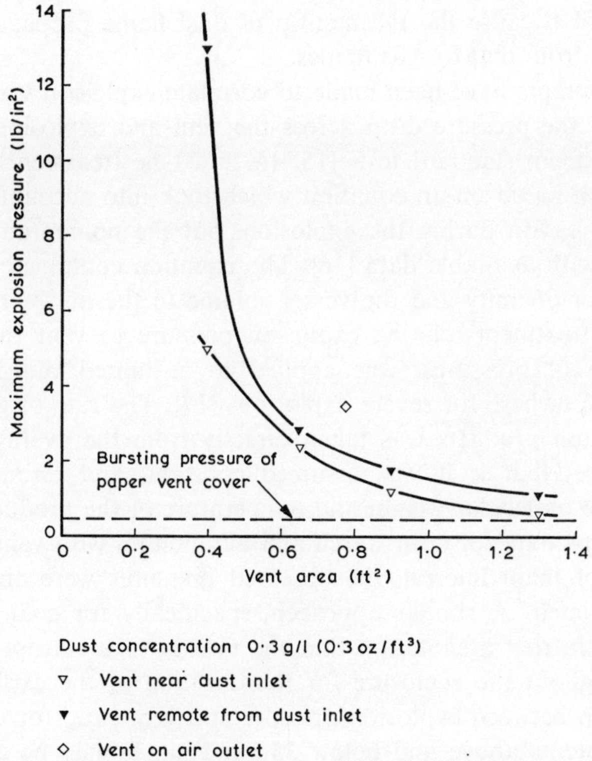

Fig. 8.9 Vented cork dust explosions in a cyclone [2 and 12]

recommended value for carbonaceous materials, Table 8.1, indicating that even with a readily explosible dust under highly turbulent conditions the explosion pressures were relatively low. One reason was probably the marked variation in dust concentration through the volume of the cyclone, so that only a part of the total volume took part in the combustion. Similar behaviour was found for several other carbonaceous dusts, of both greater and lesser explosibility than cork.

The extent to which the available data can be linked together theoretically is limited. In a comprehensive review of explosion venting [13] no unified basis emerged, although the suggestion was made that the venting requirements for dusts could be calculated from those for gases and vapours. Explosion pressure measurements in the standard test apparatus could be compared with those for quiescent gases in the same apparatus and hence the venting requirements for dusts could be linked to the well established procedures for gases. Further study of this suggestion is required, to ensure that it would still be valid because the mechanism of dust flame propagation may be different from that for gas flames.

Other attempts have been made to correlate explosion venting data in terms of the pressure drop across the vent and explosion pressure parameters from standard tests [15, 16, 17]. One treatment presented a nomogram based on an equation which took into account the adiabatic compression during the explosion, but the nomogram was not compared with available data [16]. The equation contained vent area to the power of unity and the vessel volume to the power two-thirds. A general treatment relating explosion pressure to vent ratio, for a wide range of pressures, was applied to a limited range of data published elsewhere for severe explosions [17]. The rate of generation of combustion products was taken directly from the average rate of pressure rise, that is it was assumed constant, and estimates were made of the molecular weight and temperature of the products. When tested against data for corn starch and aluminium, with vent ratios in the range of main interest, the observed pressures were about twice those calculated. A similar approach, specifically for coal dust [18], but with a further assumption that the rate of heat release was constant throughout the enclosure for the duration of the explosion led to a relation between explosion pressure and vent ratio for coals with volatile contents above and below 35 per cent. It may be applied to the design of hoppers, silos etc containing pulverized coal, provided that the ignition source is within the volume and the dust is not ignited by a powerful external source. The explosion parameters of the dusts were measured with laboratory bench scale apparatus and scaled up directly. The work may not apply to explosion flames moving at high velocity along pipelines, or to explosions in heated enclosures such as pulverized fuel furnaces.

An alternative approach assumes that the mechanism of flame propagation is different for dusts and gases. This approach is supported by known characteristics of dust explosions which have become well established by experience [2]:

(1) Dust flames are longer or thicker than gas flames. If the equipment containing the explosion is of sufficient size the dust flame thickness could be of the order of metres.

(2) In vented dust explosions the maximum pressure increases with the size of the ignition source. The pressure could be particularly high if the dust had been ignited by a large external source projected into the suspension.

(3) The maximum pressure in vented explosions also depends on the method of dispersion of the dust and on its concentration. By suitable adjustment of these variables, explosions could be graded from mild to strong.

The data of most interest are those relating to strong or severe explosions because if these conditions can be adequately designed for, the less severe conditions which may arise by chance would also be covered.

Using these observations as a basis, certain tentative conclusions can be drawn regarding the mechanism of dust explosion venting. The thick flame implies that the dust particles became ignited relatively rapidly, and subsequently burned relatively slowly. In a vented explosion in a compact vessel, initially filled with a dust suspension, the enclosure may be completely filled with flame when the pressure reaches a maximum. This would be the case particularly when the venting is generous, so that maximum explosion pressures are kept low as is required for many industrial applications, being assisted by the expansion due to the formation of combustion products with ignition remote from the vent. Because of the expansion, the majority of the dust initially in suspension in the enclosure may be ejected before it is completely burned, and would form a flame outside the vent. The emission of large volumes of flame from the vents during dust explosions is a common observation. The relatively rapid ignition of dust throughout the enclosure, and its continued burning throughout the volume, would be assisted by turbulence of the suspension which is often generated during the dispersion of the dust. In experimental dust explosions in compact enclosures the dispersion of the dust is usually by some form of vigorous mechanical action, and this can be expected to lead to an intensely turbulent suspension. In such a suspension, the existence of extensive regions of unignited mixture would not be favoured, again leading to the conclusion that simultaneous burning of the suspension throughout the entire volume of the enclosure would be likely.

For enclosures with little or no venting, so that pressures are high, mass flow to the exterior would not be dominant. Flame speeds within the enclosure, assisted by expansion of the combustion products, would be lower than for well vented explosions but the distribution of flame due to the turbulence arising during the dispersion of the dust would still be present. For small enclosures with a vigorous dispersion technique, such as for measurement of maximum explosion pressures by standard techniques (Chapter 3), ignition throughout the volume is likely to be rapid. Hence the measurements of rate of pressure rise would correspond to combustion simultaneously throughout the volume. It is likely that these conditions would not apply in larger volumes, with a relatively small source of ignition and dust dispersion from the centre of the vessel, where the duration of the explosion is approximately proportional to the cube root of the vessel volume [14].

When the size of the ignition source is increased, the dust will tend to be quickly ignited throughout the volume of the enclosure, even in large enclosures. The introduction, or generation, of a large ignition source is also likely to increase turbulence, which again favours the rapid ignition of the dust. An ignition source could be regarded as large if its dimensions were of the same order as the enclosure containing the dust suspension. A typical example would be injection of a flame from an explosion in an attached vessel. A small ignition source would be of dimensions of order(s) smaller than the enclosure, such as an electric spark or a wire coil. The ignition source could also be effectively large if the dust suspension were intensely turbulent throughout its volume. This would be the case when the dust was dispersed by the action of detonators. Where enclosures are not compact, as with ducting, some modifications of the above conclusions may be required, Even if the ducting were filled with a turbulent dust suspension, ignition from a small source is unlikely to propagate flame along the whole length of the ducting sufficiently rapidly for the assumption to be valid that combustion is taking place simultaneously throughout the volume. The assumption may be reasonable if the ducting is relatively short and the ignition source is large, or if the dust is in mass flow along the ducting.

From these tentative conclusions on the mechanism of dust explosion venting, equations were derived in which the rate of generation of combustion products was equated to the rate of discharge through the vent [14]. Two flow regimes were considered, depending upon whether the velocity of discharge of combustion products was sonic or less. The more widely applicable condition is for enclosures capable of withstanding only small explosion pressures, and in this case the flow of vented products is sub-sonic. The following basic assumptions were used.

(1) In vented explosions in small compact volumes, with vigorous dispersion, and in large compact volumes with vigorous dispersion and large ignition source, combustion would take place throughout the whole of the volume either before or when the maximum pressure was obtained, depending upon the vent ratio.

(2) The maximum rate of generation of combustion products derived from the routine test apparatus would be generally applicable where combustion was taking place throughout the enclosure volume.

(3) In explosions in large compact closed volumes with a small ignition source, combustion would take place in only part of the volume when the maximum pressure was obtained. The maximum rate of pressure rise would be inversely proportional to the cube root of the volume.

(4) For elongated enclosures, such as ducting, under vented conditions, simultaneous combustion throughout the volume would generally not be obtained, the nearest approach being ignition in a short length by a large source.

(5) Where only part of the length of the ducting was filled with dust suspension, modified calculations would be necessary.

(6) For all enclosures, where vents were large, so that explosion pressures were low, the maximum pressure was obtained when the rate of generation of combustion products reached a maximum.

(7) Where vents were small, so that explosion pressures were high, the maximum pressure may not have been obtained when the rate of generation of products reached a maximum.

(8) The main effect of turbulence of the suspension was to promote ignition throughout the stirred volume. Once ignited, the burning rate of a unit volume was calculated from that measured in the routine pressure test apparatus.

(9) The maximum burning rate of the dust suspension in the routine pressure test apparatus, per unit mass of dust, was independent of the pressure.

Using the above assumptions the following equation was derived for explosions in which the rise in pressure was small compared with atmospheric

$$(p-p_0) = \frac{2.3\rho_c p_0}{C^2\Upsilon^2}\left(\frac{V}{A}\,\frac{1}{p^{3/2}_{max}}\left(\frac{dp}{dt}\right)_{max}\right)^2 \tag{8.1}$$

where A is area of relief vent
C is discharge coefficient
$(dp/dt)_{max}$ is maximum rate of pressure rise in the closed routine test apparatus
p is maximum pressure in vented explosion
p_o is atmospheric pressure
p_{max} is maximum pressure in the closed routine test apparatus
V is volume of vessel
Υ is ratio of specific heats
ρ_c is density of unburnt dust suspension at pressure p_o

Inserting values for the constants in Equation 8.1 gives

$$(p-p_0) = \frac{10^{-3}}{p^3_{max}}\left(\frac{V}{A}\left(\frac{dp}{dt}\right)_{max}\right)^2 , \tag{8.2}$$

where V/A is measured in feet and all pressures are absolute and measured in lb/in^2.

When applied to published data on corn starch explosions in cubical vessels [5], cellulose acetate in a cubical vessel [6], cork and phenol formaldehyde in long ducts [8], and coal dust in a pulverizing mill [11], the measured explosion pressures varied approximately linearly with calculated values with some scatter. In general, the extent of the correlation between experimental data and calculated pressure gave support to the assumptions made in the derivation of the equations.

Although there is insufficient data for a thorough testing of the method a more detailed examination of the implications of Equations 8.1 and 8.2 is of interest. For a given dust, the explosion pressure varies inversely with the square of the vent ratio (A/V). The use of the vent ratio, even though it has residual dimensions, appears justified and explosion pressures would be expected to fall rapidly as vent area on a given vessel was increased. Comparison may also be made between the vent ratio requirements of Equations 8.2 and the *ad hoc* values used in practice. Typical values of the explosion parameters for dust, of moderate, intermediate, and severe explosibility are given in Table 8.3, together with the calculated vent ratio assuming explosion pressures were not to exceed 2 lb/in^2. The vent ratio increased from 1/14.9 to 1/5.9 as the explosibility of the dusts increased.

Table 8.3 Vent ratio required to give explosion pressures of 2 lb/in^2 calculated from Equation 8.2

p_{max} lb/in^2 (*absolute*)	$(dp/dt)_{max}$ lb/in^2s	*Vent ratio* (A/V) ft^{-1}
100	3000	1/14.9
110	6000	1/8.6
120	10 000	1/5.9

The *ad hoc* ratios required in practice would be 1/20, 1/15 and 1/10 respectively, giving less vent area than that calculated. In both cases at least twice as much vent area would be needed for the severely explosible than for the moderately explosible dusts. In practice other factors would be involved which would reduce the likelihood of the most severe conditions being obtained throughout the plant unit to be vented. Thus part of the volume could be occupied by deposited dust and the concentration needed for the most severe explosion is unlikely to be present, by chance, throughout the whole volume. The calculated vent ratios are thus reasonably in accordance with the requirements found from practice, but also indicate that these requirements cannot be safely reduced.

Further examination of Equations 8.1 and 8.2 shows that with a given value of vent ratio the explosion pressure varies directly with the square of the maximum rate of pressure rise and inversely with a cube of the maximum explosion pressure. As rates of rise tend to vary much more widely between dusts than do maximum pressures, the pressure in the vented explosion depends strongly on the value for the rate of pressure rise for the dust. The reason why the maximum explosion pressure appears in the denominator is that for a given rate of rise the duration of the explosion would increase as the maximum pressure rose. Additional time would be provided for the venting process, and the pressure developed would be less than in a more rapid explosion.

Although use of the equations may be justified on the available data, there is a great need for additional information. Until this has been obtained the validity of the assumptions leading to Equations 8.1 and 8.2 must remain in doubt, and in common with other theoretical treatments will be of restricted usefulness. The empirical methods, which have proved broadly satisfactory in the past, are therefore likely to continue for the near future.

REMOVAL OF VENTED PRODUCTS

Burning dusts, hot gases and flames will be ejected from vents and must be discharged safely. If the plant is indoors, ducting is usually attached to the vent to ensure discharge to the open air. The usual requirements are that the duct should be short, not more than 10 ft in length, straight, and of sufficient strength to withstand the explosion pressure, i.e. it must be at least as strong as the plant being vented (Plate 13). If the vent is normally provided with a cover, to prevent discharge of dust during normal working, this cover should be on the vent and not at the far end of the duct. It is good practice to avoid the accumulation of dust within the duct. A port should be provided in the duct, so that the vent cover can be inspected regularly. After inspection, it is important that the cover to the port is replaced and properly secured.

When a duct is fitted over the vent on a plant the maximum pressure in an explosion is increased. The amount of increase cannot at present be calculated, but information is available from experiment. Tests with cellulose acetate in a 1 ft^3 cubical gallery, to which was attached a square duct, 4 in $\times$ 4 in, over a vent of similar size, showed that as the length of the duct was increased the maximum pressure rose markedly (Fig. 8.10). A duct 10 ft in length increased the explosion pressure by a factor of about 10, which would not be acceptable for a gallery and vent of the size used. The difficulty may be alleviated by

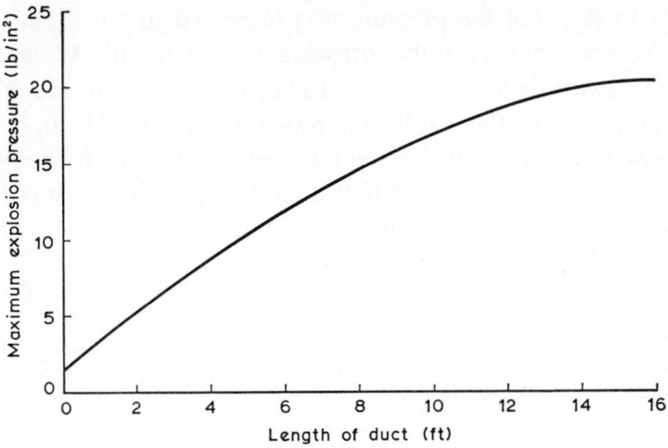

Fig. 8.10 Effect of vent duct length on explosion pressure [6].

using a duct of greater diameter than the vent [6], with a conical adaptor to attach the duct to the vent. Further tests with the same dust and gallery showed that incorporation of a 45° bend in the duct increased the explosion pressure by about 30 per cent, with short ducts, and increase in the angle caused little further change in pressure. With longer ducts the effect of bends was barely measurable, and the general conclusion was that the effect of a 90° bend about equalled that of an additional straight duct whose length was two to three times that of the bend.

In some applications, generally outdoors, it may be sufficient to deflect the emitted flame rather than to carry it away in a duct. Deflectors were fitted to the vents on a pulverizing mill [11]. When applied to a vent covered with a bursting panel, the deflector turned the vented products through 90°, close to the vent, by means of a simple elbow arrangement. The explosion pressures were increased by about 20 per cent. In other tests the vent was covered by a spring loaded plate valve and the deflector turned the vented products issuing from under the plate through 90°, by means of an expansion piece. No increase in the explosion pressure was recorded.

Ducting from vents is also required on elongated items of plant such as elevators. Tests with cork dust using a gallery 10 in. in diameter and 10 ft in length [8], with a vent at one end of equal area to the cross section, and the ignition source at the other end (Fig. 8.11), gave a maximum pressure of 1.4 lb/in². When a 10 ft length of duct was attached to the vent the pressure was increased to 4.9 lb/in². A 35 ft length of duct increased the pressure to 6.9 lb/in². When a small amount of extra dust was dispersed near the vent, with the 35 ft duct length in position, the explosion pressure rose to 11 lb/in². These results also indicated that duct lengths greater than 10 ft should not be generally used, and confirmed that dust dispersed in the duct from the vent increases explosion pressures considerably.

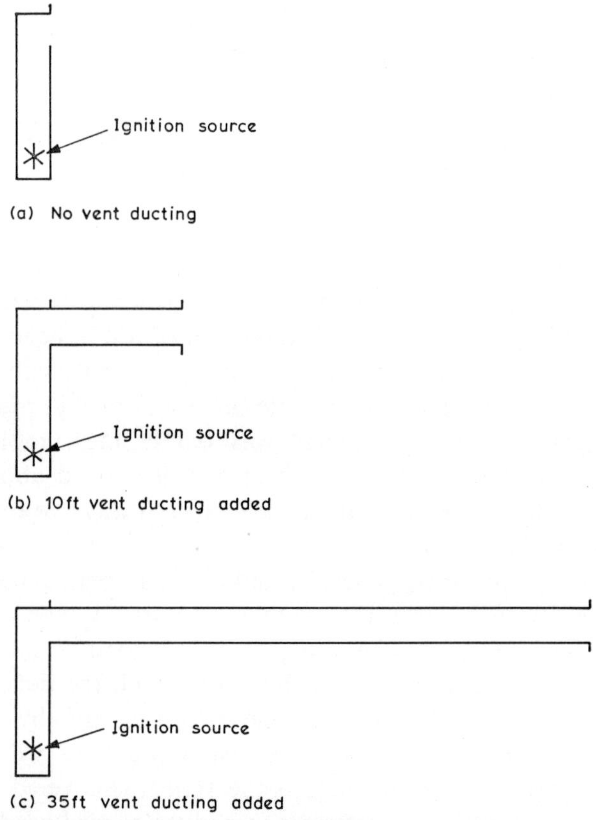

Fig. 8.11 Vent ducting arrangements on trunking [8].

Provided that the ducts are kept short, experimental evidence indicates that the explosion products can be discharged safely without greatly increasing the explosion pressure. If the plant is awkwardly sited so that a long duct would be unavoidable then the experimental evidence indicates that substantially increased explosion pressures could be expected. Thus either the system must be strengthened throughout, or an alternative method to venting of protecting the plant against pressure must be used.

VENT COVERS

Some form of cover is usually necessary for the vents to retain the dust being processed within the plant or building (Plate 14). As generally it is the plant that should be fitted with vents, rather than the building containing it, the use of vent covers for plant will be considered first. The vent cover is required to open at a present pressure reliably, and should not open accidentally. In some situations, such as when the vent cover may be exposed to mechanical shocks, it should also be robust as possible.

Common types of vent cover are:

Hinged covers or doors.

Magnetic covers.

Pop out panels.

Bursting diaphragms.

Providing the dust is not too vigorously explosible, hinged covers and doors have been found useful. They cannot be recommended when very explosible dusts are involved, say those with maximum rates of pressure rise in excess of 10 000 lb/in^{2}s, because their inertia may cause excessive pressures at the vent. As it is the inertia, rather than the weight of the cover that is important, the provision of a counterweight increases the pressure build-up rather than reduces it. A value of 10 lb/ft^2 may be taken as the upper limit for a light hinged door [9]. For accurate design calculations the weight per unit area of the door should be considered in addition to the explosion pressure for a open vent. The increase in explosion pressure may be several times that due to the weight of the vent cover.

The edges of the door should be turned to fit into a channel containing a gasket which is firmly attached to the plant (Fig. 8.12a). A good seal may be obtained by the use of suitable gasket material, fairly soft, and the edge of the door embedded into it. The hinged cover or door should not be mounted in a vertical position (Fig. 8.12b) unless a catch is provided to ensure that the door seats well into the gasket. A more convenient positioning is either angled (Fig. 8.12c) or horizontal so that the weight of the door assists in its seating. The possibility of the leakage of dust is further reduced if the plant is run at a slight negative pressure, e.g. 1 inch water gauge, but with careful design well

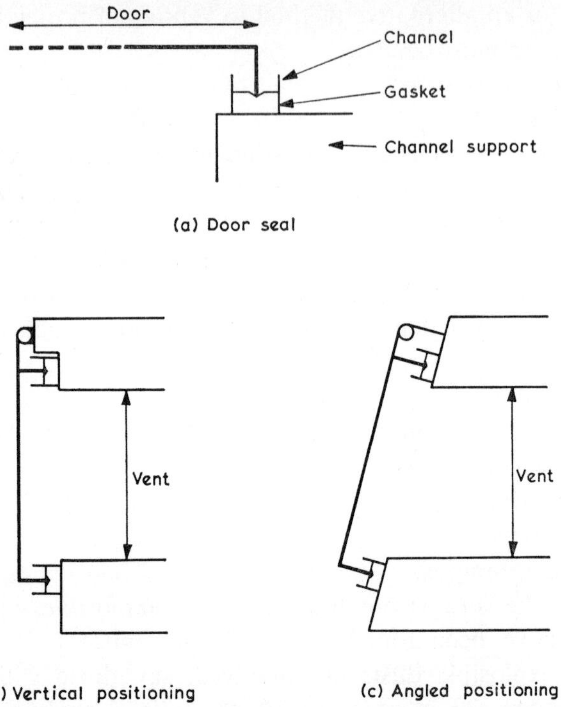

Fig. 8.12 Hinged vent cover.

seated hinged covers acting only under their own weight can be obtained when the pressure is atmospheric. Hinged covers and doors are often much less efficient when the pressure in the plant is above atmospheric, and alternative vent covers may be necessary.

The hinges should be sufficiently strong for their purpose, and it is an advantage to attach them to straps running across the door, rather than to the turned over edge. This edge should of course be on all four sides of the door to give the required sealing, and the hinges should allow the door to open and close without fouling the channel containing the gasket. The hinges must be well maintained, to provide easy opening; if permitted to deteriorate they may become stiff and thus prevent the cover opening freely. Higher explosion pressures would be the result. To prevent damage, it is useful to provide a stop so that when the cover is flung open it does not break the hinges. Such a stop can be a metal bar mounted so that when the door has opened at least 90° it is brought to a cushioned halt. There is experimental evidence that providing the door opens by at least 45° the full venting effect is obtained [9].

If the door closes within a frame, the latter should be of sufficient strength to ensure that it cannot distort and trap the door during the early stage of an explosion. In a weak structure slight deformation may occur at pressures below that at which a hinged door would open, and prevent the door opening at its design pressure.

Consideration should be given to providing a catch on the door so that when the vents is opened it remains open after the explosion. The advantage is that, on cooling, the explosion products within the plant contract, and if the vent has become sealed after the explosion a dangerous negative pressure may develop. On the other hand, if the vent is kept open after the explosion, air will rush into the plant and may stir up burning dust and cause a second explosion or may lead to a sustained fire. The procedure adopted should depend on the type and strength of the plant, and the dust contained in it, but in general it is probably advisable to allow the vent cover to close of its own accord after the explosion. Cooling of combustion products is most rapid immediately after the flame has propagated, and during this period the vent cover is still likely to be open, so that the most damaging phase of negative pressure development would be avoided. The combustion products which reduce the severity of any fire within the plant are largely retained.

On no account, after an explosion, should operatives attempt to enter or clean out the plant until all the burning material within it has been extinguished, and the atmosphere has been tested to show that it is safe to enter. Danger can arise if the oxygen level is below that in air, or if carbon dioxide or carbon monoxide resulting from the explosion are still present.

Where hinges are inconvenient, or may give unreliable performance, a modified form of cover can be obtained using a simple plate valve (Fig. 8.13 and Plate 15). This consists of a metal plate, usually steel, which is held in position over the vent by its own weight or by springs but which can be displaced along the guides by explosion pressure. The guides can be steel rods and should be of sufficient length to enable the plate to open up an area around its edge at least equal to that of the vent, and preferably about double. There should be sufficient guides to prevent the plate from jamming at an angle. The guides are supplied with stops to limit the travel of the plate, and coil springs fitted

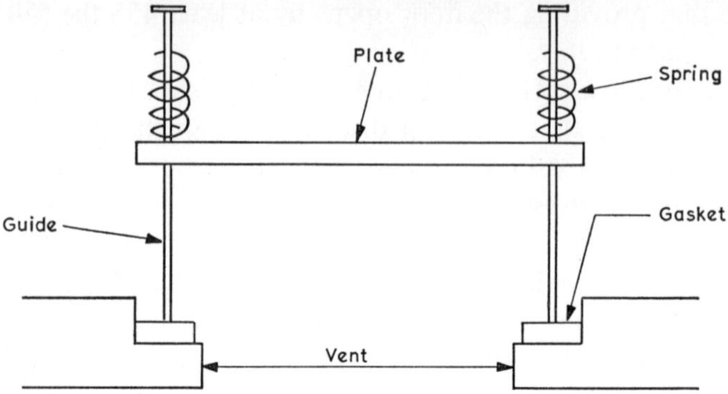

Fig. 8.13 Plate-valve vent cover.

loosely on the guides prevent a too severe impact of the plate on the stops. When a plate valve of this design, of weight 10.5 lb/ft^2, was fitted to the vent of a coal pulverizing mill, the maximum explosion pressures were the same as for open vents [11]. The plate valve reseated itself after each explosion, a good seal being obtained by means of a soft gasket (Fig. 8.13). The simple plate valve is best installed on horizontal vents and emitted flames are of course deflected sideways.

Where hinged doors or covers would be inconvenient, or too massive, they can be replaced by covers held by magnets. The principle of the magnetic cover is that the vent is closed by a light rigid panel, with good sealing around its edges, and the panel is held to the plant by magnetic force. The advantages are that light, but strong, covers can be used and the holding force can be regulated independently of the weight of the cover. Generally the weight of the cover should not exceed 10 lb/ft^2, and the weight plus magnetic force should not exceed 30 lb/ft^2. When the cover is displaced by an explosion the holding force decreases rapidly as the cover is moved away, so that increase in pressure due to the inertia of the cover is minimal. If, due to inadequate maintenance, the magnetic cover deteriorates, and the magnets lose their strength, the cover would be displaced by a lower pressure than that for which it is designed; in this way it fails in the direction of safety. There are also no hinges which could stiffen if not properly maintained.

Light but strong covers can be made from plastic, or aluminium honeycomb, covered with a lightweight material to prevent escape of dust through the cover. A suitable material is aluminium foil [19].

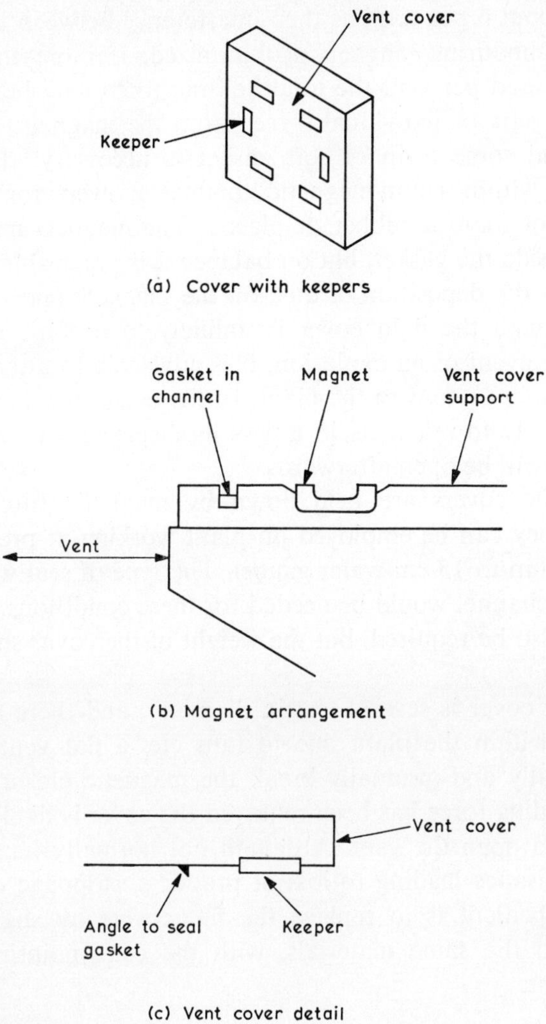

Fig. 8.14 Magnetic vent covers.

Alternatively, aluminium sheeting may be used, braced to give additional strength if necessary. The 'keepers' for magnets, usually steel plate or strips are bolted to the cover. The magnets themselves must not be attached to the cover, because they are relatively heavy. The magnets are attached to the surround of the vent, on the plant, and the 'keepers' on the vent cover matched to the magnets. A typical arrangement is shown in Fig. 8.14. It is useful if the number of magnets can be kept to about 6 per vent, as then interference between the magnetic fields of neighbouring magnets is minimized. If more than a dozen magnets are used per vent the total holding force may be appreciably less than the sum of individual forces from the magnets. To obtain a dust-tight seal some form of soft gasket is necessary, this can conveniently be within a non-magnetic channel or even, for very simple cases, strips of foamed rubber or plastic. The magnets may be either inside or outside the gasket, but on balance it is preferable for them to be outside as the deposition of dust on the magnets and their keepers is less. Although the light cover is unlikely to become a dangerous missile in the event of an explosion, it is advisable to attach the cover by a slack chain or cord to the plant. In the event of an explosion the cover is unlikely to reseat itself, unless its displacement was small, so that the vent will be open afterwards.

Because the covers are held down by magnetic force, which is adjustable, they can be employed on plant working at pressure above atmospheric (up to 15 cm water gauge). The type of seal with a gasket inside metal channel would be needed for these conditions. Extra magnets would also be required, but the weight of the cover should not be increased.

If the vent cover is several feet in diameter, and there are pressure fluctuations within the plant due to fans etc. a flat vent cover may flex periodically and gradually break the magnetic closure. Once the magnetic holding force has been removed the cover is likely to become dislodged and open the vent. Although not normally hazardous, this may be a nuisance leading to loss of products, stoppage of plant etc. A simple expedient is to replace the flat covers by slightly conical designs, from the same materials, with the cone pointing outwards from the plant.

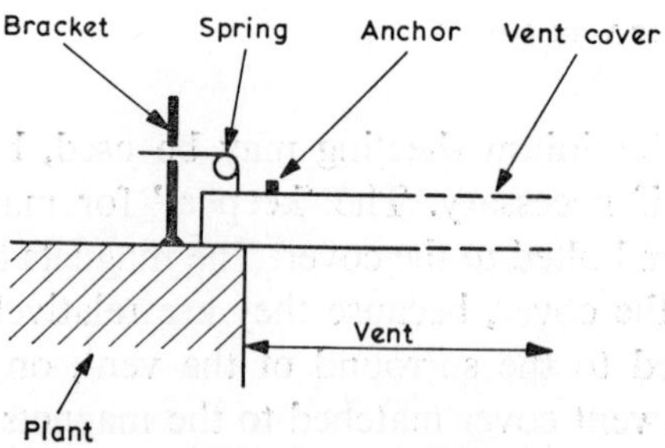

Fig. 8.15 Vent cover held by springs.

Coil springs can be used instead of magnets to provide attachment which becomes free of holding forces after slight movement of the cover. The springs are mounted with their axes parallel to the plane of the cover, and anchored to it (Fig. 8.15), and engage holes in brackets attached to the plant. As the vent cover rises during the explosion the springs twist until the engagement with the brackets is suddenly broken. This should be arranged to occur when the vent cover has travelled about 2 cm away from its seating. The springs should be of appropriate strength and number to give similar holding forces to those obtained with magnets; direct measurement of the force by a spring balance attached to the vent cover is advisable. Coil springs have the advantage of simplicity, but good maintenance is essential because if the linkage with the holding bracket stiffens, a greater force would be needed to dislodge the vent cover.

Pop out panels consist of sheets of light, flexible, material such as plastic or metal foil which cover the vents and are held in place around the edges by friction. The pop out panel is sandwiched between one frame attached to the plant and another which is screwed or bolted down onto it. The surfaces of the holding frames should be smooth so that the panel is held by uniform pressure around its edge, and the amount of pressure adjusted to ensure that the panel does not become dislodged accidentally but nevertheless is released at a low explosion pressure, possibly 0.1 lb/in^2. The weight of the panel should not exceed 5 lb/ft^2. The panel should be flexible and sufficiently strong not to shear or split before coming free. Pop out panels have the advantage that they are cheap and require the minimum of fixing equipment. Their disadvantages are that it is difficult to measure the holding force without removing the panel, so that installed panels cannot be easily checked. Because of their lightness and flexibility they are relatively weak and are very likely to give way if an operative should step onto them. To avoid this, a strong grid may be fixed to the plant, just below the panel, capable of taking the weight of an operative. The grid must never be attached outside the panel.

After an explosion the panel will have been completely removed from the vent which thereafter stays open. The panels are lightweight and are unlikely to cause damage by becoming missiles, so that anchoring chains are not usual.

Bursting diaphragms or panels are effective vent closures because dust tightness can be maintained, providing the diaphragm is correctly mounted, with either positive or negative working pressure differential to atmosphere. The principal disadvantage for plant in which explosion pressures should not exceed 2 lb/in² is that the diameter of the diaphragm may need to be large, and its thickness small, and it may thus be flimsy. It is generally impracticable to increase the area of the vent solely to allow the diaphragm to burst at a suitable pressure. The area of the vent would be estimated from the vent ratio requirement, and a suitable bursting material would then need to be found, to match the dimensional and pressure requirements.

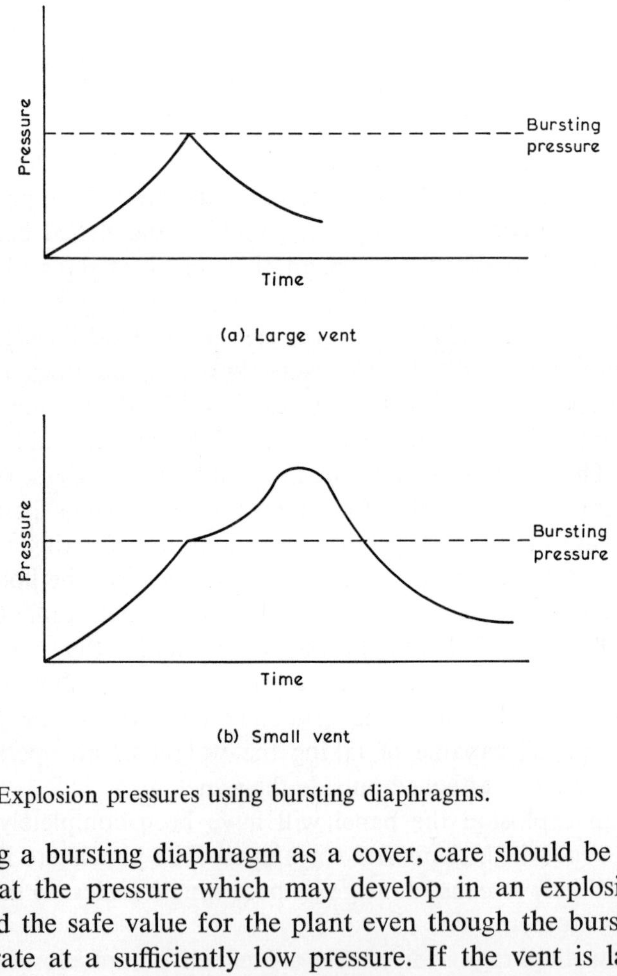

Fig. 8.16 Explosion pressures using bursting diaphragms.

In using a bursting diaphragm as a cover, care should be taken to ensure that the pressure which may develop in an explosion shall not exceed the safe value for the plant even though the bursting disc does operate at a sufficiently low pressure. If the vent is large, the explosion pressure will rise until the diaphragm bursts after which it would decay (Fig. 8.16a). With a small vent, the bursting of the diaphragm would check the pressure rise momentarily, but the pressure could continue to increase until a maximum was reached above which the diaphragm would burst (Fig. 8.16b). The bursting of the diaphragm may also induce cyclical pressure fluctuations due to vibration of the burning suspension. The pressure obtained in an explosion can be controlled either by the strength of a bursting diaphragm or panel, or by the area of the vent, and both factors must be considered in the design of plant. Where the pressures must not exceed 1 or 2 lb/in^2 for weak plant, the differential between the maximum pressure permitted and the bursting pressure of the diaphragm would be small. For strong plant, the differential could be much greater, and the importance of using the correct criterion correspondingly increased. In general a diaphragm or panel should burst at the lowest convenient pressure, providing that false alarms due to mechanical failure are avoided. For this reason the bursting pressure should preferably be at least one third greater than the working pressure of the system [20]. A similar pattern to that in Fig. 8.16 is shown by other types of vent cover, but these are usually employed where explosion pressures are kept low. Where high explosion pressures are permissible, bursting diaphragms are convenient and hence it is with them that particular care has to be taken.

Providing the diaphragm or panel bursts correctly, and does not pull out from its seating, the pressure at which it bursts increases directly with its thickness and inversely with its diameter or linear dimension. If a diaphragm is doubled in thickness, and doubled in diameter, it would still burst at the same pressure. As the bursting pressure is a function of the diameter of the vent, any sub-division of the vent would lead to higher bursting pressures.

In designing a bursting diaphragm or panel for a particular plant two approaches may be used. The bursting material may be obtained together with its support, as a commercial item. Graphite discs, which can be designed to burst at low pressures, are an example. Alternatively the diaphragm material may be selected by *ad hoc* testing, using the value obtained in a static bursting test. This test is sufficiently accurate for materials which are not appreciably elastic, but with materials such as rubber or polyethylene film which stretch readily the bursting pressure increases with the rate of pressure rise. Such materials cannot be conveniently used for design calculations. Where the method of choosing the material for a bursting panel is by *ad hoc* tests, the

Table 8.4 Approximate bursting pressures of diaphragm materials 30 cm diameter

Material	*Thickness mm*	*Bursting pressure lb/in²*
Brown paper	0.05	0.6
Aluminium foil (soft)	0.03	1
Rubber-asbestos	0.4	12 (estimate)
Asbestos millboard (for hot situations)	3	1.5

values in Table 8.4 may be used as a rough guide. Bursting diaphragms are frequently flimsy and if the pressure within the plant pulsates in normal working, the diaphragm may fatigue and burst at a pressure below the design value. Although this is failing safe, it is a nuisance and cases have been reported where diaphragms have, quite wrongly, been replaced by thicker material in order to alleviate the nuisance. Because of their flimsy nature the diaphragms may also need supporting, either by a metal grid or a wide wire mesh, and the support must always be mounted on the inner side of the diaphragm so that it does not interfere with the stretching and bursting movement during an explosion. Further information is available on the strength of diaphragm materials [4].

If it is desirable for practical reasons to use a stronger material than the unaided bursting pressure would allow, the use of a cutter may be practicable. This consists of a sharp edged device, saw-tooth or pointed or with a cutting edge, mounted in the centre of the diaphragm on the outer side. The cutter is put close to the diaphragm so that as soon as the diaphragm deflects because of explosion pressure it contacts the cutter which ruptures it, before the bursting pressure is reached. Once ruptured, the diaphragm would split readily and open up. Cutters are also used with reversed-type bursting discs, which are domed towards the source of pressure [20]. When overpressure occurs the disc reverses and contacts the cutter. These discs avoid the problem of creep. If a cutter is relied upon to cause the vent to open at a lower pressure than that at which the diaphragm would burst unaided it must of course be maintained in a good condition. Corrosion or soiling or painting of the cutter could reduce its sharpness and its effectiveness and higher explosion pressures would result. As a cutter does not fail safe in this respect its use is limited to situations where there is positive reason to assume that adequate maintenance would be carried out.

The performance of a bursting diaphragm or panel cannot be considered in isolation from its mounting. Any test carried out on the static bursting pressure of the material should be made when it is in its actual mounting. The design of mounting demands precision engineering, for example the holding frame should have a rounded edge rather than square so that the diaphragm would not be pressed against a metal edge, possibly unevenly, which may cause it to burst in an uncontrolled fashion. Designs may be obtained from the literature or may be purchased as commercial items [21, 22, 23].

So far consideration has been given to the vent closures of the plant, rather than for vented buildings. In general, the vents should be on the plant and should not discharge into a closed building. If the vented plant items must be under cover then the vents should have direct access to outside. Suitable means of providing this facility are the use of an open sided building or of ducting from vents to atmosphere. In exceptional cases the plant would vent inside a walled building because it must be situated indoors for security, toxicity, or other reasons and there may be a danger of passage of dust into the building during an explosion in the plant. The venting of buildings should never be used as the sole precaution against explosions in dust which has been deposited on the surfaces within the building. In such a case, removal of the dust is essential.

Relief venting for buildings may be obtained by providing one or more of the following:

Light roof
Light wall
Windows
Louvres

A light roof can be used for the explosion venting of single storey buildings. The principal requirements of a light roof are that it should become dislodged when explosion pressure within the building reaches a pre-determined value, and that it should be weatherproof. For ease of construction the entire roof can be provided as a unit, although the amount of vent area which opens when the roof is dislodged is generous, and likely to be considerably in excess of the vent ratio requirement for the volume as given above. Although a light roof which covered only part of the total area, the remaining fraction being a fixture, would be possible, there are usually practical difficulties in ensuring a weatherproof junction with the permanent part of the roof. It may constructed of wood, plastic, or metal, or any other convenient material. The weight of the roof would depend upon the strength of the walls of the building, and little direct investigation of appropriate weights has been reported. However if the light roof does not weigh more than 10 lb/ft^2 it is likely to be generally adequate. The roof should be pitched or sloping to encourage run off of rain water, and to discourage accumulation of thick layers of snow. It should be securely attached to the building so that it is not moved by wind pressures. The maximum wind pressures to which the roof could be exposed would depend upon the location of the building. In exposed situations, pressures as high as 50 lb/ft^2 might occasionally arise, whereas in sheltered situations about half this value might be more realistic. Consultation with the meteorological services is advisable. In any event there will be a margin between the holding force needed to avoid difficulty with wind pressure and the weight of the roof. The difference must be made up by attachments at the edge of the roof giving the required holding force. These attachments can be pins of known shear strength, or, with metal roofs, crimping the metal to the building might be practicable. The additional holding force should not be obtained by increasing the weight of the roof.

In the event of the roof being blown off, or dislodged, it must not fall dangerously or become a missile. Secure restraining chains are therefore desirable to prevent the roof falling either outside or inside the building.

If the roof is constructed in this way the maximum pressure within the building should not exceed 2 lb/in^2, and may well be under 1 lb/in^2 because of the unlikelihood of a highly explosible dust concentration being present throughout the whole volume at the time of the explosion and because of the large area of vent which opens up when the roof is dislodged. The walls of the building should be able to withstand the pressure, or else be so designed as to vent safely, and the same requirements would apply to windows and doors. Usually in a building vented in this way the door would not be expected to act as a vent and hence should be designed sufficiently strongly to withstand the explosion pressure. The total force on the door is likely to be of the order of tons and hence both the catch and the hinges must be designed to withstand this force. Normal door closures are unlikely to be adequate, and special designs are necessary.

Wherever possible plant which may give rise to an explosion discharging into a building should run automatically, so that injury to operatives is avoided. For safety the controls of the plant should be interlocked with the door closure to ensure that the plant cannot be run whilst the building is occupied.

Windows in a building with a light roof explosion relief would not themselves be required to act as reliefs and should be as strong as practicable, e.g. of wired glass. To prevent hazardous fragmentation the windows may be protected externally by wire mesh guards. In any event the number of windows should be kept to a minimum.

A light wall can be used as explosion relief in a building, instead of a light roof. The weight requirements for a light wall are similar to those for a roof. Lightweight cladding on a steel framed building is often used (Chapter 10). Polystyrene foam about 5 cm thick has been found satisfactory from the view points of lightweight and heat insulation [24], particularly when a 2.5 cm deep groove was cut around the edge of the panel, on the side exposed to the explosion. The use of light walls has disadvantages for single storey buildings as compared with light roofs. Being of light construction they are readily damaged by impact etc, and a wall is more exposed to this hazard than is a roof. Also there is the problem of security because they are fairly readily breached. The problem of wind pressures is tackled along similar lines as for roofs.

Where the provision of a light roof or light wall would be inconvenient, the use of windows as explosion relief vents can be considered. They would be practicable for many dusts, with the exception of the most explosible such as aluminium and magnesium. For convenience the explosion relief windows would be installed in one or two of the walls and should open up an area giving the appropriate vent ratio. The best arrangement is for the windows to be in frames which open at the required pressure, rather than as panes which have to be broken by the explosion to obtain relief. The breaking characteristics of window materials, particularly glass, cannot be accurately predicted and a better arrangement is to use hinged panels with weak closures whose strength can be designed and tested before or during installation. A suitable arrangement would be a transparent plastic pane in a light metal frame held closed by a spring loaded catch. The frame could be hinged at top or bottom and designed to open outwards to a sufficient angle to avoid restricting the area of vent. The window material and frame should be of relatively light weight, say up to 10 lb/ft^2, and the residual retaining force needed to prevent wind from opening the windows should be obtained by appropriate design of the spring loaded catch. Some type of retaining chain or stop is necessary to prevent the windows acting as missiles.

Windows used as explosion relief have the advantage of not detracting from the appearance of the building and of being applicable to single storey and multi-storey buildings.

Louvres are cheap, easy to install in buildings, and offer relatively little resistance to the passage of explosion gases. The total area available for venting through the louvres should be similar to that required for windows and they should of course be of the permanently open type. The disadvantages of their use include the difficulty of retaining heat within the building, their ineffectiveness in preventing the transmission of noise, and if dust should escape from the plant they will not prevent it from spreading outside the building. They give reduced protection against adverse weather conditions. They can be made mechanically stronger than light walls or windows and can be designed to give greater security. If louvres are used for explosion relief, then the walls, roof, windows and doors to buildings must be of sufficient strength not to be damaged by an explosion. Louvres can of course be used in multi-storey buildings.

Inerting

Dust explosions may be prevented by the use of inert gas or dust. In industries other than coal mining, inerting is usually done with gas, but in the specialized circumstances of mining an inert dust, such as stone dust, is used and inert gas would be impracticable.

For general industry the principle is to use an inert gas such as nitrogen, carbon dioxide or flue gas to reduce the oxygen concentration in the plant so that if a source of ignition should appear flame cannot propagate through the dust cloud. The advantages of the technique are that explosion is avoided rather than controlled whilst it is in progress. Inerting can also be used to control, or extinguish, fires which may develop in the dust. Finally, inerting may be used where venting would be awkward, due to the position of the plant or to the undesirability of discharging the dust to atmosphere. The disadvantages of inerting are that it is likely to be more costly than venting, and strict control over gas compositions must be maintained. Personnel must be protected by breathing apparatus before entering any inerted enclosure; the necessity for entry must be minimized.

The basic requirements for the inerting of plant with gas are as follows:

(1) A permanent supply of inert gas must be available.

(2) For economic reasons a closed, or mainly closed system, should be considered so that inert gas can be recirculated.

(3) Continuous and reliable monitoring of the atmosphere in the plant, usually of the oxygen content.

(4) A knowledge of the maximum permissible oxygen concentration for the dust in the plant.

A standard for inerting for fire and explosion prevention has been published [25].

A permanent supply of inert gas may be obtained by several methods:

(*a*) Commercial compressed or liquefied gas. By specifying the purity of such gas, difficulties due to the possible contamination of the dust by impurities in the gas can be avoided. Adequate storage facilities must of course be available.

(*b*) Flue gas may be available cheaply and in adequate quantities. Washing and conditioning of the gas is likely to be necessary, involving installation of appropriate plant, and special precautions may be needed to avoid contamination of the dust by constituents of the flue gas, including moisture. As storage of flue gas may not be practicable, several furnaces may need to be running simultaneously to ensure an adequate and unbroken supply of gas.

(*c*) An inert gas, or one with reduced oxygen content, may be available as a by-product from another part of the plant. Washing etc may be necessary before it can be used for inerting.

(*d*) Inert gas generators are commercially available. They commonly burn oil to produce a gas mixture of low oxygen content. The combustion gas would need cooling and probably washing, before it would be in a suitable state for use in inerting. An inert gas generator may contain a source of ignition and should be situated in a safe area and shut off if the gas produced is at an incorrect temperature, or if combustion is inefficient so that flammable vapours could be transmitted from the generator.

Whichever source of supply is installed it should be capable of delivering more than the requirement if all plants connected to the source take their maximum supply simultaneously. Where the inert gas is generated continuously, failure of one generating unit must not cause the supply to be inadequate. Either sufficient generators must be installed, or a stand-by storage capacity must be provided.

Careful design of the inert gas distribution piping system is also necessary. When inert gas is delivered to several points, non-return check valves must be used in each branch to ensure that back flows of gas are prevented. Excessive forward flow should be controlled by means of suitable valves and pressure monitoring devices. The transmission of impurities such as moisture, harmful gases, rust etc must be prevented by suitable traps and filters.

The inert gas usually needs to be introduced at several points on the plant to ensure that the oxygen concentration is maintained at a safe level. To a certain extent the necessary points can be foreseen at the design stage of the plant, e.g. where the pressures within the plant are low and hence the leakage of oxygen most likely, and the total quantity of inert gas can also be estimated from knowledge of wastage. Considerable checking of the inert gas requirements is needed when the plant has been erected, and this can be conveniently done by monitoring the oxygen concentrations at numerous points within the plant. Particular care should be taken to avoid the development of pockets of gas within the plant which are not in the main stream of flow. Where the plant is totally enclosed, and the dust is introduced and removed from it by valves giving the minimum loss of inert gas, the demand for gas would be reduced to that necessary to replace the small losses likely in the removal of the dust. In cases where the inert gas is recycled within the plant, care must be taken to remove from it the fine fractions of dust which may not have been removed by the gas cleaning equipment, such as cyclones, and which would then be re-circulated. If the fines are not removed their concentration will progressively increase and could cause a major hazard in the event of an ignition within the plant. Similarly, in a partly closed system, a certain amount of dust is likely to be carried away to exhaust and safe collection is necessary. This may require either high efficiency cyclones, or fabric filter units.

In order that the plant shall run safely it is essential that the oxygen level shall be continuously and accurately monitored. Particular care should be taken to ensure that the monitoring equipment continues to function satisfactorily over long periods even though the gas may be dusty. In the event of failure of the monitoring equipment, or of an unacceptable rise in oxygen concentration, the plant should shut down automatically. It is thus desirable that the plant in normal running should be automatic and that it should be capable of reacting safely when one of the monitors indicates that the designed conditions of working are not being obtained.

Although monitoring of the oxygen concentration would be the usual control, with dusts such as magnesium where only low concentrations could be permitted, and the usual inert gases may not be acceptable, more sophisticated monitoring arrangements may be required including continuous analysis.

In addition to the monitoring of gas compositions, other instruments should be provided to ensure that the appropriate gas flows, pressures, and temperatures are being obtained.

If the use of inert gas is to be economic, complete replacement of the oxygen in the plant may not be feasible. When some residual oxygen is to be permitted, its maximum safe level must be known in order that the plant can be safely operated. The maximum safe oxygen concentration level varies with the dust, and at the present no theoretical approach is available enabling the maximum value to be calculated. Hence recourse is made to *ad hoc* testing (Chapter 3), and to direct experimentation where appropriate (Chapter 5). When a value for the maximum oxygen level has been obtained by one of these methods, it is usual to employ a safety factor in the design of the plant. This factor would depend upon the size of the plant and the explosibility of the dust, but is unlikely to be less than a 2 per cent margin. Thus if the maximum oxygen concentration has been determined as 10 per cent, the maximum permissible level in the plant should not exceed 8 per cent. Greater margins may be needed where large plant volumes are involved, extended hot surfaces are present, or where the dust is vigorously explosible. In extreme cases the safest course could be to replace all the air in the plant by inert gas, and thus the designed oxygen concentration would be zero, possibly allowing a trace to prevent subsequent spontaneous ignition (Chapter 7).

In general the maximum permissible oxygen concentrations obtained by experiment and those published in literature, refer to plant in which the dust is in suspension in a cool environment, say below 100°C. If the dust suspension should enter a heated region then the maximum permissible oxygen concentration would be reduced, but to an extent which cannot be directly related to the temperature by calculation. If it is intended to use inerting to protect dust suspensions in heated enclosures direct determinations of the maximum safe level should be made in *ad hoc* test apparatus.

In coal mines inert dust is used to prevent dust explosions. The method has been used for a long time, and is backed by substantial investigations in many countries. The use of inert dust to prevent explosion is of minimal importance in other industries, because other preventative measures are more convenient. As the use of inert dust is not of general industrial application, a comprehensive account will not be attempted here, although for the specific application of coal mining a vast amount of information is available. The basic principles will be outlined, as these are of general interest, but details of the technique, and of the regulations applicable will not be given.

The inert dust used in different countries varies, although its function is thought to be the same. In the United Kingdom limestone is used, but alternative materials such as shale, Fuller's earth, clay, and other mineral dusts are, or have been, used in other countries. The essential function of the dust is that if an explosion is initiated, either in methane discharged from the rock strata or in coal dust, and coal dust in the vicinity is stirred up into a cloud, then sufficient inert dust shall also be dispersed to prevent the explosion from propagating. It is usually impossible to prevent some deposition of coal dust at the coal face and in the galleries or in conveyor roadways. The requirements of the inert dust are therefore:

(1) It shall be present wherever coal dust may be dispersed.
(2) It shall be of a specific fineness and readily dispersable at all times.
(3) It should be mixed evenly with the coal dust deposit.
(4) Sufficient should be present, and dispersed when necessary, to prevent flame propagation.

The requirements that the inert dust shall be present with the coal dust, and that excessive deposits of coal not mixed with inert dust shall be avoided means that the stone dusting is an economic burden to the industry. The inert dust must be well mixed with the coal because tests have shown that if it is not then although the ratio of total amount of inert to coal may be adequate, preferential dispersion of the coal may take place and explosion propagate. Even a thin layer of coal dust, not mixed with stone, but resting on a stone dust mixture can propagate an explosion. A report has been published on the problem of coal dust explosions, and the use of stone dusting, giving a survey of the present position in British mines [26].

The inert dust should not only be evenly distributed, it must also be present in sufficient quantity to prevent explosion. For coals of 25 per cent or more volatile matter content (on a dry ash-free basis) a minimum of 80 per cent incombustible is recommended for the floor, roof and sides of headings and also for the roof and sides in other roadways [26]. For coals below 25 per cent volatile matter content, a reduction of 5 per cent incombustible is recommended for each 5 per cent range of volatile matter content below 25 per cent down to 10 per cent. In order that these conditions shall be met the frequency of stone dusting must match the rate of deposition of coal dust.

Although the distribution of inert dust is the basic protection against coal dust explosions in mines, it does not afford complete protection throughout the mine. The mechanization of mining procedures has meant that coal is transported from the face by conveyor belts and quantities of fine coal dust are likely to be continuously present on the conveyor and they accumulate in relatively inaccessible spaces underneath. Stone dust cannot be mixed with the fine coal on the conveyor and to overcome this difficulty considerable attention has been given to the installation of stone dust barriers. The barrier consists basically of an array of shelves on which the stone dust is heaped and is of such mechanical design that the stone is dispersed into the air by the blast wave ahead of the explosion. When the explosion flame arrives at the barrier a dense cloud of stone dust has been formed and this prevents further propagation. An alternative approach is to detect the oncoming explosion electronically and to disperse the stone dust by independent power. With these triggered barriers a mild explosion, which might not produce a sufficiently strong blast wave to effectively distribute the stone dust can be protected against. A background review to the performance of stone dust barriers has been published [27], and research is actively in progress at present.

Although the coal dust explosion problem in mines has been recognized for many years, and the introduction of stone dusting led to a marked decline in the fatality rate, there is still need for improvement in the technique. The two aspects requiring particular attention are the need for improving the economics of the stone dusting process and the methods of distribution of the dust. Also, laboratory tests indicate that the inert dusts currently used are not the most effective (for instance see Chapter 5) and relatively large quantities are needed. The ultimate aim should be to use a dust which weight for weight is much more efficient, is economic, and does not have undesirable properties such as poor dispersability under wet conditions, toxicity, and corrosiveness. Considerable research effort is currently being given to the problem in a number of countries and significant advances should soon be forthcoming.

Automatic suppression

After ignition the flame takes a finite time to propagate through a dust suspension within a plant, and initially the explosion pressure increases relatively slowly, Fig. 8.17. In fact, immediately after ignition the pressure should theoretically increase with the cube of the time, the so called 'cube law'. This law should hold so long as the expanding flame is spherical and is not accelerated by turbulence or influenced

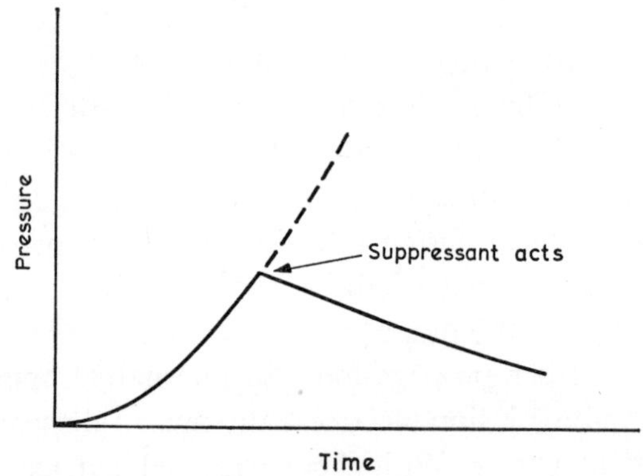

Fig. 8.17 Pressure record for suppressed explosion.

by the walls of the vessel. During the time interval before the pressure exceeds the safe value for the plant in which the explosion is occurring protective action can be taken automatically. Equipment is available commercially which is capable of detecting the incipient explosion (Plate 16) and triggering all or some of the following protective actions [19, 25, 28].

(1) Suppression of the explosion within the plant.
(2) Venting the plant automatically.
(3) Advance inerting of the plant.
(4) Isolation of the affected section of the plant.
(5) Automatic shut down of the plant.

Suppression involves the discharge of a suitable agent into the plant volume within which the explosion is developing. The composition of the agent depends upon the dust involved, and is typically a halogenated hydrocarbon or an inert gas or powder. The suppressant is contained within a sealed receptacle attached to the plant and is rapidly discharged throughout the volume by an electrically fired detonator (Plate 17). Thus as soon as the existence of the explosion is detected the control mechanism fires the suppressant into the plant and the flame is extinguished wherever the ignition may have developed, Fig. 8.18. Plant vessels of volumes up to several thousand cubic feet can be protected by this method.

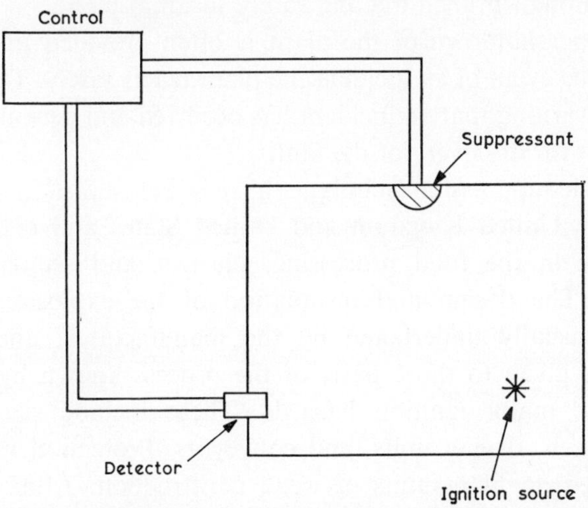

Fig. 8.18 Scheme for automatic suppression of explosion in a plant unit.

Alternatively the explosion can be automatically vented to atmosphere. When the explosion is detected a vent closure is ruptured automatically, thus providing a rapid opening of a vent. The vented explosion then proceeds as for cases in which the vents are opened by the pressure of explosion. The automatic method has the advantage that vents are opened extremely rapidly, and for very explosible dusts this helps to reduce the maximum explosion pressure. Automatic venting does not extinguish the flame, and a fire may persist after the explosion. Vented plant should not of course discharge into a work room, see above.

Advanced inerting protects plant units attached to that in which the ignition had occurred. In this way, a vented explosion can be prevented from extending throughout a plant. If an explosion develops in one vessel, it can be dealt with automatically and safely, and the use of advanced inerting insures that ducting and other vessels to which it is connected do not permit the explosion to propagate into them (Plate 18). The technique also prevents the spread of fire throughout the plant.

Isolation of plant units can be provided automatically by mechanical means. Valves, which can be rapidly closed by use of explosives, can be attached to the vessel in which the explosion is first detected and may either completely contain the explosion, or can cause it to be diverted through branch ducting safely to atmosphere.

Automatic shutdown of the plant is often provided in the system, so that in the event of an incident the plant reacts safely. The shutdown also gives warning that an incident has occurred and permits inspection of the plant for discovery of the faults.

Automatic suppression techniques have been installed in many plants, both in the United Kingdom and United States and elsewhere [28], particularly in the food processing, plastics, and animal feed stuff industries. The design and installation of the explosion protection system is usually undertaken by the manufacturer, and particular attention is given to those parts of the process known by experience to give the major ignition hazards. These include size reduction, storage vessels, drying units, and conveyors. Protection must also be concentrated on plant units of weak construction. After an incident the equipment must be reset, and regular maintenance is in any case necessary. A facility is provided on the equipment for checking that it is operational at anytime.

Automatic suppression devices are of particular use where the plant is awkwardly sited, the dust is toxic so that it cannot be freely discharged to atmosphere, or where normal working under inert gas conditions would be impracticable. If the dust becomes ignited, and an explosion develops, it is controlled and warning automatically given. Interruptions of production are thereby minimized.

References

[1] Wheeler, R. V. (1935), Home Office Report on Experiments into the Means of Preventing the Spread of Explosions of Carbonaceous Dust. HMSO, London.
[2] Palmer, K. N. (1971), I. Chem. E. Symposium Series No. 34. 142–7. Inst. Chem. Engrs. London.
[3] Dust explosions in factories. Department of Employment. Health and Safety at Work. No. 22. HMSO, London. (1970).
[4] National Fire Protection Association. Guide for Explosion Venting. No. 68. (1954). Boston, USA.
[5] Hartmann, I., Cooper, A. R. and Jacobson, M. (1950), US Bureau of Mines RI 4725.
[6] Hartmann, I. and Nagy, J. (1957), *Ind. Engin. Chem.*, **49**, 1734–40.
[7] Cotton, P. E. (1951), *NFPA Quart.*, **45**, 157–64.
[8] Brown, K. C. (1951), Safety in Mines Research Establishment, Research Report, 22.
[9] Brown, K. C. and Wilde, D. G. (1955), Safety in Mines Research Eestablishment, Research Report, 119.
[10] Schwab, R. F. and Othmer, D. F. (1964), *Chem. Proc. Engin.*, **45**,165–74.
[11] Brown, K. C. and Curzon, G. E. (1963), Safety in Mines Research Establishment, Research Report, 212.
[12] Palmer, K. N. and Tonkin, P. S. Fire Research Note 639.
[13] Maisey, H. R. (1965), *Chem. Proc. Engin.*, 526–35, 662–72.
[14] Palmer, K. N. Fire Research Note 830.
[15] Heinrich, H. J. (1966), *Modern Unfallver.*, **10**, 36–41.
[16] Heinrich, H. J. (1966), *Mitt. VGB*, 105, 380–4.
[17] Straumann, W. (1965), *Chem. Ing. Tech.*, **37**, 306–16.
[18] Shagalova, S. L. and Reznik, V. A. (1962), *Teploenergetika*, **9**, 41–5.
[19] Guide to the use of flame arresters and explosion reliefs. Ministry of Labour. Safety Health and Welfare New Series No. 34. HMSO, London. (1965).
[20] Clark, C. R. N. (1970), *Chem. Proc. Engin.*, **51**, 43–5.
[21] British Standard No. 2915. (1957).
[22] Symposium on bursting discs. *Trans. Inst. Chem. Engrs.*, **31**, 113–67, (1953).
[23] Philpott, J. E. (1963), *Engng. Mater. and Des.*, **6**, 24–9.
[24] Simmonds, G. E. and Harrison, D. A. (1968), Rutherford Laboratory Report RHEL/R 168 HMSO, London.
[25] National Fire Protection Association. Explosion Prevention Systems. 69. (1970). Boston, USA.

[26] Final Report of the Working Party on Coal-dust Explosions. Coal Industry National Consultative Council Safety and Health Committee. National Coal Board. London, 1967).
[27] Coward, H. F. (1952), Safety in Mines Research Establishment, Research Report, 51.
[28] Grabowski, G. J. (1968), *Fire J.*, **62**, 1, 21–4.

9 *Dust Fires*

Combustible dusts can cause fires, providing sufficient dust and oxygen to support the combustion are present. The dust may be in a layer, heap, or other deposit rather than dispersed as a cloud in air as is necessary for a dust explosion. There are obvious differences in the appearances of dust fires and explosions. In fires, the rate of spread is much slower than the propagation of an explosion and, because the oxygen supply to particles in a layer is reduced the burning rate of individual particles in a fire is correspondingly slower. The obvious characteristics of a dust explosion, the rapid rate of flame propagation and the generation of damaging pressure if the burning cloud is confined, are generally absent from dust fires. The quantity of dust in unit volume of a layer is far greater than in the same volume of an explosible suspension, and the total amount of heat released when the layer is completely burnt is correspondingly greater than with the suspension although the duration of the fire is much longer. In addition, certain dusts which are not explosible, such as anthracite and coke, can be ignited and cause fires.

Unlike dust explosions, there are two types of burning in dust fires; flaming and smouldering (or glowing). The type of burning depends upon the characteristics of the dust, the geometry of the deposit, and the ease of access of air. In a dust explosion, flame must have been present to cause the rapid propagation and the pressure development; a suspension of glowing or incandescent particles would not propagate into unburnt dust and could not give an explosion. But if glowing particles were dispersed into a suspension, and thereby generated flame, an explosion could be produced. In dust fires smouldering or glowing can propagate into unburnt material thereby sustaining the burning.

For present purposes dust fires involving flaming or smouldering will be considered together and the topic will be divided into the ignition, burning, and extinction of dust fires.

Ignition

In broad terms the principal sources of ignition of dust fires are the same as for dust explosions, and have been covered in Chapter 7. The sources are in the same order of importance for fires and explosions, but the amount of information available concerning the different sources varies considerably. The amount of detail given here is not a guide to the relevant importance of the source.

Dust fires can be initiated by flames, glowing particles (including those from welding and cutting), frictional and impact heating, and electric sparks or arcs, as has already been discussed for the ignition of dust clouds. The comments previously given on the avoidance of ignition, fire prevention, are applicable. The high effectiveness of flames in igniting dust clouds is also the case with dust layers, and it is not possible to establish limiting flame properties for safety, and fire prevention must be based on the avoidance of flames in contact with dust layers. Deposits of burning dust are also likely to result from a dust explosion.

Dust fires are readily initiated by glowing particles, and the ease of ignition in air is usually increased by a draught, provided that disturbance of the dust is not caused. Materials such as wood sawdust, cork, grass, coal can be ignited by a glowing cigarette end, either in still air or in draught, the ease of ignition generally increasing with finer dusts. The ignition of dust layers by glowing particles has not received a full theoretical treatment, and limiting safe conditions for practical use cannot be laid down. Apart from dependence upon the composition of the dust, the ease of ignition would also vary with the particle size, thickness of layer, ambient air velocity, amongst the principal variables. Any treatment is therefore likely to be complex. Glowing particles would of course be expected to be particularly effective with dusts which were able smoulder or glow, because the surfaces of the particles immediately adjacent to the glowing source would be in a relatively favourable position for ignition. Once these particles had commenced to smoulder the process could be expected to propagate so that a smouldering dust layer would be produced. For dusts which do not smoulder, for example because they melt, a glowing ignition source would require to distil off volatiles from the dust and ignite them in the gas phase. The dust fire would then be flame propagating across and possibly through the dust layer. Because of convection effects, the residence time of the distilled vapour near the ignition source would be limited, and the conditions for ignition may well be less favourable than for the initiation of smouldering in a dust which is capable of undergoing the process.

Ignition of dust deposits by friction or impact sparking can be readily demonstrated. In industrial processes, particularly hazardous situations are the grinding or polishing of materials on abrasive or fibrous wheels, with which extract ventilation systems are frequently incorporated to remove the dust formed. The working process is likely to produce friction sparking in dust or fibrous materials which can deposit either in the extraction system or the collector at the far end. In addition there is an air flow. The working conditions are thus favourable for the development of fire, particularly in the dust collector where the greatest amount of dust is likely to be present. For methods of fire extinguishing see below. Friction and impact are part of the process in size reduction generally and ignition of dust particles is to be expected, and dust explosions from this cause must be protected against, as discussed above. However even if explosions do not develop, the burning dust particles within the mill etc, or in the conveying or collection systems can readily come into contact with dust layers or other deposits and ignite them. The process can of course occur with dusts which are combustible, but not explosible. It is good practice to inspect those parts of the plant where dusts can deposit, or are collected, particularly before restarting after a shut-down of the plant. The purpose is to avoid the agitation of burning dust which could lead to either a vigorous fire or to an explosion, the extent of the burning of the dust having increased during the shut-down period. Where friction causes surfaces to become heated e.g. at a bearing on which a dust layer may be present, the situation is analogous to the ignition of the dust deposit on an electrically heated surface. The process is considered in more detail below, together with that of spontaneous heating.

Fires in dust layers can be initiated by electric sparks or arcs. The discharges may be produced either by a power fault, or by electrical equipment which sparks in normal operation, or by electrostatic discharge. Sparks and arcs produced by a power fault are, because of the amounts of energy released, powerful sources of ignition of dusts, or of other combustible materials in contact with the dust. Materials such as electrical insulation are obviously likely to be in the vicinity. In any event the quantity of energy released is likely to be considerably greater than that just necessary to cause ignition. The position is thus broadly similar to that of the initiation of dust fires by flames. For electrical discharges which may occur on the breaking of low energy circuits, or in electrostatic discharges, it is more relevant to consider the minimum ignition energy of a dust layer. The method of measurement developed in the United States, see Chapter 3, uses a capacitative spark of known energy, discharged from a probe onto the surface of a dust deposit. Values of minimum ignition energy for the ignition of dust layers differed from those for dust suspensions and no theory is available to link the measured values to other properties of the dust. Nevertheless knowledge of ignition energy of dust layers can be used in practice on an *ad hoc* basis and is particularly valuable in giving advanced warning of materials which are relatively sensitive to ignition by this means. Prevention of fires due to electrical discharges onto dust layers requires similar measures to those for dust suspensions (see Chapter 7).

The initiation of dust fires by hot surfaces, or dust deposits in hot environments, is an important hazard and much attention has been devoted to it both by way of experiment and theory. In broad terms three types of situation are likely to arise in practice and give dust fires.

(1) Dust deposits or heaps. Typical situations would be thick accumulations in a dryer or oven, heated material in a hopper, or a very large heap on the ground exposed to the atmosphere for long periods.
(2) Dust layers, which are relatively thin, in a heated environment. These may be found as deposits on the walls of dryers, including spray dryers.
(3) Dust layers, which are relatively thin, on a hot surface with the other surface of the dust exposed to atmospheric temperature. Relevant situations include deposits on hot pipes, light fittings, and hot bearings.

Because of the practical importance of the hazard, routine tests for the various situations have been developed and are outlined in Chapter 3. As designed, the tests give minimum ignition temperatures which may then be applied directly, on an *ad hoc* basis, to practical situations. This may be adequate for certain situations, where it is only desired to ensure that working temperatures are well below hazardous levels, but for more accurate control, further investigation is needed. In order that the relevant design of experiment shall be made, some consideration of thermal ignition theory is advisable and is summarized here.

Basically, the theory permits a relatively few determinations of the combustion properties of a given dust to be used to calculate the minimum ignition temperatures of situations 1–3 above and to permit maximum safe temperatures to be stipulated with reasonable confidence.

The minimum ignition temperatures so calculated permit the specification of maximum safe working temperatures, by incorporating an arbitrary safety margin, but in the design of plant one practical point must be remembered. The surface temperature of the plant may be the result of a balance between heat generated within the plant and heat losses to the atmosphere, and if a dust deposit is formed on a plant it will have a thermal insulating effect. This insulation can cause the surface temperatures of the plant to rise. Where plant is thermostatically controlled, or is effectively so by the action of constant steam pressures, the insulating effect of the dust would not be important; in other cases where rates of energy input are constant, on average, increased surface temperatures should be looked for.

The ignition of a dust deposit in a hot environment depends upon the balance between the heat generation produced by a reaction of the dust with the air and the heat loss arising from cooling of the dust to its surroundings, see Fig. 9.1. A similar argument applies to a dust layer on a heated surface, in cooler surroundings. In Fig. 9.1 as the temperature difference between the dust and its surroundings increases, i.e. the dust becomes heated, the rate of heat generation increases exponentially. Thus small increases in the dust temperature produce a

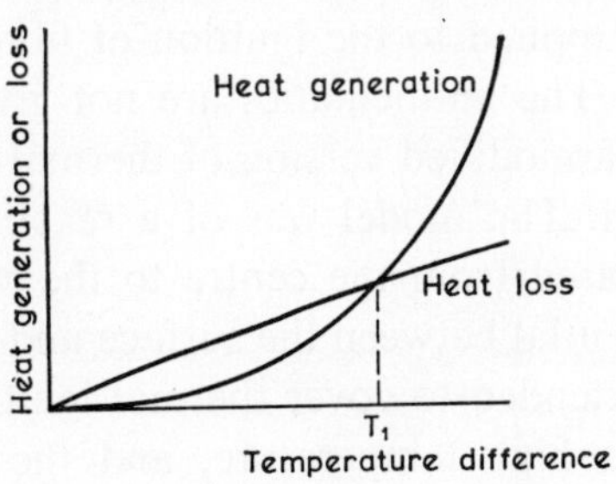

Fig. 9.1 Effect of temperature difference between dust and surroundings on heat flow.

relatively slight increase in heat generation, but larger temperature rises produce a more than proportionately greater heating effect. The situation continues until the rate of reaction of the dust is sufficiently fast to use up all oxygen available locally, and the rate of reaction then becomes dependent upon the movement of oxygen to the dusts, i.e. diffusion, rather than the rate of reaction when the oxygen has arrived. For most practical situations the ignition process is chemically controlled rather than by diffusion. The rate of loss of heat from the dust is approximately proportional to the temperature difference between the surface of the dust and its surroundings, giving a straight line relation as shown in Fig. 9.1. The condition for the dust to ignite is that the rate of heat generation should be at least equal to the heat loss, or higher. From Fig. 9.1 it can be seen that at low temperature differences the heat loss line is above the heat generation line, implying that potentially the heat loss is greater than the heat generation and thus any slight temperature rise due to reaction of the dust is immediately overcome by the cooling effect of heat loss to the surroundings. As the temperature of the dust is increased, the heat generation approaches the maximum potential heat loss. At a temperature T_1 the heat loss and heat generation are in balance, thus any slight increase in the rate of generation would not be overcome by cooling, the reaction rate would further increase, and heat generation would also increase. The situation is unstable and the further the temperature rises the more rapid is the heat generation within the dust. The reaction then 'runs away' and ignition results. The ignition temperature of the system can thus be regarded as that temperature at which the heat generation and heat loss lines in Fig. 9.1 intersect.

A detailed mathematical treatment of the processes summarized in Fig. 9.1 has been applied to the ignition of fibre insulating boards in hot enclosures [1]. The mathematics are not given here in detail, but they are based on a modified version of thermal ignition theory due to Frank-Kamenetskii. The model was of a reacting mass in which the temperature decreased from the centre to the surface, with a further temperature differential between the surface and the surroundings. The theory has been extended to cover thermal ignition in a slab with one face at a constant high temperature, and the other cooling to the surroundings [2]. Good agreement was reported between theory and the results obtained in experiments with wood sawdusts and other dusts ignited on a hot surface. [3]. The validity has thus been established of theoretical models for the ignition of dusts as deposits, heaps or thin layers, in heated environments or on hot surfaces. (Plate 19).

The range of dusts investigated has, however, been relatively narrow, and further work with dusts such as metals, plastics and chemicals would be valuable.

Although principally concerned with the ignition temperatures, the theory gives some indication of the time required for dust to ignite. These times can, however, be greatly affected by the occurrence of induction periods which postpone the onset of the principal heating mechanisms and the existence of which has to be found by experiment. The basic theory also assumes that during the time required for the dust temperature to rise to the minimum ignition temperature there is no significant loss of mass of dust. If the ignition process is regarded as a chemical reaction, this implies that the concentraton of reactant is constant.

In practical situations dusts may be contaminated or dirty. The presence of foreign matter may affect the minimum ignition temperature but recourse to experiment must be made. Contamination of coal dust with oil and grease has shown that the minimum ignition temperature is not lowered, and may be raised [4], but in order to apply the thermal ignition theory, experiments must be made with the contaminated dust.

The thermal ignition theory can also be modified to include ignition of dust coating a hot cylinder, such as a pipe [5], and also the case of dusts deposited in atmospheres of low oxygen content [6]. Tests with cubes of a sawdust in an oven showed that, for example, with a cube of side length 7.6 cm, the minimum ignition temperature rose from 180°C in air to 205°C in an atmosphere containing only 4 per cent by volume of oxygen. The burning characteristics of dusts in such atmospheres will be considered again below.

In most practical fire situations involving dusts, to which the thermal ignition theory can be applied, an adequate treatment may be obtained by assuming no loss of reactant during the ignition process. For special situations, including those where ignition arises from decomposition reactions, account may need to be taken of the consumption of reactant. Theoretical treatments have been put forward [7, 8] which are more complex than the basic theory. It is unlikely that this more advanced theory would be needed for the general case.

Although the ignition of dusts is usually considered where oxygen in the air is a reactant, other hazardous situations may arise. For instance, certain metal dusts can be ignited in atmospheres of carbon dioxide or nitrogen. Results have been reported [9, 10] for ignition of metal powders in small crucibles, a few millimetres diameter, in heated tubes containing nitrogen and carbon dioxide. Although the variation of ignition temperature with the dimensions of the crucible was not investigated, and hence the reported ignition temperatures would be arbitrary values, the ignition of the metals is of interest in relation to plant protection. The metals which ignited in carbon dioxide included lithium, beryllium (ignited spontaneously at room temperature), calcium, magnesium, aluminium, cerium, titanium, zirconium, thorium, uranium, manganese, and chromium. Some of these metals would also ignite in nitrogen, with the following exceptions: magnesium, aluminium, manganese and chromium, within the range of temperatures investigated (up to at least 1000°C). As the dimensions of the metal powder samples were small the minimum ignition temperatures would be expected to be higher than could occur with the larger samples which might be present in industrial process operations. Consequently, if metals are to be processed in gases in which they can be ignited, direct measurements of ignition temperatures would be necessary on samples of practical dimensions.

Burning rates

Dusts can burn either by flaming or by smouldering (or glowing). Flaming is the more spectacular process, and as the rate of heat release is likely to be higher than with smouldering, it will be considered first. If a suitable ignition source is present, the burning of dust layers with flame can take place with certain dusts, provided the quantity of dust is sufficient, it is distributed favourably and the supply of air is adequate. From these qualifications it can be seen that whether dust will burn with flame in a particular set of circumstances depends upon a number of factors, and these can be generalized only partially. Whether a particular dust can burn with flame so as to give a fire hazard may need to be determined by test, because it cannot be reliably predicted. Although the process is complicated certain basic statements can be made. In order that a flame can persist, sufficient vapour must be generated from the dust; the vapour may be a product of partial combustion. Dust deposits which are unable to produce combustible volatiles, or can produce only a small amount, would not burn with flame in air at ordinary temperatures. If the dust is unable to smoulder, and no other heat source is present, the supply of combustible volatiles must be generated by heat transfer from the flame to the dust surface. Thus a dust which melts when heated, and does not smoulder, may be able to burn with flame if the rate of vaporization is adequate to support the flame. If a dust can smoulder, the gaseous products produced by the smouldering may contain sufficient combustible to support a flame. In this case a flame may stabilize over the smouldering area, apparently consuming the smoke evolved by the smouldering, but the flame does not necessarily advance across the surface of unburnt dust, ahead of the smouldering zone, because insufficient volatiles may be produced there to support it. If the dust melts, and only a thin layer is present initially, it may move away from the burning region, under the action of gravity, and the flame is extinguished because of lack of fuel. Also, if the area of dust deposit is initially small, say only a few centimetres in width, heat losses from the sides of the deposit may be too great for the flame to be stable.

Although the composition of the dust, its volatile content, melting and boiling points are important factors governing flaming, other considerations are also relevant. The particle size of the dust in the layer influences whether flaming can occur and whether it would spread relatively rapidly over the surface of the layer but burn relatively slowly into the layer, usually from the upper surface downwards. If this is the case the surface spread of flame is likely to be completed before the dust under the surface becomes fully involved. Following ignition at a point, the flame would thus spread over the surface at a rate depending on whether it could propagate unaided or whether it required smouldering to propagate first and the flame would stabilize in the smoke produced by the smouldering. Under the latter conditions the speed of propagation of the flame would be governed by the smouldering rate. Once the external surface of the dust deposit had become completely involved, the flame may persist on the surface and smouldering combustion continue into the interior of the layer, the flame burning the smoke produced.

If the dust is sufficiently coarse, the flaming may be able to penetrate into the layer and propagate downwards and sideways among the dust particles. Under these conditions the dust layer is likely to be completely involved in the fire relatively rapidly and, because of the increased rate of burning, the rate of heat generation is also rapid. The fire hazard in such a situation is thereby greater than with the slower processes when flame does not penetrate the layer, and the exposure of other combustibles and structures to the fire would be more severe.

Measurements have been reported for the rate of propagation of flame over dusts formed in small layers about 15 cm long, 3 cm wide and of depth between 0.2 and 2.0 cm. The linear burning velocity in a layer 1 cm deep, in still air, following ignition either by a small flame or piece of smouldering material, ranged from 0.8 cm/min for sulphur to 5.0 cm/min for lycopodium [11]. Wood sawdust of particles up to 1.2 mm diameter only supported flaming after the dust had been oven dried. With this material the linear burning velocity increased with layer thickness from 2.2 cm/min for a 5 mm thick layer to 5.8 cm/min for a 20 mm layer. With sulphur the burning velocity varied little between layer depths of 10 to 20 mm, but increased sharply for layers thinner than 5 mm. Coal, flour, and starch did not give flaming in these experiments but it might be obtainable with different experimental arrangements, particularly with larger areas of dust.

The burning of dust deposits is strongly influenced by the presence of an air flow. There has been no systematic study of the variation of burning velocity with air movement, but the transition to flame from smouldering combustion can be made readily by the action of an air flow [12]. In layers up to 26 cm in width, ignited at one end and consisting of cork dust or wood sawdust, the transition to flaming was obtained at air velocities of 100 cm/s upwards with dusts of mean particle diameter greater than 1 mm. There was evidence that flaming could develop from smouldering under nominally still air conditions, due either to convection currents induced in the air or to ignition by radiation from neighbouring glowing particles. When the flames developed they were always present over the surface of the dust layer and caused carbonization of the dust ahead of the smouldering zone, i.e. the rate of propagation of flaming was greater than that of the smouldering. In addition with coarse dusts, of particle diameters up to 5 mm, flame could propagate through the dust layer, in the direction of air flow, again at a faster rate than the smouldering.

Until the theory of flame propagation is developed, recourse must be made to experiment for information on the flaming characteristics of dusts and the effects of the principal variables. From the practical point of view, information on whether flaming can occur, and in what circumstances, and on its rate of spread, are of importance.

Smouldering is a slow type of combustion. Its importance as a dust fire arises from its relatively easy initiation, and its ability to undergo transition to flaming, after long periods of quiescence. It provides a mechanism whereby a rapidly spreading fire can suddenly develop many hours after the original ignition of the dust. In industry it illustrates how fires may break out a long time after a factory has closed for the night, although the smouldering was initiated during working hours.

Two conditions of smouldering can be distinguished. In the first, the smouldering propagates over the surface of a dust deposit, or, if the deposit is thin, the smouldering propagates across the layer burning practically the full thickness in the process. The second condition is ignition within the deposit, the smouldering propagating through the deposit, mainly upwards, until it reaches the surface; propagation across the surface then proceeds as in the first condition. Propagation through the deposit is usually the result of an ignition source buried within the deposit, or some reaction within the deposit leading to ignition, e.g. spontaneous heating, or the ignition of the dust on a hot surface. Smouldering across the surface of a dust deposit, the first condition, is either the result of ignition at the surface or smouldering breaking through the surface from below. Most experimental work has been concerned with propagation in thin layers, and most numerical values for smouldering rates relate to this condition. It will be considered first.

Knowledge of smouldering rates is of practical importance because it enables realistic estimates to be made of intervals between ignition and the outbreak of flame or observed burning. As the dropping of small ignition sources such as matches and cigarette ends are a frequent fire hazard, the majority of experimental work on smouldering has been related to this problem. The common procedure has been for the dust being investigated to be formed into a 'train', which is a thin layer of the dust whose length is much greater than its width. The layer is usually on a horizontal surface One method of forming trains is to fill a long trough with dust, and then inverting on to a horizontal surface, finally removing the trough [12]. Alternatively, the dust may be deposited on the surface and smoothed into the required shape [11]. The train is ignited at one end, either by a small flame or glowing material, and the rate of propagation of smouldering along the train is measured. Typical smouldering rates are shown in Table 9.1, for beech wood sawdust of different particle size ranges, in still air. The smouldering rates are about 5 cm/hour and demonstrate how little dust is required to sustain a smouldering fire for many hours. The results also show that for smouldering to be sustained, a minimum

Table 9.1 Smouldering rates of beech sawdust in still air, cm/s (Packing density 0.26—0.33 g/cm^3, on a dry weight basis)

Particle diameter mm	*Moisture content per cent*	*Depth of sawdust layer cm*				
		0.30	0.80	1.00	1.65	2.65
0.48	9.4	ns*	ns	ns	1.51×10^{-3}	1.41×10^{-3}
0.27	9.8	ns	ns	ns	1.54×10^{-3}	—
0.19	9.4	ns	ns	1.38×10^{-3}	1.61×10^{-3}	—
0.14	9.8	ns	1.74×10^{-3}	1.58×10^{-3}	1.68×10^{-3}	—
0.12	9.7	ns	1.87×10^{-3}	—	1.70×10^{-3}	—

* Smouldering not sustained.

depth of dust must be present. This point is considered further below. Other dusts behaved similarly, fine cork and grass dusts gave smouldering rates of about 20 cm/hour in still air [12], and coal, cocoa, lycopodium, and pyrites gave comparable values [11]. A notable exception was magnesium, of particle diameter less than 150 microns, which gave a smouldering rate of 14 m/hour in a 10 mm deep layer [11].

The appearance of smouldering in trains was characteristic. The surface of the trains was usually darkened and the division between the burnt and unburnt portions was visible, enabling the rates of propagation to be measured directly. In still air the sides of the trains did not burn right down to the base board although in the centre of the trains the entire depth of the dust was affected. Glowing was not usually visible in still air, except for very coarse dusts, and the smouldering zones were covered by a residue which was frequently of lesser volume than the original dust and remained after the trains had burnt out. In contrast, smouldering in an air stream was frequently accompanied by visible glowing of the dusts, the brightness and extent of the glowing increased with air velocity until the division betwen burnt and unburnt portions of the trains became very marked.

In still air, the smouldering rate of a given dust was governed mainly by the particle size. As the particle diameter was increased, the smouldering rates fell, until eventually the particles were too large for smouldering to persist. The behaviour of beech sawdust, Table 9.1, is typical. The smouldering rate was practically independent of train size, implying that smouldering rates across the surface of much thicker deposits would be similar, provided that no flame was present above the smouldering zone which could radiate heat on to it and increase the burning rate. Likewise little variation was found with the density of packing of the sawdust [12]. The smouldering rate was reduced as the moisture content dust was increased, but the effect was not very marked. The smouldering rate decreased from 1.8×10^{-3} to 1.4×10^{-3} cm/s as moisture content was increased from 0.5 to 19 per cent [12]. The nature of the material upon which the dust was deposited did not affect the smouldering rate. Cork dust trains resting on asbestos and on metal smouldered at the same rates, and the minimum depth of dust layer required to sustain smouldering was also unchanged [12].

Smouldering rates can be drastically increased by the action of an air flow over the train. For air movement in the same direction as

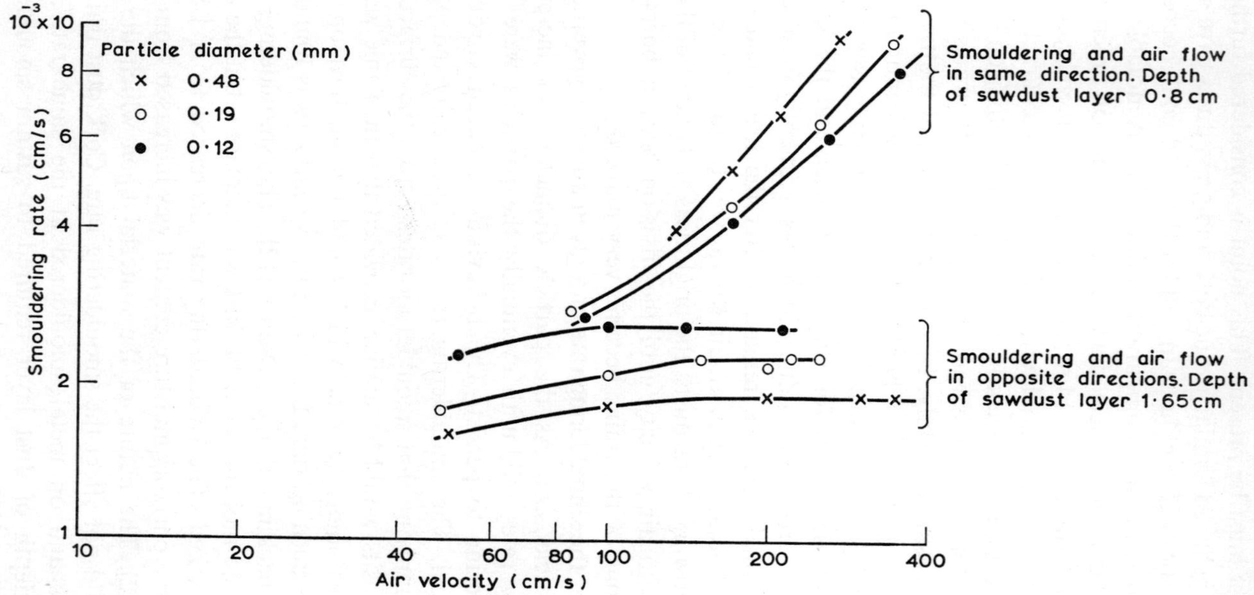

Fig. 9.2 Variation of smouldering rate of beech sawdust with air flow [12].

the propagation of the smouldering the smouldering rate may be increased by an order of magnitude (Fig. 9.2). The effect is of practical importance because it is the means whereby smouldering is most likely to be converted to flaming, and subsequent rapid spread of fire. As the air flow is increased the smouldering rate also increases, approximately exponentially (Fig. 9.2) until either flaming develops or the dust is carried away by the air flow. In practice, with wood sawdust and cork, flaming did not develop in the trains for particles of diameter less than 1 mm, and the dusts were blown away. With coarser fractions flame did develop and then spread rapidly. The fractions finer than 1 mm could give flaming if the smouldering came into contact with adjacent coarser material. This could be either dusts of particle diameter greater than 1 mm or tinder material such as wood shavings, newspaper, sacking etc. Although the fractions finer than 1 mm did not give flaming directly, they could cause it easily in other flammable material. For the greatly increased smouldering rates to occur the air flow and smouldering must be in the same direction, if they were opposite directions the smouldering rate was increased to a much lesser extent, Fig. 9.2.

Although no glowing may be visible in trains in still air, the temperature of the smouldering zones within the trains were at dull red heat. Measurements with thermocouples in trains of cork and beech sawdust gave temperatures of 550 to 610°C [12], the maximum temperature developing about 2 cm behind the front of the charring on the train surface. In an air flow the temperatures were raised, a value of 930°C was reported for cork dust of particle diameter greater than 5 mm, under an air flow of 1 m/s.

Smouldering rates can also be increased by chemical means. The addition of an oxidizing agent accelerates smouldering and there is evidence that addition of other chemicals can also accelerate the burning. A fine coal dust, 85 per cent less than 65 microns, would not propagate smouldering unaided in still air. Addition of calcium based greases enabled smouldering to be sustained and the smouldering rates were doubled by addition of sodium based greases [4].

A dust layer or train must be of minimum depth if smouldering is to be sustained. In still air, for wood sawdusts and cork dusts [12], there was an approximately linear relationship betwen the minimum depth and the particle diameter. Minimum depths ranged from 2 mm for a fine grass dust (through 240 BS mesh) to 47 mm for cork dust of mean particle diameter 1.8 mm. In an air flow the minimum depth for sustained smouldering can be greatly reduced. With a beech sawdust fraction of mean particle diameter 0.5 mm the minimum depth of layer was reduced from 13 mm in still air to 1.6 mm under an air velocity of 2.2 m/s.

The minimum depth for sustained smouldering appeared to be independent of the nature of the surface on which the dust lay, being approximately equal for asbestos, metal, and wood supports. The reason may be that the temperature at the base of the smouldering layer was low compared with the peak temperature within the smouldering zone. If it is assumed that with dust layers of the critical minimum depth the rate of loss of heat by conduction is of the same order as that transferred ahead of the smouldering zone, the minimum depth should then be directly proportional to the diameter of the dust particles [11], in approximate agreement with experiment.

The existence of the minimum depth for sustained smouldering in dusts is of practical importance because it illustrates the extremely thin layers of fine dusts that can smoulder and proves the need for general cleanliness and good housekeeping.

Instead of ignition on the surface of a dust layer, smouldering may be initiated within a deposit, the burning propagating through the deposit to the external surface, usually the top. In some laboratory experiments smouldering was initiated at the base of conical heaps or boxes of cork dust and wood sawdust, and there was no visible sign of the combustion until a short time before the smouldering reached the top surface of the dust [12]. The surface then became damp near the centre, shortly afterwards the centre of the damp patch dried out to a lighter colour, and was soon followed by charring. The buried smouldering was accompanied by a faint musty smell and the normal acrid burning smell only developed when signs of the combustion were visible. There was little change in volume of the dusts during the burning and the carbonized material formed in the cork dust heaps was sufficiently rigid to enable heaps to be cut open for inspection. There was a sharp division between the carbonized area and the unburnt dust. On exposure to air the carbonized mass, originally uniformly black, commenced to glow over its entire depth. The smouldering rates were very slow and the times required for smouldering to burn through mixed wood sawdust (through 12 BS mesh) given in Fig. 9.3, show that for a deposit 85 cm thick the time required was approximately 300 hours (i.e. 12 days). The time varied with the square of the depth of the dust, so that a 30 cm layer required just over 1 day. In these tests the smouldering was initiated in a small heap of dust in the bottom of the box, and the unburnt dust was then shovelled on top of the smouldering material. It is of considerable practical interest that burying the ignited dust below an 85 cm layer did not extinguish the burning, which continued steadily until a carbonized core had formed from the bottom to the top of the layer. It was only after 10 days' smouldering that the top surface of the dust showed signs of cracking, and further time elapsed before any evidence was obtained of steaming or smoke. Superficially the experiment gave the impression that the dust had ignited spontaneously, although this was not the case.

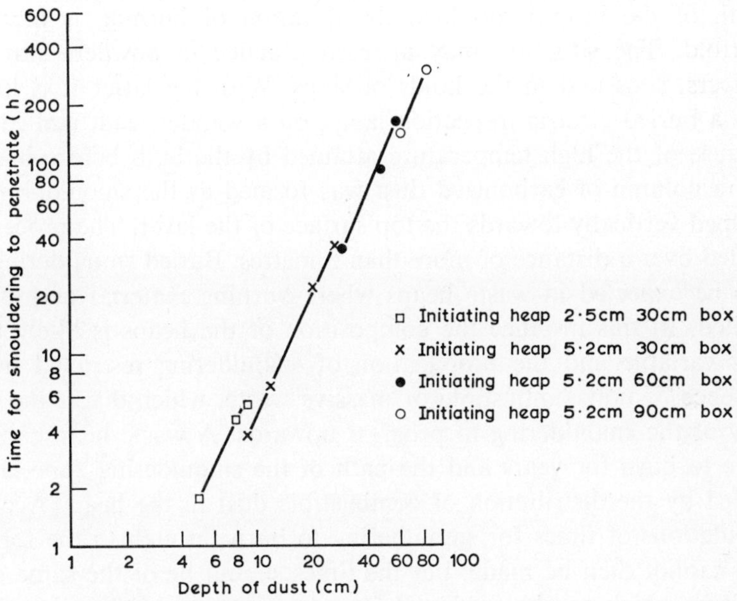

Fig. 9.3 Smouldering within deposits of wood sawdust [12].

When the smouldering had reached the top surface of the dust it propagated across the surface at a rate of the same order as that with which it burned in trains. Once smouldering had broken through to the top surface a chimney effect developed in the carbonized core which then began to burn sideways into the unburnt dust, but at a slower rate than the surface propagation of smouldering. Consequently when the whole of the exposed surface of the dust was smouldering there were zones of unburnt dust between the central core and the walls of the box. This dust was slowly consumed by smouldering advancing from the top surface and from the core. Considerable volumes of smoke were produced at this stage and flame could stabilize on the top surface of the dust.

Although these experiments were of laboratory scale, no indication was obtained that on larger practical scales buried smouldering could not be produced. The relation between time to burn through and depth of dust is therefore likely to be valid for reasonably homogeneous deposits, of depths considerably greater than 1 metre. The duration of smouldering, following ignition at the base of the heap or deposit, would be expected to increase with the square of the depth, i.e. if the depth of the heap is doubled the duration of burning is increased fourfold. The situation may arise in practice in powders stored in hoppers, silos and in the holds of ships. With the latter it is known that a buried electric inspection lamp, on a wander lead, ignited dust because of the high temperature attained by the bulb before it failed and a column of carbonized dust was formed as the smouldering developed vertically towards the top surface of the layer. The process extended over a distance of more than 5 metres. Buried smouldering can also be expected in waste heaps where burning material may be deposited. In this instance the composition of the heaps is likely to be very variable and the propagation of smouldering restricted by the presence of non-combustible or massive waste, which divert the tendency of the smouldering to progress upwards. A waste heap can continue to burn for years and the path of the smouldering zone is controlled by the distribution of combustible dust in the heap. Accurate calculations of times for smouldering to burn through to the top surface cannot then be made, but the times should be of the same order as those which can be predicted from laboratory experiment with the dust concerned.

The probable reason that the smouldering is so slow, but nevertheless persists, is that the burning is controlled by the rate of access of air. The materials which restrict the access of air also prevent the loss of heat from the smouldering zone. Thus although the rate of heat production is low, because the rate of oxygen access is low, the heat losses are also small and as a result the smouldering is able to sustain itself. An important consequence of the good heat insulation is that extinguishing the smouldering becomes difficult. If water is injected, to act as a coolant, it must be distributed throughout the mass of dust, and this is likely to be difficult. Because the flow of air to the burning zone is very restricted, prevention of access of air to the outside of the heap can only be effective if a very good seal is made. The use of inert gas, for instance in a ship's hold, to extinguish a deep seated smouldering fire can only be successful if the inert atmosphere can be maintained for a long time (certainly many days) to ensure that the heat generation within the dust has been stopped and that the burning zone has been cooled down. Failure to ensure these requirements would lead to the re-entry of air as the inert gas disperses and as soon as this air reaches the hot combustion area the smouldering can recover and continue.

Detection

There are no special types of fire detector designed solely for use against dust fires. When a dust is burning with flame it may well be emitting heat, light, and smoke. If the dust is in a building, fire detectors can be permanently installed in the building, in the conventional manner, and the detectors may be sensitive to heat, smoke, light, infra-red or ultra-violet radiation, or other types suitable for detecting fires in the solid concerned. Conventionally, the detectors are installed in the upper part of the building and are intended to give warning should fire occur at a lower level. Their installation requires specialist advice.

Automatic sprinkler systems fulfil the dual purpose of detecting fire and attacking it. They can give protection against fires in dust, provided the heat output from the fire is sufficient to activate the sprinklers. This requirement would usually mean that the dust was burning with flame, and that the sprinkler heads were directly exposed to the products of combustion of the fire. A sprinkler installation in a building has the additional advantage that if the fire should spread to other combustible materials, protection would continue to be given to the structure of the building. A sprinkler system can also be installed within a large hopper or bin, to control any fire and prevent it from breaking out into the building. Both automatic sprinklers systems and fire detectors are commercially available, and their selection and installation require specialist advice.

If a dust is undergoing smouldering combustion, with no flame, the rate of burning is usually slow and hence the heat release is also relatively slow. The quantity of smoke produced varies with the nature of the dust, and cannot be predicted without experiment, and reliable detection can be very difficult. If the detectors are set to respond to very low levels of heating or smoke generation, they may be so sensitive that false alarms will be frequent. These defeat the purpose of the system because after numerous false alarms, little notice is taken by operatives if a genuine alarm should occur. With certain dusts, particularly those containing combustible metals, the amount of smoke generated is negligible and the smouldering zone may be covered completely by a layer of oxide or ash. Even under close examination, detection of the smouldering can be very difficult. Such smouldering fires are likely to proceed unnoticed for long periods. If the smouldering is buried within a heap or layer of the dust, there is no sign of the combustion for long periods after ignition until the smouldering is nearing the upper surface of the layer. If fire is suspected within such heaps, there may be a perceptible temperature rise at the surface or side of the dust deposit, but if the burning is deep seated, it may be necessary to drive probes containing thermometers or thermocouples into the dust to confirm the presence of the smouldering. Smouldering fires within large masses of dust may be detectable by smell but this is subjective and would need independent confirmation, as by measurement of temperature rise.

Extinction

Fires in dust deposits may be extinguished either by allowing the fire to burn all the dust, or by applying extinguishing agents, or by treating the dust deposit so as to exclude oxygen from the fire. Whichever method is used, great care should be taken to avoid disturbing the dust into a cloud because burning material is present and may cause an explosion.

Allowing the fire to burn itself out may be possible for small amounts of dust, or deposits in well protected confined volumes, but it is not likely to be generally acceptable. If the method is used continuous watch must be kept at all stages to ensure that changes in burning can be dealt with. Such changes are likely to arise from sudden air draughts, the burning away of the dust below its surface, causing the dust to avalanche, and the failure of supporting surfaces under the action of heat.

The usual approach will be to extinguish the fire. Selection of the most suitable extinguishing agent will usually depend on factors such as the composition of the dust, its situation, and the presence of other hazardous material and equipment nearby. These factors will tend to be dominant, and the fact that any dust is finely divided is of lesser importance in selecting agents. The extinguishing agent most commonly used is water, but it is unsuitable for fires involving electric equipment or reactive metals such as aluminium and magnesium. Water can be used for most dusts of vegetable origin, and synthetics such as plastics, and wherever water would be used if the materials were present not as dusts but in bulk. The main point of difference is that water should not be applied from a high pressure jet to dust, but should be a low pressure spray. The action of high pressure water would be to disperse the dust as a cloud in air, thus giving rise to the explosion hazard. In contrast, a gentle spray would dampen the deposit and cause it to become more cohesive, thereby reducing the likelihood of accidental dispersion of the dust. If the dust layer is 1 m or more in thickness, water applied as a spray may not penetrate into the mass of the dust. Instead the outer layers can become waterlogged and the addition of further water only results in it running off. If this situation should arise, vigorous stirring of the dust must not be made but either it should be gently uncovered by digging and the spray application continued or the use of 'wet water' may be tried. Wet water is ordinary water to which has been added a small amount, about 2 per cent, of a wetting agent, for which detergents are often used. The wet water is more easily able to penetrate into a mass of dust, and also spreads over the surface of the dust more rapidly. The fire extinguishing properties of the water are not adversely affected by the addition of the wetting agent.

Because of the necessity of applying the water gently, and of ensuring that it penetrates into the layer, extinguishing dust fires with water can be slow. Any attempts to speed up fire fighting, by disturbing the dust, are likely to increase the hazard due to explosion and should be avoided. If the dust deposit is well spread out and of the thickness of metres difficulty may be experienced in applying the water to all parts of the surface. However because of the danger of pockets within the dust deposit burning hollow, and giving a weak crust which may collapse if loaded, fire fighters should in general not walk on the dust whilst extinguishing a fire within it. In exceptional cases, where men must traverse the dust, they should wear safety harness and should preferably walk on planks. The stability of the surface can be tested with rods pushed into the dust.

Fire fighting foams, whether of low or high expansion, are of limited use on dust fires. The foams cannot penetrate into the dust deposits, and at best can form a blanket over the surface. Although they may extinguish flaming and glowing actually on the surface, they are of limited value in controlling fires within the dust. Their action could only be to prevent access of air to the fire, but as has been shown already the air requirements for smouldering within a heap are low and air leakage due to breakdown of the foam bubbles would give an adequate supply to enable the smouldering to persist for long periods, possibly days or weeks. As the foam blanket will not retain its integrity for this time unless continuously reinforced these agents are of limited use for dust fires. Clearly if the fire fighting agent needs continuous attention over long periods, alternative agents which give direct attack on the fire are more attractive. Foams should not be used where water cannot be used.

Dry powder fire extinguishing agents are used on dust fires only in special circumstances. Their main use is for fires involving reactive metals, where the burning dust can be covered by the dry powder, usually by manual application with long handled shovels to avoid disturbance of the dust. Special dry powders are available commercially for metal fires, and traditional powders such as sand, graphite, common salt are still currently used. As these powders are used on reactive metals, which are of severe explosibility, particular care must be taken to avoid stirring up the burning dust whilst applying the extinguishing agent. Dry powders can also be used with dusts in contact with electrical equipment. The requirement of manual application means that only small fires can be tackled; and an extensive fire involving metals may in practice be impracticable to fight. Instead, the tactics should be to allow it to burn itself out, causing the minimum of damage to its surroundings. Attention should be given during a fire to protecting the neighbouring plant, buildings etc., against the radiation and any burning material from the fire itself.

Vaporizing liquids such as halogenated hydrocarbons (chlorobromomethane, BCF, TBM) have a limited use in fighting small scale dust fires. They are likely to be of most use in fighting fires associated with electrical equipment, but must not be used with reactive metal dusts. Where automatic explosion suppression equipment is installed the suppressant, which is a vaporizing liquid, is capable of extinguishing subsequent fires as well as suppressing the explosion. In using a vaporizing liquid fire extinguisher on dust care should again be taken to avoid disturbance of the burning material.

Gases or vapours such as carbon dioxide, nitrogen, or steam may be used as extinguishing agents for dust fires, apart from reactive metals etc., again providing that the disturbance of the dust is avoided. Their action is basically the exclusion of air and hence their effectiveness as extinguishing agents depends upon their being present for a sufficient time to allow the dust to cool sufficiently to prevent re-ignition when the dust is again exposed to air. These agents will thus be particularly effective for dust which is contained within a relatively gas-tight volume, e.g. burning dust in a hopper, silo, or hold of a ship. The gases have the advantage that they are able to penetrate relatively easily into the dust deposit and can thus be used in fighting fires in large volumes of dust. However, in such large volumes the heat insulation is good so that natural cooling of the dust is very slow. To ensure complete extinguishing of the dust fire the gas must be maintained for long periods of the order of days or weeks. For many reasons this would be impracticable either because the plant is required to be available for operation in a short time, or because of the expense of maintaining the inert blanket, or because the ship containing burning dust has reached port. The procedure should then be to open up and commence to unload the dust from the top surface of the hopper or hold, continuing until either burning material becomes uncovered or the burning has become sufficiently vigorous for further work to be impossible. At all times careful check of the atmosphere must be maintained, to ensure that adequate oxygen is available for safe working, particular care being taken after opening up but before operatives are allowed to enter. The wearing of breathing apparatus may be necessary. When work has to be temporarily suspended, an extinguishing agent, water or other appropriate medium, should be applied to the dust until the burning has subsided. A further session of digging out should then be started and continued until conditions are too arduous. Further addition of extinguishing agents is then made. These cycles of operations must be continued, once started, until the main volume of burning material has been uncovered. The possibility of the dust burning hollows within the deposit, mentioned above, is again relevant and periodic checks may be necessary. Mechanical aids to unloading are to be encouraged, provided that they do not cause serious disturbance of the dust and that hollow pockets in the dust do not give rise to hazard.

In fighting a dust fire with inert gas there will inevitably be some leakage of the gas, requiring a continuous supply to replace that lost. Because of practical difficulties the usual procedure is to tend to displace all the air from around the dust deposit, rather than to attempt to dilute it to a controlled extent. The relative efficiencies of carbon dioxide, nitrogen, or steam in controlling and extinguishing dust fires do not then have to be taken into account athough some differences can be detected in laboratory experiments under strict control. The difference in performance of most importance in practice is that when the supply of gas is stopped, steam will condense to water, giving a great reduction in volume, when the temperature has fallen to 100°C. Because of the large change in volume on condensation, air is drawn into the dust deposit to replace the volume formerly occupied by the steam. Thus on turning off the steam supply and allowing the dust to cool a rapid resurgence of the fire can occur because most of the space between the particles would be filled with incoming air. With carbon dioxide and nitrogen the situation is different because some time is required for the gas to be displaced from within the dust by air, and this process would be much slower than the inrush of air due to condensation of steam. The resurgence of the fire when carbon dioxide or nitrogen are used is therefore slower than with steam.

A variation of the inert gas method is to use liquefied carbon dioxide or nitrogen as the extinguishing agent. These must be introduced to the dust through piping which should penetrate a little way into the dust so that the cooling effect as the liquid boils can assist in removing the heat from the fire. Gas introduced in liquid form is likely to be a more effective extinguishing agent, but a closer approach to the fire may be needed by operatives in setting up the system.

Fires in dusts particularly in large masses such as tip heaps, can be fought by the exclusion of oxygen with a solid barrier. The method involves covering the burning heap completely with a non-combustible layer, such as earth, which has then to be compacted down to a more impervious layer than the underlying dust. Typically, a metre or so of earth would need to be applied over the top surface and sides of the heap and compacted down mechanically. This is in itself a hazardous proceeding because of the steep slopes which are likely to be present and also the danger of the burning away of hollows within the heap which would collapse as the machine was compacting earth above it. A more positive procedure is to inject a waterborne slurry through fissures in the heap, usually a cheap non-combustible material such as limestone or cement would be pumped into the heap from above, in water under pressure. The purpose is twofold; the water is pumped down into the heap, and some of it at least may reach the burning zone and cause partial extinguishing of the fire, the slurry material then dries out and tends to close fissures etc. which act as channels for the ingress of air. If the slurry treatment is followed by covering and compaction, the fire in the tip may be eventually extinguished. This is likely to take a long time, of the order of weeks or months because of the slow rate of loss of heat from the interior of the tip to the atmosphere. The method should be used with caution if the underlying ground is steeply sloping; the injected slurry might then act as a lubricant and cause the tip heap to avalanche. If quicker extinguishing is needed, as may be the case if a building on made-up ground is threatened by a fire approaching from below, the usual solution is to dig. Channels should be cut through the heap, sub-dividing it down to its base, and further digging undertaken to attack the burning material. Such a procedure is time consuming and expensive and, quite often, the material of the heap is of little commercial value for salvage. Digging out should thus only be undertaken as a last resort, where circumstances make it essential.

References

[1] Thomas, P. H. and Bowes, P. C. (1961), *Brit. J. Appl. Phys.*, **12**, 222–9.

[2] Thomas, P. H. and Bowes, P. C. (1961), *Trans. Farad. Soc.*, **57**, (467), 2007–17.

[3] Bowes, P. C. and Townshend, S. E. (1962), *Brit. J. Appl. Phys.*, **13**, 105–14.

[4] Hardy, V. O. (1956), Safety in Mines Research Establishment, Research Report, 136.

[5] Thomas, P. H. and Bowes, P. C. (1967), *Pyro.*, **5**, 29–37.

[6] Bowes, P. C. and Thomas, P. H. (1966), *Comb. and Flame*, **10**, 221–30.

[7] Thomas, P. H. (1961), *Proc. Roy. Soc. A.*, **262**, 192–206.

[8] Adler, J. and Enig, J. W. (1964), *Comb. and Flame*, **8**, 97–103.

[9] Rhein, R. A. (1964), *Comb. and Flame*, **8**, 346–8.

[10] Rhein, R. A. (1965), *Pyro*, **3**, 161–8.

[11] Cohen, L. and Luft, N. W. (1955), *Fuel*, **34**, 154–63.

[12] Palmer, K. N. (1957), *Comb. and Flame*, **1**, 129–52.

10 *Application to Industrial Processes*

So far separate consideration has been given to the behaviour of dusts in explosions and to factors relevant to plant design, such as ignition sources and explosion protection. The separate treatment enabled the subject to be explored in detail, so that limitations of knowledge were made apparent. For the purposes of plant design, however, unit processes are of basic importance. Thus, a series of plant units is usually required in a manufacturing process and it is convenient to consider dust explosion and fire problems for each of the units in turn. The procedure will be to consider a number of principal unit operations in dust handling plant, to give a brief account of the main aspects for each of these units, and to refer back to the appropriate chapters where these aspects were covered in greater detail.

The present chapter is a selection of important aspects concerning dust explosions and fires for various operations, and covers the basic methods of protection. Variations may be necessary because of local conditions, either arising from special properties of the dusts or from particularly hazardous siting of plant, and so the present considerations cannot be inflexible or comprehensive. However, they may be used as a guide, providing they are not followed blindly, and will enable the basic protection to be designed. In many instances the projected design will need to be submitted for approval to some authority (see Chapter 1) and it is useful if a scheme with firm proposals is available as a basis for discussion.

The operations in a dust handling plant have been arbitrarily divided up into the following, which will be considered in turn: dust manufacture, conveying, separation, drying, screening, and storage and disposal. Not all processes will be represented in all plants, the selection and order depending upon the material being handled. Other operations, for specialized processes, may not have been covered here and for these reference should be made to earlier chapters, drawing on the appropriate information. The absence of a unit operation from consideration in the present chapter does not necessarily mean that explosion or fire hazards do not arise with it.

Consideration will also be given to the buildings that house plant. Important factors are the separation of individual structures, and their construction.

Dust manufacture

DESIGN OF GRINDING MACHINES

Size reduction of solid material to powder or dust is a necessary process in numerous industries, because the finely divided material is either the product or is used in the makeup of the product. Obvious examples are flour from grain, pulverized fuel from coal, and animal foodstuff from oil seed cake. The size reduction may be carried out either in one stage, using a single machine, or in a series of stages involving different types of machine. The choice usually depends upon the requirements of the industry, rather than any consideration of dust explosion or fire hazard. The types of machinery range from crushers, kibblers, grinders, mills, to micronisers, and from slow to high speed. Higher grinding speeds tend to be used for the more finely divided feed stocks.

It is a general characteristic of grinding machines that they are massive and of high strength construction, to withstand the mechanical forces involved in the process. Where the delivery from the mill is of finely divided material, so that a dust explosion hazard can arise, the mechanical strength of the mill is an advantage. The strength, coupled with the entry and exit apertures in the mill which can give some explosion relief, may ensure that even if an explosion should occur within the mill, structural damage to it will not result. Information on the strength of the mill may be obtained from the manufacturer or from the results of static pressure tests. Adequate strength is more likely to be obtained with mills of small volume, but even those of internal volume of several hundred cubic feet (tens of cubic metres) such as are used in coal pulverization, can be designed to withstand pressures as high as 350 kN/m^2 (50 lb/in^2). Further protection can be obtained by the provision of explosion relief vents on the mill, Chapter 8, or by the installation of automatic explosion suppression equipment. A hazardous explosion in a mill or other grinding plant may be avoided in this way, provided the emission of flames and burning materials from vents is safely handled (Chapter 8).

Although the explosion hazard of the mill itself can be readily dealt with, serious explosion hazards can arise where the product of the grinding is conveyed pneumatically to further units of plant. The ducting from the mill should preferably be able to withstand the full explosion pressure but the dust collection units are often of much weaker construction than the mill itself, and quite unable to withstand full pressure. Explosion protection of such units is essential, and is considered further below.

Where size reduction takes place in stages, the explosion risk is low for those crushers delivering product of a few centimetres (1 in) diameter, particularly if the machine is slow speed. A fire hazard may be present, and appropriate fire protection measures should be considered, see Chapter 9. Care should be taken to ensure that the small proportion of dust or powder produced simultaneously with the breakdown of the massive solid does not escape and accumulate within the workroom. This material is hazardous both because of its potentiality for explosion and also because of causing rapid spread of flame in a fire. The product from this preliminary crushing, or original feed stock of similar or lesser size, is on comminution likely to give a product which presents an explosion hazard. Explosion precautions should be concentrated on this stage of the process unless legislation requires otherwise, but fire precautions are necessary throughout.

SOURCES OF IGNITION

Ignition of dust in grinding and pulverizing machinery is particularly likely because of the presence of friction and hot surfaces arising from the energy used in the comminution process, sometimes accompanied by the injection of heated air if the feedstock is wet. These sources of ignition were considered in detail in Chapter 7.

Friction or impact ignition can be caused by the presence in the feed stock of foreign material such as tramp metal, stones etc. Protection against these materials can be obtained by installing separators, magnetic or pneumatic, which are available commercially. Magnetic separators give protection against ferrous metals, whilst pneumatic separators are intended to remove material of greater density than the feed stock. As the foreign material may be present within large lumps of the original feed stock, the separators should be installed after these large lumps have been broken down, but before the material passes into machines delivering dust or powder. Pneumatic separators frequently use the air current induced by the mill, and are mounted in ducting attached to the input of the machine. Magnetic separators can be installed likewise or, if the feed stock to the mill is delivered by conveyor belt, may be mounted just above the surface of the feed on the belt. Alternatively an electro magnetic detector could be installed which would stop the conveyor belt if ferrous material was passing. The foreign material would then have to be removed by hand, and the conveyor subsequently re-started.

Foreign material collecting in separators must be removed regularly before the separator becomes choked. If ferrous material collects on a magnet, and is not properly removed, a stage may be reached at which the magnetic force is unable to continue to support the load. The ferrous material is then likely to fall back into the feed stock and be delivered to the mill in a mass, causing a severe ignition hazard. If electromagnetic separators are used, provision must be made to divert the feed stock from the mill in the event of a power failure, otherwise previously collected material will be delivered to the grinding machine.

Hot surfaces may develop in a mill if the rate of supply of feed stock is far from the optimum. Fire and explosion hazards can arise whether the mill is overloaded or underloaded. In the design of a grinding unit the optimum output from the mill should be balanced with the delivery of feed stock, and the unit powered by a motor, usually electric, designed to match the optimum load. Safe working can then be assisted by installing an overload trip on the motor, so that if the power required to run the mill becomes excessive, indicating imbalance in the mill, the motor will be shut off. If the power of the motor has not been matched to the grinding process, but is much greater, then it could continue to run even though the grinding machine was overloaded. Likely causes of overloading are either blockage of delivery from the mill, or excessive feed into it. In either case the feed stock will be resident in the machine for a longer time than usual and may overheat unless the motor is switched off. Ignition of the dust can then result and, unless the system has been closed down and inspected to ensure that the dust has not ignited, burning dust may be delivered into the subsequent units of plant.

If the supply of feed stock to the mill is less than the optimum throughput, or is interrupted, the interior of the mill becomes relatively empty of dust. This has two consequences. The concentration of dust in suspension in the mill will be lower than under optimum conditions, and is more likely to be in the explosible range. The free volume within the mill, within which the explosible suspension can be present, is increased. The probability of explosion within the mill is increased and causes an unnecessary hazard. Where hot air is introduced into the mill, in order to dry the feed stock whilst it is being milled, departure from the optimum working conditions is particularly to be avoided. If the supply of feed stock to the machine is interrupted, as may happen accidently due to bridging in a wet material, the operating temperature within the mill will rise because of the reduced delivery. Any deposits of dust within the machines are thereby exposed to higher temperatures than are designed for, and can ignite. A fire in the mill results, and may be accompanied by explosion, and the burning dust can be delivered from the mill into subsequent units.

In designing grinding units to avoid dust fires and explosions, the optimum capacity of the mill must be matched to the required throughput of feed stock, the motive power should also be matched, and the feed should be regulated. These requirements are the basis of a safe design.

PROTECTION AGAINST EXPLOSION AND FIRE

Protection against explosions in mills, if the unit itself is of insufficient strength to withstand the full pressure of an explosion, can be obtained by using either relief venting, automatic explosion suppression, or by circulating inert gas, see Chapter 8. Fires in mills should be extinguished by the appropriate agent for the dust concerned, Chapter 9, and access for fire fighting is not usually difficult.

Some difficulty may be experienced in deciding the procedure if a dust explosion or fire occurs in a mill. In many cases the solution is to shut the mill down immediately, extinguish any burning material, remove it, and inspect for signs of mechanical damage in the machine. When the position is clear, the process can be restarted. If burning dust has been delivered from the mill, before it is shut down, then there are subsequent hazards. If the dust is conveyed to the succeeding unit by air current from the mill, and the mill is then shut down, burning dust will deposit. Even though the fire in the mill is extinguished and all burnt material removed from it, on restarting with fresh feed stock the burning dust in the other units will be thrown into suspension and can cause a serious explosion. Where this hazard is possible, the entire system must be inspected for burning dust before restarting. In the particular case of a pulverizing mill for coal delivery to a furnace the option arises, in the event of a fire in the mill, of continuing to run the plant and delivering the burning material to the furnace, where it is dealt with safely. For this to be carried out successfully it is usually necessary to increase the suppy of coal to the maximum, decreasing both the free volume and the likelihood of explosion within the mill. If successful, all burning coal is fed safely to the furnace. If fire persists in the mill, recourse may need to be made to stopping the coal supply to the mill but to continue the air supply until all dust has passed from the ducting. This procedure involves the risk of explosion in the ducting as the concentration of burning coal decreases. Another method of obtaining protection is by the introduction of inert gas into the pneumatic system.

From the above it is clear that detailed consideration must be given both to the design and the method of operation of grinding units bearing in mind both the hazards involved and their effect on subsequent units in the plant.

Where the material to be ground has a low ignition temperature, such as sulphur, ignition within the mill can be prevented by the use of inert gas during normal working. Reference can be made to Chapter 8 for details of the technique. Inerting can be obtained by using nitrogen, carbon dioxide, flue gas, and the usual precautions are necessary to control the oxygen to a safe level. Indication of whether the use of inert gas should be seriously considered can be obtained from values of the minimum ignition temperature of dust suspensions and dust layers, Appendix 2.

In order to control gas compositions when inerting is used, the entry and delivery points of the grinding unit are likely to be enclosed. In the more usual circumstances, where the mill contains air, and dust may ignite, protection must be provided against the emission of flame from the entry and delivery points of the unit. Such discharge of flame, as in the case of explosion relief vents, could cause injury to personnel and damage to surrounding plant. Adequate protection can be obtained by installing some form of choke, e.g. rotary valves, at the input and delivery points. The performance of the chokes should be such that the supply of feedstock to the mill should neither overload nor underload it.

Even if adequate precautions are taken to minimize the risk of explosion and fire in a grinding unit, complete protection cannot be obtained. Also, even though good housekeeping is stringently enforced, the possibility remains that flammable material can escape from the unit and deposit in the work room. The noise emitted by the machine may also be a problem. The siting of grinding units in separate enclosures is sometimes recommended for fire, explosion and noise protection. Good housekeeping within the enclosure should be stringently enforced. The walls of the enclosure should be fire resistant, for example 14 in brickwork, and the interior should be separated from the workroom by a strong door.

For dusts of severe explosibility and for others of less hazard but which must be situated near other plant or operatives, explosion protection of the enclosure should be provided. For severe dusts the enclosure should be outside the main workroom, on the outside wall, and this wall and the two side walls of the enclosure should be strong. The outward facing wall and the roof should be of lightweight material. Entry to the enclosure is from the workroom, through a strong door opening inwards, and so interlocked with the machinery that operatives cannot enter whilst it is running. For less severe dusts, the enclosure may be within the main building, and all walls should be strong and fire resistant. Under no circumstances should feedstock awaiting grinding be stored loose, or in bags, in the enclosure with the grinding unit. Stored materials would greatly increase the fire and explosion risk from a minor mishap.

Where units are manually loaded or unloaded, dust is liable to escape during normal working and even good housekeeping cannot ensure that no deposits would be formed temporarily. It is then good practice to install local exhaust ventilation at the points of escape. The ventilation should run continuously and be connected to a dust collection unit in a safe area.

An alternative method of obtaining safety in the milling or grinding of hazardous solids is to use a non-flammable liquid as the dispersing medium, instead of air. The product is then in the form of a paste or slurry, and a dust fire or explosion is only likely if it should dry out. If a flammable liquid has to be used as a dispersing medium, and particularly if its working temperature is above its flashpoint, explosion and fire hazards are increased. The ignition energies of vapours are lower than those of dusts, being below 1 millijoule, and the movement of the solid during the grinding process generates static electricity

(Chapter 7). A spark produced by static can then ignite the vapour, causing an explosion or fire involving both liquid and dust.

DUST AS A BY-PRODUCT

The presence of grinding and similar machinery usually indicates that the dust is being generated as a product, and not as an unwanted by-product. Dust as a by-product is more likely to be produced by other types of machining, such as sawing, sanding, granulating or the breaking down of compressed material into separate fibres. Industries where dust is produced as a by-product in this way include woodworking and furniture, leather, rubber, polished metal, and textiles. The manufacture of dust in these industries differs from that where grinding is involved because the rate of production of dust is likely to be less, the process may not be in a strong machine as in grinding, and may be under very open conditions, and the dust is of relatively little value.

To minimize the fire and explosion risk the process should be provided with local exhaust ventilation, with a dust collection unit. Operations such as sawing and polishing are usually under open conditions, so that the risk of explosion of dust at the point of production can be disregarded. Ignition of the dust can however occur and the burning dust would be transported and eventually arrive at the collector. There is a dust hazard in the collection system and appropriate measures should be taken, see below. In the instances where the dust is being produced in an enclosed volume, which is unlikely to be designed to stand the full explosion pressure (including the attached dust extraction system) provision for explosion relief venting may be necessary, see Chapter 8.

In general, processes generating dust as a by-product can ignite the dust at the point of its production, leading to explosion in attached units and if housekeeping is not adequate there is a hazard of secondary explosion in buildings housing the process. Ignition of the dust at the point of production can also lead to a fire, and appropriate measures against dust fires should be taken, Chapter 9. As in the case of grinding machines it is good practice to avoid the storage of combustible material adjacent to processes which can produce ignited dust; provision of fire resistant separating walls is advisable.

Conveying

MAIN METHODS

After the dust has been manufactured from the feed stock, or if the feed stock itself is a dust, it frequently requires to be conveyed to a dust collection system, pending further processing. Three main types of conveying can be specified: manual, mechanical and pneumatic. The system is usually designed to meet the requirements of the plant, rather than to obtain the optimum protection against dust explosions and fires, but where a choice is possible increased safety may thereby be easily obtainable. For both economic and safety reasons the extent of conveying of dusts in a plant should be minimized, but long lengths of ducting may have to be accepted particularly if vulnerable and hazardous plant units, e.g. dust collectors, are to be in a safe place. As a conveying system connects various plant units, it provides opportunity for the containment and separation of units, which is one of the methods of preventing an explosion from propagating throughout an entire plant, Chapter 8. Conversely, those components which isolate the individual units can also prevent the entry of explosions originating within the conveying system. When mechanical or pneumatic conveying must of necessity run within a building the casing of the conveyer should be either of sufficient strength to withstand the full explosion pressure, or should be given adequate explosion protection by methods described in Chapter 8. Use of the first alternative is particularly attractive if the conveyer is attached directly to a unit and its attached ducting can be considered as an entity.

MANUAL

Manual conveying is only likely to be economically attractive where batch process is involved, with limited conveying requirements, unless labour costs are extremely low. It can also be involved at 'bagging off' points, where quantities of dust are limited. An example is the removal of an unwanted by-product from a bag filter unit and disposal of the dust on another part of the site. Manual conveying often involves the transport of the dust in bags or sacks, for which purpose paper sacks are to be preferred to those from textile as leakage of dust through the walls of the sack is prevented more easily. As the quantity of dust that can be conveyed manually is restricted by the weight that the operative can manipulate the method is only suitable for limited quantities of dust. When material is being removed from or transferred to the plant by manual conveying the dust can escape to atmosphere and may deposit within the workroom, incurring the risk of a secondary dust explosion. Local ventilation at bagging off and tipping points is advisable, coupled with particularly good housekeeping along the conveying route. There is also a fire hazard associated with the dust during manual conveying and good supervision is needed. Because of the relatively high cost per unit weight of dust moved, manual conveying is only normally economic for small scale and sporadic dust movement.

MECHANICAL

Mechanical conveying has the advantages that the throughput can be high, continuous operation is possible, and the dust may readily be confined within the plant. The method involves transport of the dust in bulk, and not as a suspension in air, using methods such as screw, drag link, belt conveyers or bucket elevators. In the ideal mechanical conveyer there would be no free volume in which a dust suspension could form, and no ignition sources. The absence of free volume can be obtained with the screw conveyer, which has already been mentioned (Chapter 8) as a means of separating plant units providing it is fully loaded with dust. Screw conveyers may occasionally act as a source of ignition, by generating friction which can lead to ignition of vulnerable dust Chapter 7; the hazard may be minimized by installing an overload trip on the motor driving the screw. Drag link conveyers generate less friction in the dust, but the return leg has free volume in which a dust suspension could form and become ignited. The free volume can be reduced by designing the carrying links of the conveyer to fold up when empty, and the return can thereby be made narrower than the delivery leg. Some explosion protection is desirable in the return leg, either by ensuring that the conveyer casing is sufficiently strong to withstand the full explosion pressure, or by explosion relief venting to the outside of the building, Chapter 8, or by a judicious mixture of both.

The belt conveyer is very simple but is unsuitable for conveyance of dust unless enclosed. The free volume within the enclosure is likely to be much greater than with the drag link conveyer and the explosion hazard is thereby increased. The drag link type should be used in preference. If a belt conveyer is chosen, then full protection against explosion should be provided (Plate 20).

In all the mechanical methods of conveying dust, there is likely to be generation of static electricity. It is essential that all metal components of a mechanical conveying system shoud be securely bonded to earth, Chapter 7. Effective bonding ensures that sparks will not be drawn from the plant to earth, although residual charge may persist in the dust. Mechanical conveyers will also have bearings, which may overheat as a result of failure or faulty maintenance. All such bearings should be external to the dust carrying components, and preferably should give an alarm if overheating occurs. Belt conveyers can also give overheating either of a jammed idler roller or, if the belt jams, as a result of drive rollers continuing to run. The belt can also generate static electricity, and should therefore be of anti-static material.

Bucket elevators are liable to cause explosions and fires in dusts, and numerous examples have occurred in the past, particularly with vegetable dusts, for which these elevators have customarily been used (Plates 4, 21, 22). The reasons are that, because of its design, dust clouds are likely to be continuously present during working, particularly in the head and the boot of the elevator. Secondly, the buckets are regularly subjected to impact and the belt supporting the buckets can slip on the pulleys and generate frictional heat. As a result, a source of ignition and a dust suspension can be present simultaneously, causing explosion or fire. Modern high capacity elevators, with separate delivery and return legs, have a reduced risk because of the reduced volume per unit weight of dust conveyed. In the general case, with explosible dusts, other types of elevator are preferable, and are particularly necessary for dusts of more severe explosibility, e.g. those giving maximum rates of pressure rise in excess of 5000 lb/in^2 s (35 000 kN/m^2 s) in the small scale test apparatus (Chapter 3). Use of the elevators should also be avoided for dusts known to be readily ignited by friction, e.g. sulphur.

Where the use of bucket elevators is unavoidable, their positioning should be carefully specified, and regular maintenance is essential. The elevators should be mounted outside the building, e.g. supported by an outside wall, and the intake and delivery points should preferably be isolated from the rest of the dust handling plant by means of chokes (Chapter 8). The elevator casing should be of fire resisting construction, sufficient to retain a fire, and should be dust tight and of sufficient strength not to disrupt in the event of an explosion. To meet the strength requirements the casing should be provided either with automatic suppression or with explosion relief at the head and the boot, with a vent ratio appropriate to the dust concerned, Chapter 8. For example, with grain dusts, the required vent ratio is usually 1 ft^2 per 20 ft^3 of volume of the whole elevator. Long elevators, say more than 6 m (20 ft), may require additional relief at intervals along the casing to ensure that no point is too remote from a vent. Particular care should be taken to ensure that flame, burning dust, etc, discharged from the vents during an explosion cannot injure operatives or damage nearby plant. Provision of ducting or deflectors over the vents may be required, Chapter 8. Where a bucket elevator is used inside a building, instances of which should be discouraged, it is desirable for the internal pressure to be slightly below atmospheric to minimize leakage of dust. Discharge of combustion products from vents to the outside of the building is essential, and the design requirements of ducting from vents should be observed, Chapter 8. The need for dust tightness and adequate casing strength and fire resistance should be rigorously met.

Steps may also be taken in the design of bucket elevators to minimize the generation of sources of ignition. These include the provision of strong fixing for the buckets and strong bearings for all shafts, external to the casing, provided with detectors for overheating. The main drive to the elevator should be external to the casing. Belt slip within the casing can be detected by belt speed meters, and anti-runback devices can also be installed. Development of friction within the casing can thereby be reduced.

Exhaust ventilation can be applied to the casing of bucket elevators. Ventilation assists the removal of suspensions of dust in air, which would contain those dust fractions of smaller particle size, and also would give a slight negative pressure relative to atmosphere, within the casing. The ventilation system should be connected to a dust collection unit in a safe area, and should be provided with explosion protection on a similar basis to that in the elevator casing itself. Care should be taken to ensure that the air flow is sufficient to prevent deposition of dust in the ducting of the collection unit, and also sufficient to permit the collector to function efficiently. Dust collection systems are considered in more detail below.

Automatic explosion suppression can be used to protect bucket elevators when explosion relief is not practicable, often because of the situation of the elevator on plant within a building. The design of the suppression system is usually undertaken by its manufacturer, Chapter 8.

PNEUMATIC

Pneumatic systems differ from mechanical and manual systems in that the dust is transported in suspension, usually in air. Two types of pneumatic system are available:

low volume/high pressure air

high volume/low pressure air

The low volume/high pressure systems are a comparatively recent development and are applied to the transport of dust within plant and also to delivery of dust, e.g. from road vehicles to storage hoppers in plant. The basis of the method is that a high concentration of dust is transported rapidly through a narrow pipe with a relatively high air pressure. A typical pipe diameter would be 10 cm (4 in) and an operating air pressure of 50 kN/m^2 (7 lb/in^2). Systems can also be designed to operate at negative pressures of about this magnitude. The weight of dust/air will vary with the nature of the dust, but can be of order 40 to 1. This ratio is well above the upper explosible limit of all but the most exceptional dusts, such as those containing their own oxygen for combustion, and for most materials the mixture being transported is not within the explosible concentration range. Pneumatic conveying has the advantage that both horizontal and vertical travel is readily obtained and the system is self cleaning and does not cause dust to accumulate in the pipe runs; the linear velocity within the pipes will vary with the material being transported but is often about 25 m/s (about 80 ft/s). Because of the high flow velocity and the density of the suspension there may be abrasion of the pipe walls, particularly at bends, and the design of installation should take this into account. The pipe is relatively strong and by the use of suitable connections, could be readily made sufficiently strong to withstand a full explosion pressure should one develop. Because of the normal working pressure within the system, all hoppers and collectors incorporated must be sufficiently strong to withstand the air pressure but they may not be sufficiently robust to withstand a dust explosion.

Where the system works only at pressures above atmospheric, the air movement may be generated by a blower which forces clean air into the system in the vicinity of the material inlet. The pressure then drops throughout the system and reaches atmospheric where clean air is delivered at the end. Alternatively, the whole system can be at less than atmospheric pressure, in which case the air drawn through the system is progressively lowered in pressure and either the fan unit must be subjected to the dust laden suspension or a final filter unit is subjected to the maximum pressure differential. There is also a hybrid system in which the blower is situated at an intermediate point, part of the system is then below atmospheric and part above. The blower is again subjected to the suspension and because of other complications the system is often more expensive. The system actually installed in a dust handling plant is likely to be selected on the basis of the economics of running, with the materials to be transported, and explosion and fire hazards are then considered subsequently.

The conveying system can conveniently be considered in the following parts: feed inlet, conveying pipes, dust separators and delivery units. At the feed point the air, or other conveying gas, comes into contact with the dust. The air must be taken from a safe area, and be free of ignition sources. For systems working under positive pressure the blower, fan, pump, or compressor should be supplied with an air relief valve and an overload trip on the motor to ensure that in the event of blockage of the system air will not be delivered at excessive pressures or temperatures. A non-return valve can also be fitted to the air delivery line to prevent damage to the blower should a pressure surge arise from an explosion downstream in the system. The dust inlet should be of strong construction, preferably capable of withstanding full explosion pressure, and with the minimum of moving parts. It is important that burning dust should not be fed into a pneumatic system and also that the dust feed to the system is maintained at the optimum level. The dust should be supplied to the system from a storage hopper, road tanker, etc., at a controlled rate to ensure an adequate supply and to prevent choking. Where the dust is produced by grinding or similar processes, some form of choke, e.g. a rotary valve, must be installed to prevent explosion propagating from the grinding unit through the storage hopper into the pneumatic system. The choke valves may not prevent the movement of smouldering dust through the system, and with dusts which smoulder readily, the use of an inert conveying gas instead of air should be considered.

The generation of static electricity is to be expected at all parts of the system because of the movement of the dust suspension. Appropriate precautions should be taken, Chapter 7, and bonding of all units to earth is essential. The use of electrically non-conducting flexible links in the pneumatic system is undesirable, even if these are bridged by bonding, and flexible metal parts are preferable.

At the feed point the occurrence of a dust explosion is unlikely, because of the high concentration of dust in the suspension, but if the dust is burning before introduction to the system, then it is likely to continue to do so as it moves away from the feed point.

The pipe system conveying the dust is customarily strong, and may be capable of withstanding the full explosion pressure, and providing it is non-combustible, is then unlikely to be damaged by fire or explosion occurring in any part of the system. When running under the optimum conditions, the concentration of dust in suspension is high and explosion is not then expected, but burning dust may continue to burn during transit. The most hazardous periods are at start-up and shut-down, when the dust concentration may not be optimal, or at any other time when the supply of dust decreases due to bridging in the hopper, the hopper becoming empty, or failure of supply of dust to the system. The hazard should be minimized by reducing the period for which the dust suspension is not at optimum concentration. On start-up, the air supply should be established, in the clean system, and the dust then fed in at its optimum rate as rapidly as possible. On shut-down, the dust supply should be stopped quickly and air continue to run until all dust has been removed from the pipe system. A serious hazard could arise if the dust is not completely removed before shut-down of the air and if it is burning. On subsequent start-up, ignited dust could be thrown into suspension at less than optimum dust working concentration and explosion could ensue. This may not damage the pipe system, because of its mechanical strength, but could cause serious explosion in the separating and collection units.

Sight glasses are generally not required in pneumatic pipe systems, but if they are installed should be of equal mechanical strength to the pipe to which they are joined, and should be able to withstand the heat from burning dust. The sight glass must not interrupt the electrical bonding of the pneumatic system.

On arrival in the delivery area the dust must be separated from the conveying air and the concentration of dust in suspension inevitably varies, and will be within the explosible range at some points. It is at the separation stage that the dust explosion hazard is most serious in the low volume/high pressure air systems. The separation may be made either by delivering the suspension into a large volume, e.g. a silo, and using it as a settlement chamber. The air escapes via a filter unit, direct to atmosphere and the dust accumulates at the bottom of the silo. Alternatively, the dust suspension may be fed into cyclone units, probably followed by bag filter units. Whichever method of separation is used, the units must be protected against the effect of dust explosion, usually with relief venting, Chapter 8. The opening of an explosion relief vent may cause the flow of conveying air to be reduced, or may allow the discharge of the dust suspension. More details and consideration of the protection of dust collection units are given below. Where cyclones or filter units are used, the dust is delivered subsequently into storage bins or hoppers, whereas with the settlement chamber the dust accumulates in the separating unit. Apart from the explosion hazards, there is also a fire risk because of the residence of the dust over a period permitting the burning to spread throughout a mass and also because of the disturbance of the air associating with the pneumatic system. A small fire is thereby likely to be fanned into a large fire relatively quickly. It is advisable for the dust storage units to be of non-combustible construction, although plastics materials are beginning to be used.

If an explosion should occur in the separating units, the dust stored in the unit can be expected to ignite and the whole system should be subsequently emptied to remove all traces of burning material.

Delivery of the dust from the separation and storage units presents problems that are common to all dust delivery processes and will be considered in further detail below. At this point it is emphasized that the delivery of burning dust from hoppers, silos etc., is extremely hazardous because such dust is likely to be thrown into suspension during the loading and operatives are likely to be in close attendance. The dispersion of the burning dust can cause the sudden production of volumes of flame and serious injury to operatives is likely.

The low volume/high pressure pneumatic conveying system produces dust explosion and fire hazards related to those involved in the fluidization of powders. Fluidization is used in the conveying, storage, and mixing of powders and involves the introduction of air, or other gas, at the base of settled dust in plants thereby causing the powder to acquire some of the properties of a fluid. As with the low volume/high pressure pneumatic system the concentration of dust in the fluidized layer is high and in most cases would be outside the explosible range. Above the surface of the fluidized layer the dust concentration decreases sharply and there is likely to be a zone within the explosible concentration range. In addition finer particles are more likely to be carried above the fluidized layer than coarser materials, and so the explosible zone can contain a greater proportion of fine particles than the average. Where the fluidized material is delivered into a storage vessel or pipeline the concentration of dust in suspension will change, usually being reduced, so that larger volumes of dispersion whose concentration is within the explosible range are to be expected. The explosion and fire hazards associated with fluidization are particularly important in fluidized bed dryers, because of the presence of a heating source. These dryers will be considered below.

The precautions to be taken against explosion and fire in fluidized bed units are basically similar to those for the low volume/high pressure pneumatic systems. It is of equal importance that sources of ignition, or burning dust, should be prevented from entering the system, and the generation of static electricity is to be expected and appropriate precautions taken.

High volume/low pressure pneumatic systems have been used frequently for the conveying of dust between plant units and also for the removal of by-product dust from its point of manufacture. Conveying within the plant may be exemplified by the use of the air entrained by a grinding machine to blow the reduced material into a collector. In this situation no fan is required because the machine generates sufficient air movement by design.

Where the system is used for the removal of dust generated by cutting or polishing machines, or for exhaust ventilation generally, a fan is necessary to move air and this should be of good design to ensure that the air velocity is sufficient to prevent deposition of dust in the exhaust ducting. The following general points should be considered in the design of such a system; these points are additional to any others which may be required specifically for the process.

1. The average dust concentration in suspension should be less than 30 per cent of the minimum explosible concentration, as established by tests, Chapter 3. This requirement is only a partial protection because of variations in the rate at which the dust can enter the system, the possibility of dust being temporarily deposited and then redispersed, uncontrolled variation in the particle size and other properties of the dust which would effect its minimum explosible

concentration. The requirement does however ensure that on average the dust concentration would not be in the explosible range.

2. Provision should be made in the design of ducting to facilitate regular cleaning. If dust or tarry deposits are permitted to accumulate in the ducting, they will form a highly flammable layer which, if ignited, would allow flame to spread rapidly because of the air flow. Thin metal ducting of the usual design has little fire resistance and flame is liable to break out of the ducting and ignite nearby plant and the roof of the building.
3. Where the design requires that the extraction system should be served by a centralized dust separation unit, which should be outside the building, long runs of ducting are likely. The ducting should either be of sufficient strength to withstand explosion or be provided with explosion protection, either relief venting or automatic suppression equipment, Chapter 8. The simplest solution may be to provide strong ducting, of circular section, preferably welded seams, and with flanged joints. Relief vents should discharge to outside atmosphere, not into work rooms, and their fitting should not promote dust deposition within the ducting. It is good practice to avoid a large number of ducting systems being serviced by a few dust separator units. If an ignition should occur in one of the separators, the explosion could propagate along all ducting attached to the unit, and blow back flame as far as the dust entry points. The number of ducts attached to a single separator unit should therefore be minimized. In the same way, the number of branches in the ducting within the building should be minimized because explosion originated in one branch can be transmitted back into others. Although the provision of separate exhaust and collection systems for individual machines or groups of machines entails a higher initial capital expenditure than with a common system, advantage is obtained should an explosion occur in one of the systems, in that the companion systems can continue to run after the incident and a total shut-down of production is avoided.
4. Consideration should be given to the provision of an individual dust collection unit for each machine generating dust. Although unit collectors would be installed within the workroom, and fairly close to operatives, the quantity of dust in suspension is likely to be much less than with a common extraction system and any

ignition is localized to the machine producing it. The local system is necessary for severely explosible dusts, such as the light metals, where the manufacturing process may need to be in an isolated cubicle, and long runs of ducting are not desirable. Unit collectors are restricted as to the quantity of dust handled, and are mainly suitable for conditions of light loading.

5. Where ducting runs to central dust separators the fan is a potential ignition source if it is in the dust stream, Chapter 7, and explosion protection must be provided either by relief venting or by automatic suppression, Chapter 8. Placing the fan at this part of the system permits the dust separators to be emptied easily because they will be at slightly above atmospheric pressure. If the fan is in the clean air downstream of the separators, air interlock valves are necessary to enable dust from the separators to be removed whilst the plant is running. Emptying of the separators to ensure that they do not become overfilled is essential.
6. In branched ducting, dampers may be provided to balance the system. They should be correctly adjusted and then locked in position, otherwise dust may deposit in the ducting.
7. The design of an extended ducting system should cater for the possibility of a fire or explosion which, if adequate protection is given should not lead to disruption of the ducting, but which may produce deposits of burning material. Facilities should be provided for inspecting the ducting throughout its length, and for enabling burning material to be readily removed after an incident. On no account should the system be restarted after a fire or explosion until all residual burning material is known to have been removed. The extinguishing of fires in deposits in ducting, with inadequate access, can be difficult. Even if the fire is detected quickly, and the extraction fan is shut off, air will continue to move for possibly several minutes unless stopped by a damper. This air movement would intensify the fire. It also removes extinguishing agents injected at the entry to the ducting, or at points along its length, so that vaporizing liquid or dry powder agents may not have sufficient residence time to extinguish burning deposits. Injection of water is difficult in long and complicated duct systems, and may not be permissible if the dust can react with it, Chapter 9. Injection points for extinguishing agents can be useful in long ducting. In the event of an inaccessible fire developing, extinguishing may only be

possible by breaking into the ducting and, as this is frequently running overhead, damage to underlying plant is probable.

8. Where the extraction system is taking off waste dust from a machine manned by an operative, attention should be given to the human element. If smoking is forbidden on the premises, under penalty, the extraction system forms a convenient method of disposing of contraband smoking materials. As a cigarette butt can continue to smoulder for several minutes after being discarded, and a partly smoked cigarette would last longer, smoking materials introduced into the exhaust system are likely to be still burning on arrival at the dust collector. A dust fire can then ensue which, because of the air flow within the collector, can rapidly increase. It may result in an explosion. The characteristics of smouldering fires have already been described, Chapter 9.

The high volume/low pressure air systems can allow the propagation of explosion through them and explosion protection is likely to be needed, e.g. by relief venting or automatic suppression, Chapter 8. If venting is used the appropriate vent ratio is essential, ensuring that vents are situated near any bends, junctions, or fans in the system. Where the ducting is open to atmosphere, as when attached to a collection hood, the end of the ducting can be regarded as equivalent to a vent. Where the ducting enters a dust collector, a vent should be installed near the point of entry, either on the ducting or in the wall of the collector. Such vents should be designed so that they do not discharge directly into a work room nor, although they discharge into the open air, should they be liable to cause injury to operatives or passers-by.

Separation and collection

TYPES OF UNIT

When a dust has been conveyed pneumatically it must usually be taken from suspension in order that further operations on it can be carried out, e.g. storage, sieving, blending, "bagging-off". The type of dust separator installed will usually be decided by the functional requirements of the system, the relevant considerations including type of dust, quantity, particle size, subsequent plant operations and the overall legal requirements. Apart from legal requirements, e.g. those for magnesium, some flexibility in the design of plant in relation to explosion hazards may be possible, particularly if a choice of separator and collector is feasible.

In some cases the choice of plant is severely restricted by functional needs, and explosion protection has to be considered for all of the common types of separator and collector. These include the following: settling chambers (including silos), cyclones, bag filter units and wet scrubbers. In the majority of applications adequate efficiency of separation is obtained by the use of a single separator or two in combination, e.g. a cyclone and bag filter unit. To minimize explosion and fire hazards the design of dust separator and collector should be on the basis that the volume of dust suspension within the unit is minimal, ignition sources or processes liable to give ignition are absent, the dust is rendered non-hazardous or removed from the unit rapidly, and the amount of dust passing through the unit without being collected is low. In contrast, for mechanical and other reasons, the volume of the unit may have to be large in order to cope with a high throughput of dust, and retention of the dust within the unit may be an essential part of the process, e.g. a silo. Recirculation of air which has passed through the separator, back to the inlet of the conveying system, is not recommended because of the possibility of a gradual accumulation of fine dust which is not separated.

In addition to the explosion and fire protection of dust separation and collection units, careful consideration must be given to the siting of those units for which explosion protection is required. Particularly where relief venting is used, Chapter 8, the separator and collector units should be situated where they should not cause injury or damage due to discharge of flame from vents. Where the separator must be in the workroom, automatic explosion suppression should be considered, Chapter 8, and in particularly difficult cases a rebuilding of the plant may be necessary in order to obtain a satisfactory arrangement.

SETTLING CHAMBERS

The principle of the settling chamber, as a method of dust separation and collection, is regularly used with low volume/high pressure pneumatic systems where the delivery from the conveying pipe is directly into the silo. The air passes from the silo to atmosphere via a subsidiary separation unit, often a bag filter, which empties collected dust directly back into the silo. In such a system both the silo and the subsidiary separator should be protected with explosion relief, Chapter

8, particularly as the free volume within the silo is likely to contain a dust suspension. A more refined method is for the conveying system to discharge into a cyclone, from which the separated dust is discharged into the silo whose atmosphere is relatively undisturbed. The air from the cyclone is likely to contain residual dust which is removed by a bag filter unit also on top of the silo. The cyclone, bag filter unit, and silo should all be provided with explosion relief. Air displaced from the silo by the entry of the dust should pass through the filter before discharging to atmosphere. The volume of dust suspension in the silo can be further reduced by arranging that the dust from the cyclone does not discharge directly under gravity from roof level but can pass into the silo from a vertical duct within the silo having an aperture along its length enabling dust to move sideways into the silo.

The simplest form of settling chamber consists of an enclosure having a cross-sectional area much greater than that of attached ducting through which the dust suspension flows, and the dust falls out of suspension within the chamber, under gravity. Such chambers should be restricted in use, when explosible dusts are being transported, because of the relatively large volume of suspension present within the chamber, its relative inefficiency for dusts finer than 70 microns and because alternative separators, e.g. a simple cyclone, would be more compact. During use dust collects within the settling chamber, and there is an airflow through it, so that the possibility exists of redispersion of dust. If a settling chamber must be used, it should be provided with explosion protection, probably relief venting, and should be situated where an explosion in it would not cause a hazard. Other types of collector and separator are more widely used, and are to be preferred.

CYCLONES

The simplest settling chamber relies on gravity or momentum to remove the particles from suspension, and rapidly diminishes in efficiency as particle sizes are reduced. A more efficient method is to use centrifugal force for the separation, and the cyclone has been designed for this purpose. The geometry of a cyclone should be designed in relation to the characteristics of the dust to be separated, the flow rate, and the efficiency required. Where the dust to be separated contains a wide range of particle sizes, either multiple cyclones may be required, taking progressively finer fractions of the suspension, or one cyclone taking the majority of the suspension followed by a further unit, such as a bag filter, to remove the residual fines. In any event, the geometry of the cyclone will be conditioned by normal working conditions, and explosion protection will need incorporating when the basic requirements have been met.

When a dust suspension is fed continuously into a cyclone, the separated dust should flow continuously to the base, and be removed forthwith. Any deposit of dust in the base of the cyclone is liable to be entrained by the cleaned air passing through the vortex pipe and the efficiency of the cyclone is greatly reduced. As the separated dust must be removed continuously from the cyclone, safe and efficient means must be provided. A valve is usually necessary at the base of the cyclone, and a rotary choke is often employed for this purpose, (Chapter 8). Where the cyclone is at pressure below atmospheric, because of a fan downstream drawing air through it, an air valve must also be installed at the base of the cyclone to prevent ingress of air. The rotary choke valve may discharge into a hopper, or at a bagging off point, or to a conveyor. For enclosed units, such as a hopper, explosion protection should be provided, e.g. relief venting, and at bagging off points local exhaust ventilation is required to prevent dust accumulation. The starting up procedure for the plant should be such that the rotary valve is energized before dust begins to be separated in the cyclone, to prevent initial build up of dust within the cyclone.

In efficient operation the gas within the cyclone should be free of dust at the centre of the unit, and the dust concentration will increase steadily as the wall is approached, and at the wall dust will no longer be in suspension but moving down towards the base of the cyclone, for removal. An explosible concentration of dust must exist within a band, which may be narrow, probably near the cyclone wall, irrespective of the concentration of dust suspension in the inlet ducting. Any burning material entering the cyclone is liable to pass into the band having an explosible concentration, and may give an explosion (Plate 12). Although ignition of dust may originate at some other part of the plant, an explosion is particularly likely to develop in the cyclone. For example, material fed from a grinding machine through ducting, to a cyclone separator may be ignited in the grinding machine, but because of the high mechanical strength of the mill the effects of explosion may only be apparent in the cyclone. This is the origin of the belief that cyclones are particularly susceptible to explosion, and are inherently dangerous, whereas the true position is that the cyclone contains part of its volume as an explosible suspension and that, being structurally much weaker than a grinding machine, is more likely to be damaged by explosion.

Cyclones handling explosible dust must be given explosion protection. The two frequent methods are relief venting and automatic explosion suppression, Chapter 8. There is evidence from experiment that the area of venting appropriate to the vent ratio for the dust being handled, Chapter 8, gives adequate relief area, using the volume of the cyclone itself, without attached ducting, in the calculation. The vent is best placed near to the inlet ducting on the cyclone and must be dust tight even though the cyclone may be working at a positive pressure relative to the atmosphere. The vent cover should be robust because of the abrasive action of the suspended dust as it passes by but not unduly massive, Chapter 8. The vent can be installed on top of the cyclone, or on the top of the attached inlet ducting, to minimize the effects of abrasion. If there is little free area on the top of the cyclone, the only position for the vent may be the top of the vortex pipe. (Plate 13). Explosion pressures, with the vent in this position, are higher than when the vent is near the inlet, Chapter 8, and the length of vortex pipe also acts as a restriction to the free discharge of combustion products. The vortex pipe must be designed so that it has sufficient strength to avoid being collapsed by the pressure differential between the interior of the cyclone and the atmosphere in the event of an explosion, otherwise the venting area is reduced or lost. The design must also ensure that the flat top of the cyclone is capable of withstanding an explosion pressure of 14 kN/m^2 (2 lb/in^2) which is the pressure that can be expected during an explosion even with the recommended vent area.

Where a further separator is attached downstream of the cyclone, e.g. a bag filter unit or another cyclone, protection is not usually provided to prevent transmission of explosion. In any event, an explosion within the cyclone will disturb the suspension pattern and burning dust is very likely to be expelled through the vortex pipe into further ducting. Ignition and explosion in downstream units are therefore to be expected.

The cyclone contains no moving parts, apart possibly from an impellor at the top, and hence ignition of dust due to bearing friction etc., is not likely. The rapid movement of dust within the cyclone will generate static electricity, and appropriate precautions should be taken, Chapter 7, all metalwork being bonded to earth. It is essential that no hot work is carried out on the exterior of a cyclone which may contain dust in suspension or deposited, and safe routines for maintenance must be laid down.

The positioning of cyclone separators is important, because of the possibility of dust explosion. Wherever possible, cyclones should be installed outside buildings (it is frequently convenient to place them on the roof) and arranged so that discharge of combustion products through the vent cannot harm passers-by or neighbouring plant. The installation should be sufficiently accessible for maintenance to be regularly carried out, particularly to ensure the good fit of vent covers. Where a cyclone is installed within a building, vented combustion products must be discharged via ducting to the outside atmosphere, Chapter 8, and preferably only cyclones of small volume should be accepted. Automatic explosion suppression is useful for cyclones in buildings, where venting is not possible, but wherever a new design is being laid down the cyclone separators should be outdoors. Where the cyclone is within a workroom the efficiency of dust collection must be high, which may involve further separators being attached, to ensure that there is no leakage of dust into the building. The cleaned air should not be re-cycled through the plant.

Where the cyclone is installed outdoors, and is separating a waste product of little value, e.g. wood sawdust from woodworking machines, the cyclone often discharges directly to atmosphere, even though a certain amount of fine dust is released. When deposited in the open air, e.g. on the outer surface of the cyclone, such dust is unlikely to be a severe explosion hazard, but it represents a fire hazard because it is readily ignited. In conditions where the dust is not adequately dispersed by the wind, periodic removal of accumulation is advisable.

For most dusts periodic cleaning of the interior of a cyclone is unnecessary because the scouring action of the dust suspension on the walls prevents a build up of deposited material. Where the dust suspension also contains tarry material this may not apply, and deposited tar and dust should be removed regularly particularly from the underside of the roof and the exterior of the vortex pipe. Otherwise the deposit can cause a fire, the flames from which are likely to be blown from the cyclone down further ducting to any subsequent separator, thereby causing unnecessary damage.

Where cyclone and other separators are installed in sequence an explosion in one can propagate into neighbouring units because the pressure developed is sufficient to reverse the normal working air flow. Further, the ignition sources presented to the neighbouring units will be large, and this increases the explosion pressure, Chapter 8. Explosion in one separator can cause more violent explosions in neighbouring units either due to the presence of finer dust fractions or to the increased size of ignition source.

If an explosion or fire has occurred in a cyclone, the plant should not be restarted until all burning material has been removed. A detailed inspection of the interior of the unit is necessary.

FILTER UNITS

A dust suspension consisting of a range of particle sizes can be largely separated by a well-designed cyclone, or set of cyclones, but there is usually a serious drop in efficiency of collection with the fraction of small particle diameter. With many dusts, a cyclone system can be designed to be capable of collecting the majority of particles of diameters as small as 10 microns, but where the dust contains such fine material a subsequent collection unit may be required. A bag filter, or fabric filter unit is often provided downstream of the cyclone system to take out the remaining fine particles before discharging the air from the system. Where the dust suspension is known to consist only of fine particles, then a filter unit may be acceptable, without preceding cyclones.

In the filter unit the air carrying the dust passes through fabric, often in the form of a sleeve, and the dust is deposited on the fabric and the clean air passes out of the unit to atmosphere (Plate 24). As the dust accumulates on the fabric the resistance to flow increases, and the unit must be designed to accommodate the increasing resistance. Periodically, the dust must be removed from the fabric, and the design of the unit depends on whether the airflow through it can be temporarily stopped to allow for cleaning, or whether the plant process is such that continuous air cleaning must be maintained.

If continuity is essential, the most efficient method is to have multiple filter units which can be cleaned singly. The method requires that the filter unit shall be relatively large, and correspondingly expensive, but it is preferable to attempting to clean the system at the same time as further dust suspension is being introduced into the unit. Where the dust separation can be interrupted whilst the filter unit is cleaned, the problems are simplified and the installation is more economic.

The two principal methods of cleaning of filter units, as opposed to replacement, are mechanical shaking and reverse air flow. With mechanical shaking the dust is dislodged from the fabric and falls into a collection hopper at the base of the unit. During this action a dust suspension is inevitably formed within the filter unit and if any source of ignition is present an explosion is likely. The concentration of dust is not controlled in any way, and there is no relation to the average concentration of dusts supplied by the airflow to the unit during normal working. Because of the shaking action, the fabric is subjected to mechanical stress which can distort the apertures within it, which may facilitate the passage of deposited dust through the fabric to the clean air side and also lead to a long term weakening of the fabric. These disadvantages can be largely overcome by the substitution of reverse air cleaning, in which the fabric is not disturbed but the dust is dislodged by briefly blowing air from the clean side through the fabric. The stress on the fabric is considerably less than with the mechanical cleaning and enables inherently weaker materials to be used, which may be more effective in filtration. As with the other method of cleaning, a dust suspension of uncontrolled concentration is formed within the filter unit, which can lead to an explosion in the presence of an ignition source.

The most likely source of ignition within a filter unit is burning dust which has been conveyed to it from units upstream in the plant. When deposited on the fabric, burning particles are likely to come into contact with other unburnt dust and extended burning is likely. In addition, there is an airflow through the filter unit which accelerates burning, Chapter 9. Other possible sources of ignition are mechanical heat produced by the fabric cleaning operation, arising from a malfunction causing friction, and the unauthorized carrying out of hot work on the casing of the filter unit, Chapter 7. Dust is likely to be deposited on the interior of the casing, because of the lack of a positive scouring action during the working of the unit, and is a serious hazard if heat is applied externally to the casing.

The dust on arrival at the fabric is likely to carry a static electricity charge which can be transferred to the fabric. Attempts are sometimes made to remove static electricity by incorporating metal wire or thread in the fabric construction. The metal is connected to earth via the structure of the filter unit, all the metal components of which should be bonded to earth, Chapter 7. The presence of static electricity can interfere with the cleaning of the filter fabric, as well as cause an ignition hazard. Incorporation of conductors within the fabric will facilitate the flow of static electricity to earth, but may introduce other disadvantages particularly those of causing wear to neighbouring threads in the fabric and the possibility of the metal thread fracturing and giving unearthed metal within the structure of the fabric. This could spark to neighbouring earthed threads and, because the dust is in close proximity, could become hazardous. If undue wear of the fabric occurs, holes will develop and dust will pass through to the clean side of the filter. An alternative method of removal of static, when the filter fabric is in sleeve form, is to install an earthed expanded metal sleeve on the clean side of the fabric against which it rests lightly.

The casing of filter units is usually of sheet metal, and may involve spans of several metres. The unit is mechanically weak and should be provided with explosion protection, and also situated in a safe place. As with cyclones, the best position is outdoors, away from other plant and from easy access by personnel. Even if relief venting with the appropriate vent ratio is provided, Chapter 8, extra bracing may be necessary across large flat surfaces to enable a pressure of 2 lb/in^2 (14 kN/m^2) to be withstood safely. The relief venting should be distributed round the casing in such a way that all parts of the internal volume of the filter unit have direct access to a vent, without being obstructed by the presence of the fabric. If automatic explosion suppression is used, a number of suppressant heads may be necessary to ensure that all parts of the volume of the unit, including both sides of the fabric, can be easily reached.

Although adequate explosion protection is given to prevent damage to the basic structure of the filter unit, it is unlikely to be operational after an explosion because the filter fabric will probably be destroyed or holed. Before attempting to replace the fabric, all burning dust and residues must be removed from the unit and neighbouring units, because flame from a filter unit could propagate against the normal working flow and could ignite dust in neighbouring units. If a fire or explosion develops in a fabric filter unit, the air cleaning process must be shut down until the unit has been repaired. If the unit serves a number of machines this may mean a considerable loss in production. With this type of plant, consideration should be given to having several filter units so that, in the event of an explosion, interruption to production is minimized. Alternatively, unit dust collectors on each machine may be preferable, particularly where amounts of dust are modest (Plate 23).

A common type of unit collector consists of a short length of extraction ducting, not more than a few metres, from the point of production of dust to a box casing incorporating a fan. The dust suspension is drawn by the fan to filter sleeves within the unit and clean air is discharged back into the workroom. Large particles of dust fall directly into a dust collector at the bottom of the unit, and the finer particles are retained by the fabric. The fabric is cleaned either by manual mechanical action or automatically each time the unit is switched off. As the volume of a unit collector is usually less than 20 ft^3, explosion protection is usually not required, Chapter 8, but sound construction is essential to prevent leakage of dust into the workroom. As the length of extraction ducting and the volume of dust suspension within the collector are both minimized, the units have a reduced dust explosion hazard as compared with a centralized system with long extraction ducting runs.

A fabric filter may be installed directly on to a grinding machine, where the feed stock is supplied pneumatically to the grinder. The conveying air is discharged from the mill through fabric filters to atmosphere and if the mill is of sufficient mechanical strength, it may not be necessary to provide it with explosion protection, see above. The fabric unit is much weaker and is best protected by enclosing it in strong steel tubing which should be straight and discharge directly to atmosphere outside the workroom. If the dust explodes in the grinding machine the emitted flame will then pass to the atmosphere, but the fabric is likely to be destroyed. The plant must then be shut down until all burning material has been removed and the fabric replaced.

SCRUBBERS

With fabric filter units and cyclones explosible concentrations of dust can be expected at some points and because ignition sources cannot be entirely prevented, Chapter 7, the basic requirements for the development of an explosion are liable to be present. Numerous dust explosions have occurred in both types of collector; explosion protection and correct siting are necessary. Whereas these disadvantages are inherent in dry dust collectors, some can be eliminated by the use of wet collectors or scrubbers.

The principle of the wet scrubber is that the dust suspension is washed by a liquid, usually a water spray, and the dust is removed from suspension and collected as a sludge in the liquid, the cleaned air being discharged to atmosphere. The use of wet scrubbers is mandatory in the United Kingdom in the grinding of magnesium and its alloys, Chapter 1. The main use of wet scrubbers is to collect dust which is an unwanted by-product, such as from polishing and sanding operations, but they could also be used in cases where the production of small quantities of a wet product would be acceptable. However, wet scrubbers are basically for the collection of moderate or small throughputs of dust, often at low concentrations (Plates 25, 26).

Because of the absence of ignition sources and of considerable volumes of dust suspension, which would be capable of giving a serious dust explosion hazard, wet scrubbers can be installed in work rooms and, indeed, should be situated as near as possible to the point at which the dust suspension is generated to avoid the necessity for long runs of extraction ducting. The scrubbers should be installed so that the air and water flows are at their operational levels before any dust suspension is fed into the collecting system. The safety of the scrubber relies on the air and water flows being effective and a high standard of maintenance is required. Otherwise, if the collection is inefficient, dust may either settle out before being wetted, or may escape from the collector into the workroom.

The sludge should be removed from the collector at regular intervals, to avoid overloading, and in special cases such as when magnesium is present extra precautions are necessary because of its reaction with water. This reaction liberates hydrogen which is both flammable and easily ignited, but providing the designed air and water flows are maintained the risk of a hydrogen explosion may be neglected. When the scrubber is switched off, and ventilation either ceases or is much reduced, the magnesium sludge should be removed from the scrubber, otherwise hydrogen may accumulate during the shut down period. The hydrogen could cause a serious gas explosion during starting up operations, particularly if there is a fan in the air ventilation system.

Although the sludge is relatively safe when wet, it can become much more hazardous if it should subsequently be allowed to dry out. As the dust which formed the sludge is often of small particle diameter it would, after drying, give a serious fire hazard and if dispersed in air and ignited a very serious dust explosion hazard. The sludge removed from scrubbers should thus be disposed of in a controlled manner and not just dumped outdoors. It should be regarded as a hazardous industrial waste material and treated accordingly.

ELECTROSTATIC PRECIPITATORS

Where very high efficiency of separation is required, because of toxicity problems or air pollution or the large quantity of air discharged, an electrostatic precipitator as a final collector is sometimes used. These are particularly effective in trapping fine dusts at low concentrations, well below 0.01 g/l, but have disadvantages where the dust can give an explosion or fire hazard. The concentration of dust entering the precipitator is usually below the minimum explosible concentration, if the precipitator is to be run under its conditions of design, but the precipitated dust accumulates within the unit. Fires can develop in the precipitator either as a result of electrical discharges resulting from dust accumulations reducing clearances within the unit to below the voltage breakdown distance or because of particles ignited outside the unit being collected and involving further dust. The units are often cleaned by rapping mechanically and burning dust dispersed in this way could cause an explosion. As high efficiency collection can be obtained with other separators already mentioned, and as these types are in any case necessary to remove the majority of dusts before the airstream could reach an electrostatic precipitator, efficient collectors of types other than electrostatic precipitators are often preferred for hazardous dusts.

Dryers

GENERAL REQUIREMENTS

The drying of dust may be carried out at any stage in the manufacture of the product and should always be regarded as a potentially hazardous operation. Drying involves the application of heat to the dust which may be either a deposit, as in a tray dryer, or a suspension, as in a pneumatic dryer, or both, as in a rotary dryer. The hazard is compounded if the liquid to be dried off is flammable. The burning of the fuel generating the heat for the drying process is also potentially hazardous.

The basic requirements for the safe design and operation of dryers for dusts will be considered first, in general terms, following which brief consideration will be given to individual classes of dryer. The information must of necessity be in general terms, and for a specific installation more detail may well be required. Some of this can be obtained from earlier chapters, but other are outside the scope of the present work e.g. design of flame failure devices for the heat supply, explosion protection for solvent vapours, and detailed advice must be obtained elsewhere.

In selecting the type of dryer for a particular plant the choice may be restricted by factors connected with the economic running of the plant, rather than from the aspect of dust explosions and fires. Where options are open the selection will be based on the properties of the dust, particularly its minimum ignition temperature, the volume of the dust suspension within the plant, the amount of dust deposited within the plant at any instant, and the design of the heat supply.

If the liquid being dried off is flammable, the principal hazard of the process is that of a vapour explosion. Detailed discussion of the appropriate protection should be sought elsewhere, but, in brief, the primary protection is to supply sufficient ventilation to ensure that the vapour concentration does not exceed 30 per cent of the lower flammability limit concentration of the vapour measured at the working temperature of the dryer. In addition, the plant should be provided with explosion protection, either relief venting or automatic suppression. The vapour/air mixture should preferably not be recirculated through the dryer but if partial recirculation is required for economic reasons then the mixture composition must be continuously and accurately monitored to ensure that the permissible concentration is not exceeded. In the event of a vapour/air explosion the dust is likely to be ignited, causing a fire, and if circumstances are favourable for the dispersion of dust by the vapour/air explosion, a dust explosion may also be generated.

Drying can take place in plant units other than dryers, simultaneously with another operation. An example is the use of hot air to dry a wet feed stock which is being milled, see above, and many of the hazards associated with dryers, particularly those with dust in suspension, also apply to the milling/drying operation. The explosion protection necessary for both milling and drying operations need to be blended together in such instances, to ensure a sound approach.

A fundamental requirement in the safe design and operation of dryers is that the plant should run with the optimum loading of wet material, which should be in such a form that it meets the specification for the plant. Failure may give rise to immediate fire and explosion hazards because of overheating, excessive residence time in the dryer, erratic feed and delivery, and uneven drying. In addition the process is likely to be of less than optimum efficiency, with poor economic return and lower quality product. If a dryer is to be used for a range of materials of differing physical properties and forms, it should be so designed from the beginning, in consultation with its manufacturer, that adequate control can be maintained over a range of required working conditions.

Explosions and fires in dryers are particularly to be expected during start-up, shut-down, and emergency operations. During these operations the process is not in equilibrium, and it is more likely that control by the operator will be lost. The difficulties arise particularly with dryers of continuous operation where the simultaneous introduction of wet feed stock and heated air is required. If the flow of feed stock is insufficient, the thermal load on the drying medium, air, is reduced and high temperatures will become established, causing ignition directly. If the flow of feed stock is too high, deposits of partly dried product may adhere to the internal surface of the dryer and have an extended residence time. The delay may enable layers to ignite spontaneously, Chapters 7 and 9, causing fire which can lead to explosion. Any material which has been inadvertently left in the dryer since its previous use would be dry, and the introduction of heated air can again lead to an immediate ignition in pneumatic dryers. Comparable hazards arise during the shut-down procedure, where again it is necessary to retain a balance between the heated air and wet feed stock input, to ensure that all product is removed from the dryer before complete shut-down, and to avoid the material being present on the next start-up operation.

The control of the process during start-up and shut-down is of great importance and if some emergency should arise during the procedure, the operator must be well trained to ensure that he can cope with the unexpected situation. Emergencies can arise because of either some failure of equipment of the plant, e.g. a thermostat, which could lead to local overheating, requiring immediate remedy, or to failure of delivery of feed stock into the dryer or of product out of it due to blockage or bridging of the material. These failures would result in a reduced load on the drying medium, requiring an immediate cut-back in the heat supply, and possibly to excessive residence time of the product within the dryer leading to spontaneous ignition. Attention must be given in the design of the dryer, particularly those of continuous working, to automatic temperature control, the balance between the flow of feed stock and drying medium, and the detection of holdups within the dryer.

In general, indirectly fired dryers are preferable to those with direct firing. With indirect firing gases from a combustion chamber, or an alternative heating medium, pass through a heat exchanger and the drying medium also passes through the exchanger but direct contact between the two streams is avoided. Hence flame and burning particles from the fuel in the combustion chamber cannot enter the dryer and directly ignite the dust. The dust must be prevented from entering and settling on the heat exchanger, and regular inspection is required to ensure that such accumulations do not occur and also that the heat exchanger has not developed leaks between the two circuits. With both indirect and direct firing it is essential that the combustion air should not contain flammable dust and the air intake should be designed to draw air from a non-hazardous area. The combustion chamber itself should preferably be in a non-hazardous area, but if this is not possible, it should be in a separate enclosure, of fire resisting construction, having at least 2 hours fire resistance. The combustion chamber should be provided with flame failure devices, if it is fired by gas or oil, and especially if the dryer is direct fired. Loss of ignition can lead to the discharge of unburnt fuel into the combustion chamber, and possible further through the plant, and the fuel/air mixture is likely eventually to become ignited either by nearly burners or heated surfaces, giving an explosion in the chamber capable of disrupting it.

After the dried products have been separated from the drying medium the latter may still contain dust in suspension, including an unrepresentative proportion of the finer fractions. The suspensions should not be discharged directly to atmosphere nor, where appropriate, recycled and should be passed through a dust separator. If it is desired to add the separated dust to the product this should be done at a point downstream of the dryer, and so that the fine dust, which is already dried, does not pass again through the dryer.

The dried product is frequently delivered to some form of collector, such as a hopper or silo, and where it is conveyed mechanically from the dryer there should be rotary choke valves, or equivalent (Chapter 8), to isolate the collector from the dryer. Additional explosion and fire protection may be necessary if the dryer or collector are in hazardous areas. Fires may still develop in the collector due either to the transmission of burning material from the dryer or to spontaneous ignition of the material in the collector. Delivery of burning material from the dryer may be minimized by the precautions already described, and other specific precautions for various dryers, but transmission of burning material may occur particularly if the products smoulders readily, Chapter 7. Warning of incipient trouble may sometimes be observed, with light coloured products, from the occurrence of black specks of material in the collector. Fires in the collector may also arise due to the presence of a relatively large bulk of heated dried products. Whereas the heated material in shallow layers, as in a tray dryer, or in suspension, as in a pneumatic dryer, can be designed to be at well below the ignition temperature for its geometry, this protection may be lost if the dried product is deposited in a collector of much larger volume. The ignition temperature decreases as the thickness of the dust deposit is increased, Chapter 9, and the ignition temperature may be reached in the collector whereas it is not reached in the dryer. The product, delivered to the collector, then heats up instead of cooling down. If conditions are favourable the heating can lead to ignition. Appropriate fire protection measures are either to allow the product to cool further before collection, or to reduce the thickness of the dust in the collector, or to periodically well stir the contents. In the event of fire the extinguishing techniques described in Chapter 9 are relevant.

The required design of the fire and explosion protection varies between different dryers and will be considered in relation to the individual classes of dryer. The types commonly used in industry may be classed as follows:

(1) Tray
(2) Drum
(3) Rotary, including rotary-louvre and screw conveyer
(4) Fluidized bed
(5) Pneumatic
(6) Spray
(7) Vacuum

Other types of dryer are not commonly used for dust and powders; these include dielectric and infra-red, but where these are to be used for specialist applications the fire and explosion hazards and methods of protection can be inferred without difficulty. Specialist advice might be necessary in particularly unusual cases.

TRAY DRYERS

The basic process is that the wet dust or powder is spread relatively thinly in trays which are then stacked in some form of oven. Air may be blown over the trays by means of a fan. Tray dryers involve batch drying, with a limited throughput and relatively high fuel and labour costs. They are used for the drying of relatively expensive or thermally sensitive materials such as certain chemicals, dyestuffs, etc, or for drying on pilot plant scale.

Providing the liquid being dried off is not flammable, and solvent vapour explosion cannot occur, the risk of a dust explosion is small because a suspension is unlikely to be formed unless there is a failure leading to considerable increase in the air circulation through the dryer.

Materials that are sufficiently explosive to initiate explosion in the form of a layer e.g. unstable chemicals, may require less ardous drying methods and expert advice should be sought.

Combustible dust in a tray dryer is likely to present a fire hazard if process design is at fault. Care should be taken to ensure that the dryer temperature is low enough to avoid ignition of the dust in the tray, at the thickness of deposit being dried. The residence time of the dust in the dryer must also be considered, and deposits, or spillages of dust onto the walls or floor of the dryer, must not accumulate. Information on safe drying temperatures, residence times, and deposit thickness can be obtained from tests and calculations based on thermal ignition theory, Chapter 9.

After the product is dried, it may require to be cooled before being placed in some form of collector of increased dimension compared with the thickness of the layer on the trays, see above.

In the event of fire in a tray dryer, the air circulation should be stopped immediately. If the dust requires air for its combustion the door of the dryer, and any other openings, should be closed and the flames are then likely to rapidly extinguish. Smouldering may however persist. Where fire fighting must be undertaken, because of the situation of the dryer, or of the need to extinguish smouldering combustion, or because the dust contains sufficient oxygen to support combustion, the method of attack, using a suitable fire extinguishing medium (Chapter 9), should be as gentle as possible to avoid disturbing dust from the trays. If the fire has been controlled by keeping the door of the dryer closed, care should be taken on subsequent opening of the door because of the likely presence of flammable vapours and residual burning. Personnel should not stand in front of the dryer in direct line, in case flame should be emitted.

DRUM DRYERS

When the material is in the form of a paste, or slurry, it can be dried by applying it to a revolving metal drum, on which it is dried by conduction. The product is then scraped from the drum after a fraction of a complete revolution and may be conveyed mechanically to a collector. Drum dryers give continuous working and have the advantage that the product is exposed to the drying temperature for a restricted, and controlled, time.

Apart from the hazard of solvent vapour explosion, if flammable liquid is used, the dust explosion risk is small. Ignition of dust on the dryer is unlikely unless very high drying temperatures are used, because of the short residence time. The possibility of dust overheating in the collector should be considered.

ROTARY DRYERS

This class has been taken to include rotary louvre and screw conveyer dryers as well as the simple rotary type. The basic design is that of a horizontal cylinder or trough, which is either stationary or rotates on its axis. In the latter case the cylinder may be tilted. The wet material is introduced at one end of the cylinder and passes through either because of the action of a screw conveyor or, if the cylinder rotates, due to the action of combined mechanical and gravitational forces. The drying medium, usually heated air, may be either co- or counter -current. Where applicable, the screw conveyer may also be heated. With the rotary louvre dryer the air flow is not axial but is introduced between louvres attached to the cylindrical shell which rotates on its axis, and the drying medium passes through the material being dried.

All types of dryer in the class can be used for continuous drying and, because of the throughputs possible, are suitable for large scale operations. At any instant most of the material being dried will be a deposit within the dryer but, because of the rotary motion imparted to it and the movement of the drying medium some dust will be in suspension in the dryer. The finer fractions of the dust, which may be produced during its manufacture or by abrasion due to the rotary motion of the dryer, are more likely to be in suspension giving an explosion hazard. The dust will remain in suspension as the drying medium passes from the dryer, and the medium cannot thus be discharged directly to atmosphere. The dust in suspension should be removed by a collector, such as a cyclone, and the collected material may be returned to the product, downstream of the dryer. Explosion protection, usually relief venting, is needed on the collector. Because of the large throughput with this class of dryer, explosion and fire protection should be installed in the conveying and collection units attached to the delivery side of the dryer. The information given above, under these headings, would apply.

A fire hazard can arise in the dryer itself due to material being overheated or present for an excessive residence time. This is more likely to occur if the material being dried is sticky, and other classes of dryer should be considered if the physical state of the material prevents its regular passage through the dryer. In the event of a fire in the rotary dryer the mechanical movement should be stopped, and the drying medium cut off. The fire is then likely to subside, but may continue smouldering, and can be dealt with accordingly. All burning or burned material should be removed from the plant before the drying process is resumed.

FLUIDIZED BED DRYERS

The material to be dried is fluidized and the heating medium passes upwards through the fluidized mass providing efficient heat and mass transfer. In this class of dryer the dust is in suspension. However, in the fluidized bed itself the concentration of dust is usually so high that it is above the upper explosible limit and propagation of explosion within the bed is not likely. If a fire should start in the fluidized bed it is likely to burn rapidly because of the good contact between the dust and air inherent in the design of the dryer. The motion of the dust during the fluidization promotes the generation of static electricity and all metal components of the dryer must be bonded to earth, Chapter 7.

The dust concentration above the fluidized bed decreases rapidly, and the dust in suspension is likely to contain a relatively high proportion of fines, either present in the original material or produced by abrasion during the fluidizing process. A dust explosion may propagate in the zone above the fluidized bed and explosion protection should be provided. In some dryers the dust is prevented from escaping by a filter installed horizontally over the bed, any particles escaping upwards being caught by the filter and eventually dropping back. If no filter is present the heating medium should be cleaned by passing it through a dust collection system provided with explosion protection. The dust so collected can be added to the dried product which should pass from the dryer via a rotary choke valve or worm conveyer, Chapter 8.

Relief venting is often provided on fluidized bed dryers for explosion protection, and, because it is relatively easy to install, the vent can be at least half the cross-sectional area of the dryer. In a single storey structure the vertical walls of the dryer can be continued upwards through the roof, so that any vented explosion products discharge directly to atmosphere.

In the event of explosion or fire the fluidizing medium should be turned off, and residual burning dealt with. Extinguishing of fire may be facilitated by installing a water spray system over the fluidized bed. All burning and burned material must be removed from the dryer and attached dust collection plant, before the process is restarted.

PNEUMATIC DRYERS

Pneumatic or flash dryers operate by drying the dusts in a heating medium, frequently air, in which it is simultaneously conveyed. The process is continuous and the dust is in suspension whilst it is passing through the dryer. Because of high heat and mass transfer rates the drying time is short. The dryers vary in complexity from the simple single pass of dust and heating medium to more complex systems where parts of the product and of the stream of heating medium are recycled. Efficient dust collection units must be provided for the dryer.

The need for recycling of product arises because of coarse particles requiring a longer period than finer particles, and so the partially dried material is separated and recycled, or because the feed stock is in a difficult state to handle and the mixing in of a proportion of dried material improves the handling characteristics. Recycling of the heating medium may be necessary because of fuel economy or, where direct heating is used, to maintain a reduced oxygen concentration in the gas.

Pneumatic dryers should be installed outside workrooms, as should the associated dust collection equipment, unless they are of sufficient strength to withstand the full explosion pressure. For weaker plant, explosion relief venting or automatic suppression is customarily used, Chapter 8, and chokes should be fitted at the feed delivery points of the dryer.

In the operation of pneumatic dryers, it is essential that both the distribution of dust and the gas temperature should be well controlled, both during steady running and during start-up and shut-down operations. In particular, the formation of dust deposits within the dryer must be prevented, by adequate gas velocities, the minimizing of horizontal runs of ducting, and control over particle size.

Where tarry materials may accumulate on the walls of the dryer, and could lead to a build up of dust deposit, they should be removed at frequent intervals. The dryer should be so designed that this operation can be carried out without the need for operatives to enter the plant; if smouldering material is dislodged during operations flames can be produced which are hazardous to life and which may lead to explosion.

The dryer should be fitted with thermostats, near the entry of the heating medium and also at downstream points, particularly if recycling of the dust or medium is involved, to cut off the supply of heat if safe temperatures are exceeded. With direct heating the precautions mentioned above for the control of oil or gas to the heater apply. The temperature at any point in the dryer will depend on the flow rate and entry temperature of the heating medium and also on the nature, flow rate and moisture content of the feed stock. As the heating medium is also a conveying gas, its flow rate cannot be decreased so much that dust falls out of suspension, see above under conveying, and if the supply of feed stock is reduced, due to bridging, or its moisture content is reduced because of variation in quality, both the flow and the temperature of the heating medium should be adjusted, otherwise overheating of the product may result.

Where the feed stock is a slurry or paste, or is otherwise difficult to disperse, mechanical force may require to be provided at the inlet to ensure that an adequate dispersion is formed. For difficult materials the feed stock entry may need to be through some form of centrifugal motion, e.g. a slinger, or even a disintegrator. The inlet can be with advantage in a vertical long leg of the dryer so that the material has the maximum time to disperse before meeting a horizontal surface on which it could deposit, and on which further deposits would form by subsequent impact of the suspension. If agglomerates of particles can enter, sufficiently large not to be carried by the flow of the heating medium, they will fall back against the flow and must be prevented from travelling back to the heater. Some form of trap, vertically underneath the inlet point, which can be cleared regularly is desirable. If difficult materials are introduced in a horizontal leg of ducting they may just fall and deposit under the inlet point and not be dispersed at all. Providing the feed stock is reasonably dispersed at the inlet, as drying proceeds the dispersion is likely to improve. Particular attention should still be paid to ensuring that the product cannot deposit at bends or other fittings in the dryer because once a deposit has formed it is unlikely to disperse itself again.

During start up and shut down operations, the conditions within the dryer are varying and satisfactory control requires adequate instrumentation for the operator, coupled with the maximum of automatic control. An adequate procedure may require the permanent presence of a substantial stock of wet feed, which is available for rapid addition to the system if the heating is increasing too rapidly. On the other hand the load of feed stock in the heating medium must not reduce the velocity of the heating medium too much, and the temperature should not fall below the dew point in any part of the dryer. Where part of the heating medium is recycled, having passed through the dust collection system, its velocity and humidity must be such as to avoid deposition of dust particularly of fine material. During shut down operations it is essential that all dust is cleared from the dryer; otherwise there will be a serious explosion risk on restarting.

The instrumentation of pneumatic dryers, particularly complex ones, requires much detail at the design stage. In addition the operators must be adequately trained to enable them to use the information provided by the instrumentation, and also to appreciate the hazards involved in the process.

If a dust explosion should occur in a pneumatic dryer, then either it will be completely contained, if the dryer is of sufficiently robust construction, or will be relieved to atmosphere or automatically suppressed, if the plant is of the usual weak construction. The dust separation and collection system should be similarly protected. After an explosion there may well be residual burning dust within the dryer, and the atmosphere is likely to be vitiated so that personnel should not enter. The explosion may also be transmitted to the collection system, although the presence of rotary chokes etc should minimize this hazard. Extinguishing a fire in the dryer and attached units, or dealing with the effects of an explosion, will require adequate access from the outside of the plant for the insertion of the fire extinguishing medium, Chapter 9. An emergency shut down procedure for the dryer must be laid down in advance, because to shut off all parts of the plant instantaneously may give additional hazard due to the trapping of combustible material in overheated parts of the plant. When the fire has been extinguished all burned material must be removed from the dryer and associated units before restarting the process.

SPRAY DRYERS

A rapid drying process is obtained by atomizing a solution or slurry containing the dust in a stream of heating medium which may be co- or counter-current. The drying is continuous and large throughputs of feed stock can be used. The dryer consists of an upper cylindrical portion attached to a lower inverted conical portion and the feed stock usually enters near the top of the dryer. In co-current drying, the more usual system, the heating medium enters near the atomizer and the particles of dried material collect at the base of the dryer, the apex of the cone, from where they are conveyed to other units. In this type of equipment the heating medium at its highest temperature meets the feed stock with the initial moisture content. In counter-current drying the heating medium is introduced towards the base of the dryer so that the final stages of drying of the product proceed rapidly and the earlier stages relatively more slowly.

With both forms of spray dryer the inlet temperature of the heating medium will be part of the initial design, depending on factors such as the geometry of the dryer, the throughput of medium and the water content of the feed stock. The temperature is commonly in the range 200–300°C for combustible materials. The outlet temperature will be designed to be as low as possible, in the interests of fuel economy, but ensuring that the temperature does not fall below the dew point and is not sufficiently high to cause deterioration of the product. The outlet temperature is commonly in the range 80–100°C.

Although the majority of the product is designed to fall to the base of the dryer for collection, the finer particles can be expected to accompany the heating medium as it is discharged from the dryer and a separation unit, such as a cyclone, would be required. Product separated may be returned to the line downstream of the dryer.

Rotary choke valves, or worm conveyers, should be fitted to the outlet for product from the spray dryer and also to the cyclone, and any hopper beneath it, to ensure that explosion in the dryer or cyclone will not be transmitted directly into the product line.

If the liquid being dried off is a flammable solvent, and not water, protection against vapour explosion must be installed. When the dryer is for the removal of water, sufficient steam may be produced to reduce appreciably the oxygen concentration within the dryer, but not necessarily sufficiently to prevent a dust explosion. On the other hand the dust suspension is in an environment well above ambient temperature, so that explosion hazards are greater. Protection against dust explosion must be installed and the dryer should preferably not be operated in a work room. Similar considerations apply to the product collection and dust separation units attached to the dryer. Relief venting is frequently used for explosion protection, with the vents on the roof of the dryer. The appropriate vent ratio for the dust and the volume of the dryer is advisable, Chapter 8, although if sufficient roof area is not available the structure should be strengthened to withstand higher explosion pressures, for estimating which Equation 8.2 may be used. The unquantified safety factors involved are the reduction of oxygen concentration, the unlikelihood of an explosible dust concentration being present throughout the entire volume of the dryer, and, particularly in dryers with a spiral path for the drying material, the considerable reduction of quantity of dust in suspension in the conical part of the dryer. If the dryer is outdoors the vents may discharge directly upwards to atmosphere; if the unit is indoors, in a single storey building, the ducts from the vents can be continued upwards through the roof giving direct discharge to atmosphere.

Explosion may be caused in a spray dryer either by a source of ignition entering from outside, e.g. with direct heating, a passage of a spark from the furnace in the heating medium, or by ignition within the dryer itself. Although overheating or sparks due to mechanical or electrical causes is unlikely, except bearing failure of a rotating disc atomizer, the walls of the dryer and the gas included by them are heated. Ignition can occur if dried material accumulates on the wall of the dryer and a sufficient thickness is present for a period long enough for spontaneous heating to develop. The usual sequence is that some of the atomized feed stock is flung against the wall of the dryer, at about the level of the atomizer, or even on to the roof above it and a layer of dried product builds up. The hazard is increased if deposits can form above the level of the atomizer, in a region where the temperature and oxygen concentration of the incoming air have not been reduced by steam from the drying process. If the dryer temperature and the time interval are favourable the layer can ignite and smoulder. The combustion may then either convert to flaming, Chapter 9, or the adhesion between the deposit and the wall of the dryer may be reduced so that the burning material falls off. It can then disperse in the turbulent heating medium and ignite the product suspension. Information on whether the ignition of deposits on the wall of the dryer is likely, and of the time required for ignition, may be obtained from small scale laboratory experiments, as described in Chapter 9. Where hazardous deposits can form, remedial action is essential. This consists of redesigning the atomizing process to prevent throw of the feed stock, adjustment of the flow of the heating medium, and regular cleaning of the dryer. Installation of an automatic cleaner which continuously scrapes the wall of the dryer is not recommended with combustible products, because of the friction inherent, but jets of heating medium or air may be effective. Alternatively the drying process should be stopped at regular intervals to enable direct cleaning to be undertaken; this may be necessary in any case for foodstuffs and other products on hygienic grounds. The ignition of deposits within spray dryers is known to have caused fire and explosion.

The product delivered from the dryer is hot, and care should be taken to ensure that if delivered to a collector or bin the dimensions are sufficiently small to prevent spontaneous heating, see above. Where conveyance is pneumatic, some cooling of the product will be obtained in the pipeline and attached collector unit. This unit, cyclone or filter, and hopper, should be provided with explosion protection.

In the event of fire or explosion in a spray dryer and associated separation units, the dryer should be shut down quickly as possible and any residual burning extinguished, Chapter 9. The process should not be restarted until all burned material has been removed.

VACUUM DRYERS

The use of vacuum techniques enables drying to be carried out at a lower temperature than that required at atmospheric pressure and is particularly useful for products which deteriorate at high temperatures or are chemically unstable. The dryers are of tray or rotary type, or may be freeze dryers in which the feed stock is frozen and the water removed from it by sublimation. In the latter type the feed stock is not exposed to elevated temperature. The tray and rotary vacuum dryers principally present fire hazards, as with corresponding types working at atmospheric pressure, but to a lesser extent because of the generally lower temperatures involved. The risk of disturbing the dust by a sudden release of the vacuum has to be overcome for efficient normal working, and so the risk should not enhance dust explosion hazards.

Screening

The processes of screening, classifying, and mixing or blending, involve the generation of unwanted suspensions of dust in enclosed volumes, but without deliberate size reduction or application of external heating. Mechanical power must be supplied and can involve the generation of friction and impact, and also static electricity from the movement of the dust.

The plant should be enclosed and preferably operated under a slight negative pressure to prevent escape of dust into the workroom. Any necessary apertures in the casing should be provided with local exhaust ventilation. Vigorously explosible materials, such as light metals and other dusts having a maximum rate of pressure rise in the standard apparatus, Chapter 3, of more than 70 000 kN/m^2 s (10 000 lb/in^2 s) can with advantage be processed in isolated cubicles, in a similar way to grinding machines for these dusts, see above. Explosion protection should be given to collection bins attached to the screens.

Because of the vigorous movement of the dust in the machines, and the possible generation of static electricity, Chapter 7, all metal components must be bonded to earth. The screening surfaces, whether of gauze, perforated plate or bars, should also be earthed if metal, but residual charge is likely to persist on the dust. For dusts of low minimum ignition energy provision of anti-static footwear and clothing for operatives is desirable, Chapter 7.

In the event of explosion or fire all burned dust must be removed before the unit is restarted.

Because of the severe mechanical stresses to which the units are subjected they should be of strong construction and all components subjected to particularly severe stress, such as bearings, should be outside the unit and kept dust free, to prevent ignition of dust in the event of mechanical failure.

Storage and disposal

In most plants the dust will terminate in some form of storage bin before removal. Storage vessels, ranging from small bins to large silos, should be of fire resisting construction and, if large, situated outdoors. The storage unit should be enclosed to prevent escape of dust, including the period when dust is being put into the unit and dusty air is thereby displaced. The exhaust should be cleaned before discharge to atmosphere. Consideration has been given above to methods of minimizing the generation of dust suspensions in silos, see under separation and collection units.

Explosion protection should be provided on storage vessels and is usually explosion relief venting or automatic suppression. The appropriate vent ratios for the dust and the volume of the vessel should be used, Chapter 8.

If electric lighting is required, to obtain visibility within the storage unit, permanent installation should be provided external to the unit, Chapter 7. Portable lead lamps should not be used, but battery operated hand lamps suitable for the purpose are acceptable.

Where delivery from the storage unit is a batch process, e.g. at a bagging off point, spillage of dust and subsequent accumulations must be guarded against. This is particularly important because an operator may well be standing near the point. Local exhaust ventilation is advisable, as this ensures that any spillage is removed and dusty air from the bag or drum does not escape. Local exhaust ventilation may also be necessary at an automatic take off, e.g. at a weighing machine, where because of the greater quantities of dust likely to be processed stringent precautions against leakage and spillage are needed.

In some instances the dust is an unwanted by-product and is removed from storage for the purposes of burning. Where the removal is automatic and the dust is fed directly to a furnace, precautions are essential to prevent blow back from the furnace and spillage in the furnace room, see below.

Where smaller quantities of waste materials are involved, it may be handled batchwise and burned locally using manual feeding. It is hazardous to throw dust onto an open fire, or into an incinerator, by means of a shovel, so that a dust suspension is inevitably formed in the region of burning material. A flash of flame can result which can injure the operative, even though pressure effects are insufficient to cause any damage to structures. A more satisfactory procedure is to bag off the waste material, in paper sacks, and the filled sacks are then put onto the fire or into the incinerator without being emptied. The efficiency of combustion as a result of manual feeding is likely to be low where dusts are being handled, and consequently there may be a smoke emission problem. In such cases an automatic process is preferable. If the burning is done manually on an open fire, care should be given to its situation, because of the likely dispersal of burning particles by wind. Some form of simple incinerator is advisable, as this ensures that the burning is carried out in an approved place. It is good practice to ensure an adequate water supply for fire fighting if the burning should become too vigorous. Material to be burnt should be kept well separated from the material being burned, bearing in mind the likely dispersal of burning particles from the fire, as a result of wind.

Heating units

Dusts are used as fuels for boilers, furnaces etc. and may either be manufactured specifically for the purpose, as with pulverized coal, or may be by-product material for which disposal by burning is economic, e.g. wood waste. Brief consideration has already been given to the design of heating units for dryers.

With pulverized fuel, a principal hazard arises from loss of ignition of fuel on introduction to the furnace, because of instability in the flame, leading to the build up of a volume of unburnt dust which is subsequently ignited by some means. Rapid burning in the hot enclosure of the furnace can lead to a pressure rise sufficient to cause structural damage and the emission of flame. Protection is obtained by ensuring control of the dust feed and of its burning.

For this purpose it is necessary with a material like wood waste that an adequate stock of dust is available from a hopper or bin, from which it can be fed steadily to the burners by means of a worm or other choke conveyer. As the production of dust may fluctuate with time, direct passage to the burners is not advisable and the use of a hopper to maintain a steady flow at the optimum rate is desirable.

The hopper should be fitted with high and low level alarm indicators. Whether the dust is delivered directly to the burner, or added to another fuel, such as coal, some form of flame detector or sensor is necessary to ensure that if the flame should extinguish, even momentarily, the supply of fuel is automatically cut off. Oil or gas pilot flames, continuously burning, may be necessary if instability of the burner flame can develop, and the pilot flames should also be provided with detector or sensor.

The flame detector or sensor may be sensitive either to the ultra violet of the flame, or to the infra red in which case it usually senses the flickering of the flame, thereby distinguishing the infra red radiation emitted from the walls of the furnace. As electric sparks generate ultra violet radiation, a suitable control circuit is necessary where automatic spark ignition is used with ultra violet sensors, or the sensors should be installed so that they do not receive the ignition signal either directly or by reflection. As the safety of the burning is dependent on the sensing mechanism it is common practice to employ both types of detector in a furnace.

In the event of uneven temperature distribution developing in the furnace due to partial malfunction, there is a possibility that partly burned dust may escape from the flame zone and be cooled so that combustion is not completed. Such material can form a pocket in the furnace which could re-ignite explosively. To deal with this hazard a thermocouple may be installed at the entrance to the flue so that the reduced temperatures developing from the partial malfunction can be detected and the supply of fuel stopped.

In the event of a shut down due to some malfunction being sensed by the flame detectors, the controls of the boiler or furnace should be so designed that the complete starting up procedure must be gone through to restart the combustion. Short cuts by introducing extraneous sources of ignition should not be attempted. Access doors to the furnace should be so controlled, e.g. by interlocking, that they cannot be opened whilst fuel is being blown into the furnace.

Secondary combustion air must be drawn from a dust-free area, otherwise ignition of the dust suspended in the air by the furnace can cause a blow back. The air should preferably be drawn outside the building housing the furnace, and away from the regions where dust is manufactured or stored or the air supply can be interfered with. The safety of the combustion process depends upon adequate air being delivered to the furnace, so that in the event of a complete or partial failure of the combustion air flow, the fuel supply to the boiler or furnace should automatically shut down. Failure to provide adequate air may lead to loss of ignition or to inefficient combustion and the possibility of explosion in the furnace or the flue.

Where dust which is a waste material is being burned, the combustion should be as efficient as possible, even though the economic advantage may be minimal. Failure to achieve complete combustion may result in the emission of unacceptable quantities of smoke, and of burning particles from the flue and chimney which may be carried by the wind and lead to fire in surrounding buildings and materials. A spark guard on the chimney is unlikely to afford complete protection.

Pulverized fuel furnaces, burning coal dust, are a specialized risk. A standard arrangement is for bulk coal to be reduced in a grinding mill, which may be heated to dry the coal, and the resulting dust is conveyed pneumatically to burners in the furnace where it is supplied with secondary air for the combustion process. The fire and explosion hazards of the grinding and conveying processes have been considered above, and they may be particularly severe because of the large scale of the plant. For instance, mills delivering 1 ton/min of pulverized coal are currently in use. Ignition leading to explosion may arise either because of friction or overheating in the grinding mill, see above, or as the result of a flashback from the burner which reverses the conveying air stream. Protection of such installations is highly specialized and involves both adequate temperature, air flow, and dust feed monitoring equipment as well as use of flame sensors. The starting up, running, and closing down procedures must be rigorously specified and so controlled that difficulties which can arise because of irregularity in the fuel supply, or variation in composition, can be safely handled. Further references are given in Chapter 11.

Furnaces or boilers should preferably be in a separate building from manufacturing or storage processes, or inside a fire resistant enclosure. Collection and storage units should be outside the building or enclosure, and the plant and its surroundings kept free of dust deposits.

Spraying processes

Industrial operations in which explosible dusts are dispersed in the presence of an ignition source should be given particular attention. The spraying of metals is one such operation, and involves the feeding of a metal such as zinc, in the form of a wire, through a gun on which a premixed gas flame is stabilized, e.g. propane / air, and the molten metal is dispersed and transported by a blast of compressed air to the work being coated. Provided the gun is operated as designed the process can be carried out safely and in a controlled manner. Some of the sprayed metal is not collected on the work, becomes overspray and cools to a fine powder. Unless constantly removed the powder deposits are an explosion and fire hazard. Adequate exhaust ventilation should be provided, with explosion protection for any dust collectors, and the process should be situated so that it is not hazardous to operatives or neighbouring plant.

The electrostatic spraying of powders, often plastics, is another operation requiring careful design. The powder is sprayed from a gun at a high voltage, up to 100 kV, towards the work which is electrically earthed. The electrostatic forces ensure that the amount of overspray is minimal, but an exhaust system with dust collector is necessary. The gun should be so designed that, in normal operation, the powder spray is not ignited if the gun is brought close to an earthed object and even if a spark passes. The dust collector should be given explosion protection because dust passing into it is likely to be electrically charged, and all metal components of the collector should be earthed.

Buildings

CONSTRUCTION

Dust handling plant should ideally be situated in separate open structures, so that if a fire or explosion occurs the flame and pressure can be readily dispersed, and escape of operatives is simplified. The structures may consist only of a roof or, for increased protection against weather, may be supplied with walls generously fitted with louvres. The design of such structures presents few problems, particularly if the plant can be operated from ground level so that means of escape in case of fire or explosion are readily incorporated.

Where circumstances do not permit this approach in full, partial application can be made by installing plant units that are structurally weak and are liable to undergo explosion, e.g. dust collectors, in the open type structure, with the remainder of the plant units in a conventional industrial building. If the risk is divided in this way, the plant units in the open structure can be at ground level or can be situated on the roof of the main building, in which case only a minimum of weather protection for the units may be necessary. This approach is commonly used for dust collectors protected by explosion relief vents where, in the event of an explosion, substantial volumes of flame and burning material may be ejected and should be contained by any surrounding structure to a minimum extent. Fire and explosion damage to the structure would result if it hindered the dispersion of combustion products ejected from the plant. The design of such structures again requires little comment.

For dusts of severe explosibility, such as light metals, a separate building for the entire plant may be the optimum solution in terms of safety and economics. As far as possible the plant should run automatically and should interlock with the door so that it cannot be started whilst operatives are within the building. The building should be of frame construction, steel or concrete, clad with lightweight panels, such as corrugated metal or asbestos. The roof should also be of lightweight construction. If brickwork is required for the lower parts of the walls, this should not exceed half the total height and should be at least 14 in thick, with reinforcing. The attachment of the lightweight members to the framework should be sufficient to prevent detachment of the panels by wind pressures. The maximum wind pressure depends on local climatic conditions, but often a suction pressure of 0.5 lb/in^2 on the walls and the roof, may be taken. A building from which lightweight panels are designed to blow off in the event of an explosion should be situated so that flying debris cannot cause damage to neighbouring buildings or passers-by, particularly if the lightweight material is liable to fragment. Metal sheeting can be controlled by retaining chains and would be preferable in more congested situations. The door and its surrounding framework should be strong and attached to the building frame so that it does not become detached in an explosion. Severe injury can be caused by a door which is suddenly blown open.

Where a hazardous operation has to be carried out in the proximity of other operations, segregation in a strong cubicle may be desirable, see above under grinding units. The wall containing the door, and the attached side walls, should be strongly constructed, e.g. of 14 inch reinforced brickwork, or equivalent and the fourth wall and roof of light construction, such as corrugated metal or asbestos. As described special care should be taken in the design of the door and in the interlock system to prevent operatives being in the cubicle whilst the machine is running. The area of ground in the vicinity of the lightweight wall should be fenced off, to prevent injury to passers-by in the event of the panel being displaced, and the building of a simple shield wall would give some protection against flying debris.

The most difficult cases are where dust handling plant must of necessity be inside a building and cannot all be vented directly to outside atmosphere or provided with alternative explosion protection. Some discharge of flame into the building may then be possible, but it must of course be arranged that operatives are not put directly at risk to the flames. To prevent damage to the building structure, explosion relief may be necessary and can either be a lightweight roof or wall panel construction, or suitably designed windows. Windows which open to give the venting are preferable to those which would need to break in order to relieve the pressure because of the considerable variation in the strength of glazing (Plate 27). The windows should open outwards and it is convenient to hinge them at the bottom and to provide a retaining chain so that they do not become missiles. To reduce the mass of the window, transparent plastic can be substituted for glass. The windows should be secured so that they cannot be opened by external wind pressure, but nevertheless will open under internal pressure at a value well below 1 lb/in^2. Distribution of the vent area is advantageous, to minimize the distance between the initiation of the explosion and the vent. 1 ft^2 of vent area per 80 ft^3 of room volume should be provided Chapter 8, the vents being widely distributed around the building. The recommended areas of vent are only an approximate guide because they are derived from a combination of practical experience and experimentation on a smaller scale of construction, and are not based on the result of comprehensive full scale trials.

The venting recommendations should keep the explosion pressures within the construction down to safe valves, probably about 1 lb/in^2. In the design of new buildings information may be required on the strength of various wall constructions that may be exposed to explosion forces, and relatively little detailed information is available. One of the reasons is that with brickwork, for example, the strength of a wall panel would depend not only on its dimensions but also on the amount of restraint imposed, including vertical loads, the quality of workmanship and materials in the bricklaying, and its age. It should be stressed that for present purposes, the behaviour is required of walls exposed to explosion forces from within the structure; the relatively large amount of information available on the behaviour of structures subjected to external explosion pressures gives only general guidance and should be used with some caution if calculations are to be attempted.

In designing a building in which a dust explosion may occur, the problems are simplified if the building is of frame construction. Loss of cladding or of panel infillings, in the event of the explosion being too powerful for the venting provided, does not result in the collapse of the building (Plates 27, 28, 29). The floor slabs should be securely anchored to the building frame so that they will not be displaced. The strength of the brickwork in such a construction is variable, as outlined above, but as an approximate guide a 10 ft × 10 ft panel, 9 in thick, should withstand an internal pressure of 1 to 2 lb/in^2. If required, a several-fold increase in strength can be obtained by supporting the panel on all four sides, e.g. by keying them into the webs of steel girders. The strength of brick walls would be expected to increase proportionally with thickness, and to vary inversely with its length, width or height. Concrete panels are likely to be at least 4 to 6 inches in thickness, such panels would have at least comparable strength to 9 in brick walls of the same length and width and installed in the same manner.

If the building is to be of brickwork, but not of framed construction, additional problems arise. Past experience has shown that brick walls with little restraint at the top can fail by being pushed over rather than by bursting. The top of a high wall may need to be moved only a few feet for the whole wall to overbalance and collapse. Support to all the floors within the building is then lost and they, together with the roof, can descend, leading to a complete collapse of the building, loss of contents and hazarding the lives of occupants. The strength of such a structure, depends much more upon its geometry, and the strength of internal ties, than in the strength of the walls themselves. Another form of construction that can lead to difficulties is the use of wall panels as loadbearing members. An explosion within the building can lead to the loss of one, or several, panels and unless detailed attention has been given to this possibility in the design a substantial collapse can again occur. In general a framed structure is likely to give the most economic and simple solution.

The design of the building should also take into account the need to construct the interior to minimize dust accumulations, facilitate removal of unwanted dust, and provide safe lighting and heating. Prevention of dust accumulations is assisted by good design of plant and storage facilities and also by ensuring the absence of wall cavities, suspended ceilings or unnecessary horizontal surfaces. Ledges, sills, beams, light fittings etc. should be constructed so that their upper surfaces are at an angle to the horizontal, about 60° commonly being used. Provision of angled surfaces should not be made in such a manner that cavities are formed behind the surfaces, in which dust may collect. Unwanted dust can be removed by providing local exhaust ventilation at dust handling points and by mobile or permanently installed cleaning systems. Removal is facilitated by provision of smooth interior surfaces to the building. Lighting and heating equipment should be suitable for use in dusty environments, and particularly should not have exposed surfaces at sufficient temperature to ignite dust which may deposit on them.

FIRE REQUIREMENTS

In addition to those aspects of design specifically related to explosion, the requirements for fire and other regulations in the design of buildings must be met. These requirements are applicable to industrial buildings generally and include important aspects such as:

- means of escape for occupants;
- separation from neighbouring buildings;
- fire resistant separation from adjoining buildings;
- enclosure of escape staircases and lift shafts;
- provision of fire alarms and fire protection equipment e.g. sprinkler installations;
- avoidance of unprotected communicating openings between floors in multi-storey buildings;
- provision of adequate illumination;
- access for fire fighting and provision of necessary equipment.

Consideration of these requirements is beyond the scope of the present work and reference should be made to authorities as appropriate.

Operation and maintenance

The safety of industrial processes involves both the correct design of the plant and also the provision of clearly defined running procedures and supervision of maintenance operations. The provision of these procedures is the task of management, and includes supervision to ensure that they are correctly followed. A summary of normal and emergency operation procedures will be considered first, followed by the special provisions required in maintenance operations. Only a broad outline can be given, and it should not be regarded as comprehensive. Local conditions, the proximity of other hazards, and the characteristics of the material being processed must also be taken into account, and inspection of the site and familiarization with the layout are necessary. Where local conditions clearly raise problems not covered by the broad outline, some assistance may be obtained by referring to earlier chapters where the specific hazards of industrial processes and their operating procedures have been described.

RUNNING PROCEDURES

If a plant is to be run economically and safely the operatives must be adequately trained, and this training should include an outline of any fire and explosion risks and the procedures to be taken in the event of emergency. Instruction in fire drill, and means of escape in an emergency, should be included. With more complex plants economic and safe running may be assisted by the incorporation of complete or partial automatic control in which the process variables in the plant, e.g. flow rate, temperature, pressure, oxygen level, are automatically monitored and recorded and equipment installed to correct any deviation from the pre-set conditions. For both economic and safety reasons the positioning of monitoring, sampling and other devices, should be given careful consideration in the design of the plant to ensure that they are at their optimum situation. If the sensing devices are situated wrongly, a delay may ensue between the pre-set level being lost and the sensor picking up the appropriate signal. Because the development of fire and explosion can be very rapid the plant may depart so far from its optimum level during the delay associated with the sensor that automatic correction cannot be made and the plant goes out of control. In less serious cases, the plant may continue to be held under control but because of the delay in the system the plant does not revert to its pre-set condition but 'hunts' about it. This situation is unsatisfactory because the plant may be relying on safety devices to hold it within control during normal working, and these devices may be designed only for operation in infrequent emergencies. Also the economics of running the plant would not be optimal.

In the majority of instances the automatic control of a plant is designed to cope with normal working and to control relatively minor deviation from it. It will include the start-up and shut-down procedures which may be more hazardous than steady running because the plant is not in equilibrium and a large number of interrelated factors may be varying simultaneously. In an emergency some control of the plant is likely to be lost and the automatic installation may not be capable of dealing with it. Some manual overriding control would then be required and this can lead to further difficulties. One of the reasons for installing automatic control is that the manpower requirement is reduced, but if fewer operatives are employed on the plant then in an emergency less assistance can be quickly brought to the spot. In addition automatic control often requires a more complex monitoring system for the plant, and this adds to the difficulty of operatives who may be required to take over in an emergency. Both the design team and the operating management should give advance consideration to these problems, to ensure that operatives on the plant are adequately trained and instructed for emergency conditions.

The ways in which emergencies can arise in plants are too numerous to consider individually, but they may be broadly grouped into those involving failure of mechanical, electrical, or control systems, but before any outbreak of fire or explosion. Secondly, an emergency may arise because of outbreak of fire on a plant, and thirdly the emergency may be an explosion and subsequent fire. In all cases the emergency procedures should be aimed at preventing the development of a more hazardous situation, which would endanger life and property, and involve decisions on whether a partial or complete shut-down of the plant is required. Consideration of the possibilities must be given during the design of the plant because the appropriate control must be installed at the time of construction. The extent of an emergency shut-down often depends on the balance of conflicting interests: to prevent further damage or deterioration but also to continue running to clear the plant of hazardous material, or give temperature time to fall, and thereby prevent further hazard developing. Each type of plant requires individual consideration, in such a way that all the factors are given due weight. Where fire or explosion has occurred the plant may be under only partial control and the requirements for fire fighting may overide consideration of the first favourable method of shut-down. Even if damage is superficial, after a fire or explosion, the plant should only be restarted after inspection by a competent authority and usually after removal of all burned residues.

After fire or explosion in a plant the oxygen concentration in the atmosphere within process vessels etc, may be low, and noxious gases may be present, so that operatives should be instructed not to enter until given permission, which should be preceded either by adequate purging or by checking of the composition with suitable instruments and detectors. During and immediately after an emergency operatives do not have the time, or the inclination, to read and understand emergency procedures, so they should be trained in these procedures in advance. Management should ensure that this has been done.

Special care in instruction should be given where operatives are expected to take over manual control from automatic in an emergency. The combination of more complex instrumentation, because of the automatic control, with the unfamiliarity of assuming full charge of the plant combine to make detailed instruction essential, and this must be done for emergency conditions that can arise.

Management has a key role in the safe running of the plant, particularly in the training of operatives, and also in foreseeing hazards which may arise due to some lessening of the control of the plant, e.g. at changes of shift, or during the night when supervision may not be as strict as during the day time. The role involves not only instruction in procedures for running the plant, but also explanation.

MAINTENANCE PROCEDURES

The importance of control by management is increased further when maintenance work is necessary on the plant. The first consideration is that the staff who carry out the maintenance may be brought in from other areas where they would not be familiar with the hazards on the plant or they may be brought in from outside the organization altogether, in which case they may have misunderstandings of the process hazards involved. It is the task of management to ensure that the plant is in a safe condition for work to be carried out, that the methods of work to be used are suitable, and that the maintenance staff are closely supervised. Even so, maintenance staff unfamiliar with the hazards may carry out operations which management has not foreseen.

Before maintenance is begun, management must ensure that the plant is in a safe condition to work upon. Where flammable or explosible materials have been present within the plant, care must be taken to ensure that these materials have been removed and that legal obligations in this respect have been met. Extra care should be taken where any form of hot working is to be used because the presence of a film or thin deposit on the walls of an enclosure can lead to explosion if sources of heat are applied. Where arrangements can be made, it is preferable to remove from the plant to a safe area those individual items requiring maintenance. Cleaning is still essential. The maintenance operations are then carried out in safety, and the items of plant subsequently replaced. This method is particularly to be advised with highly flammable or explosible dusts, where hot working is necessary. In whichever situation the maintenance work is carried out, control of the operation is assisted by using a work permit system. A work permit is issued by the individual in charge of the plant to the individual in charge of the maintenance staff, or, if only one maintenance man is involved, directly to him. The permit should state what the job is, and include instructions on the safe method of carrying out the work, together with any conditions or limitations necessary. On receipt of the permit, the individual in charge of the maintenance staff should ensure that they also understand the conditions laid down. The work permit should specify a time limit of validity, which should not extend into another shift or work day; on change of shift a new work permit should be issued. On completion of the work, or at the end of the shift all permits should be returned to the person who issued them or to a nominated alternative. This person should then ensure that all staff working under permit have left the plant, equipment brought in for the work has been removed, safety procedures including fire and explosion protection systems have been reinstated, and where the maintenance has been completed the plant is safe for normal operation.

Strict use of a work permit system should ensure that the plant operator and the maintenance supervisor co-operate, and that hazards arising during maintenance work are minimized. If the work permit system is not used, conflict can arise between the plant supervisor who wishes to maintain production as far as possible, and the maintenance supervisor who may be working to a restricted timetable and who wishes to expedite the work.

Although the hazards of maintenance are particularly acute during repair operations, they are still present during maintenance which is aimed at verifying that equipment is in good order. Much fire and explosion protection relies on mechanical and electrical apparatus and equipment being in good working order and a regular maintenance and inspection schedule must be provided by the management. A work permit system for these operations may also be desirable, to clarify the areas of responsibility.

An essential part of maintenance procedures, in the case of dusts and powdered materials is to ensure that accumulations do not build up, but are removed on regular cleaning schedules. The organization of housekeeping schedules is again the responsibility of management. Removal of dust should be by means which do not raise dust clouds, and may require portable or permanently installed equipment depending on circumstances. For small spillages and accumulations, portable or mobile vacuum cleaners are suitable, care being taken to ensure that the exhaust air does not stir up dust suspensions. For more serious accumulations, over prolonged periods, installation of a fixed vacuum system is advisable. The fixed system ensures that facility for removal of any spillage is to hand permanently, whereas a portable cleaner may require to be fetched from a distance and is therefore less likely to be used. The dust collection unit from a permanent installation should be in a safe position. Good housekeeping can be encouraged at the design stage of a plant by minimizing the area of horizontal surface in buildings and on fittings, and by ensuring easy access for cleaning to all parts. It is essential for the safe running of dust handling plant, and may be a legal requirement.

11 *Specific Dusts and Processes*

In the design of plant, considerable assistance may be obtained if a Code of Practice or guide is available as a basis. In this chapter lists are given of available codes and guides which may be for specific dusts, specific industrial operations, specific fire and explosion precautions, or may be generalized.

Care should be taken in following codes which should not be accepted blindly; if used with caution, however, they are helpful. They may not have legal authority, but are likely to be taken as examples of good practice and any serious departure from them in design or operation of plant may be questioned.

In general, a particular code is only applicable to the types of dust and plant operation envisaged when it was drawn up. Where a code is not available for a particular dust, a code for another dust should not be used unless an expert technical opinion is obtained to agree to its use. Factors that would have to be borne in mind include the explosibility and fire properties of the dust involved and whether the methods and scale of working and the precautions are applicable and adequate. In addition, technical changes in industry often result in different operations being carried out on the dust, possibly with increased scale of operation, and expert advice should again be sought to ensure that the code covers the new situation. In cases of serious doubt, rather than attempt to use a code which may not be applicable, the design should be based on the technical information available in earlier chapters. This approach will, in any case, have to be made for all applications where no specific code is available.

Where a particular code for the design of the plant is available for a specific dust, a change in the composition or size characteristics of the feed-stock being processed may give rise to the need for different plant design and protection. In such cases a thorough investigation must be made before the feedstock change is permitted.

Codes for specific dusts are listed in Table 11.1 and cover a range of the more common types. In the case of coal, the codes apply to its utilization in surface industries and not to the mining or winning of coal. The fire and explosion hazards associated with coal mining are specialized, and not found generally in surface industry, and because of the long history of coal mining detailed requirements have been laid down by legislation in many countries. These requirements are outside the scope of the present book and, indeed, would require lengthy treatment.

Codes are also available for specific process operations, and are listed in Table 11.2. If a code is to be followed, care should be taken to ensure that the dust being processed has suitable fire and explosion properties for the operation. Reference back to earlier chapters, listed in Table 11.2, can assist but if there is doubt expert advice should be taken. Codes are not available for all the common dust handling operations, but some recommendations can be taken from the general codes listed. In assessing whether the procedure to be adopted is correct, there is no substitute for a sound knowledge of the technical background to the hazard involved, and knowledge of the properties of the dust and the hazards associated with the plant processes to which it is to be subjected is essential.

In many of the codes published by the National Fire Protection Association a warning is given concerning the use of powder or cartridge operated tools. The warning states that 'gun-type tools using powder or cartridges for driving pegs or pins into concrete, brick, steel, etc, shall not be used where flammable dust or dust clouds are present. When the use of this type of equipment becomes necessary, all dust producing machinery in the area shall be shut down, all equipment, floors and walls shall be carefully cleaned, and all dust accumulations removed. A careful check shall be made to be sure that no cartridges or charges are left on the premises where they could enter equipment or be accidentally discharged after operation of the dust producing or handling machinery is resumed.'

The reason for the warning is that with these tools there may be a risk of hot spots developing where the projectile is driven, and this could ignite dust clouds or layers. Also, unspent or spent cartridges might find their way into the plant where if they came into contact with moving machinery they could cause ignition of dust. Providing precautions such as those quoted above are taken the use of the tools can be considered. In cases of doubt, as where dusts of severe explosibility or low ignition temperature are processed, or where cleanliness inside and outside the plant cannot be relied upon, the tools should not be used until the advice of authority or an expert has been obtained.

Table 11.1 Codes for specific dusts

Type of dust	*Dust*	*References*
Vegetable origin	Flour	1, 2
	Grain	3
	Provender	1
	Starch	4
	Sugar	5
	Confectionery	6
	Cocoa	5
	Wood	7, 8
	Spice	9
Carbonaceous	Coal	8, 10
Chemicals	General	11
	Sulphur	12
Plastics	General	13
	Celluloid, pyroxylin	14
Metals	Aluminium	15, 16
	Magnesium	17, 18
	Titanium	19
	Zirconium	20

Table 11.2 Codes and guides for process operations

Plant or process	*Relevant chapter in this work*	*References*
General	2	21, 22, 23, 24, 25, 26, 27, 28, 29
Explosion relief venting	8	29, 30, 31
Inerting, explosion suppression	8	32
Electricity, mains	7	29, 33, 34
Electricity, static	7	29, 35
Pneumatic conveying	10	36
Dryers	10	3, 10, 29

References

[1] Flour and feed mills, No. 61C, (1962), National Fire Protection Association, Boston, USA.

[2] Dust explosions in flour mills and bulk flour containers. Incorporated National Association of British and Irish Millers Ltd, London, (1967).

[3] Standard for the prevention of fire and dust explosions in grain elevators and bulk grain handling facilities, No. 61B, (1970), National Fire Protection Association, Boston, USA.
[4] Starch factories, No. 61A, (1962), National Fire Protection Association, Boston, USA.
[5] Standard for the prevention of dust explosions in the production, packaging and handling of pulverized sugar, and cocoa, No. 62, (1967), National Fire Protection Association, Boston, USA.
[6] Standard for the prevention of dust explosions in confectionery manufacturing plants, No. 657, (1967), National Fire Protection Association, Boston, USA.
[7] Woodworking and wood flour manufacturing plants, No. 664, (1962), National Fire Protection Association, Boston, USA.
[8] Pulverized fuel systems, No. 60, (1961), National Fire Protection Association, Boston, USA.
[9] Spice grinding plant, No. 656, (1959), National Fire Protection Association, Boston, USA.
[10] Coal preparation plants, No. 653, (1959), National Fire Protection Association, Boston, USA.
[11] Guide to fire prevention in the chemical industry. British Chemical Industry Safety Council of the Chemical Industries Association Ltd., London (1968).
[12] Standard for prevention of sulphur fires and explosions, No. 655, (1968), National Fire Protection Association, Boston, USA.
[13] Standard for the prevention of dust explosions in the plastics industry, No. 654, (1970), National Fire Protection Association, Boston, USA.
[14] Storage, handling and use of pyroxylin plastics in factories, No. 42, (1967), National Fire Protection Association, Boston, USA.
[15] Standard for the processing and finishing of aluminium, No. 65, (1970), National Fire Protection Association, Boston, USA.
[16] Code for the prevention of dust explosions in the manufacture of aluminium powder, No. 651, (1967), National Fire Protection Association, Boston, USA.
[17] Standard for explosion and fire protection in plant producing or handling magnesium powder, No. 652, (1968), National Fire Protection Association, Boston, USA.
[18] Standard for the storage, handling and processing of magnesium, No. 48, (1967), National Fire Protection Association, Boston, USA.
[19] Standard for the production, processing, handling and storage of titanium, No. 481, (1961), National Fire Protection Association, Boston, USA.
[20] Guide for fire and explosion prevention in plant producing and handling zirconium, No. 482M, (1961), National Fire Protection Association, Boston, USA.
[21] Dust explosions in factories, Health and Safety at Work No. 22, Department of Employment and Productivity, HMSO, London, (1970).
[22] Fundamental principles for the prevention of dust explosions in industrial plant, No. 63, (1964), National Fire Protection Association, Boston, USA.
[23] Dust explosions in factories, Pamphlet SHW 830, Department of Employment and Productivity, HMSO, London, (1968).
[24] Dust explosions analysis and control, Factory Insurance Association, Hartford, Connecticut, USA.

[25] Prevention of dust fires and dust explosions, Verein Deutscher Ingenieure 2263, Dusseldorf, (1969).
[26] Explosible dusts – the hazards, FS6021, Fire Protection Association, London, (1971).
[27] Explosible dusts – control of explosions, FS6022, Fire Protection Association, London, (1971).
[28] Explosible dust – elimination of ignition sources, FS6023, Fire Protection Association, London, (1971).
[29] *Handbook of Industrial Loss Prevention*, Factory Mutual Engineering Corporation, 2nd Edition, McGraw Hill, New York, (1967).
[30] Guide for explosion venting, No. 68, (1954), National Fire Protection Association, Boston, USA.
[31] Guide to the use of flame arresters and explosion reliefs. Safety, Health and Welfare New Series No. 34. Ministry of Labour. HMSO, London, (1965).
[32] Explosion prevention systems. No. 69, (1970), National Fire Protection Association, Boston, USA.
[33] Electrical code, No. 70, (1968), National Fire Protection Association, Boston, USA.
[34] Verbandes Deutscher Elektrotechniker No. 0165.
[35] Recommended practice on static electricity, No. 77, (1966), National Fire Protection Association, Boston, USA.
[36] Standard for pneumatic conveying systems for handling feed, flour, grain, and other agricultural dusts, No. 66, (1970), National Fire Protection Association, Boston, USA.

12 *Procedure in Plant Design*

In many instances where dust explosion or fire protection is to be specified for a plant, the basic design of the plant may already have been fixed in considerable detail, e.g. because it is already in existence and there is to be a change in the type of material being processed. In such cases the procedure to be adopted in installing explosion and fire protection is largely pre-determined and reference may be made directly to the appropriate chapters. In other instances, and particularly where a new plant is being considered, there may be little advance commitment and a number of alternative approaches to the design may be possible. Consideration will now be given to procedure in plant design, to ensure that the basic necessary requirements are met and are dealt with in logical manner. Although the approach applies particularly to new plant, it will also be applicable to existing plant where a substantial proportion of it has to be redesigned to meet the new conditions. Whether the plant is new, or is a redesign, the basic process, materials involved and plant units are likely to have been specified in terms of plant performance and product manufacture, largely irrespective of dust explosion and fire consideration. The design procedure for fire and explosion protection thus involves ensuring that the plant is adequately and economically protected within a framework of conditions laid down for production and other reasons.

The procedure in designing for dust explosion and fire protection may be divided into ten stages as summarized in Table 12.1, which includes reference back to appropriate chapters. This approach can be used as a basic guide, but special local conditions must be taken into account and these could substantially alter the approach. Basic generalizations in design have limitations, in that local conditions or neighbouring hazards could cause substantial redesign, which require individual consideration. In cases of doubt expert opinion should always be sought.

The first stage is to establish whether the materials being processed are, at any stage of the process, capable of causing explosion or fire. Consideration should be given both to run of the mill material, to by-products, and particularly to any fines which may be generated or processed and which may separate out at stages in the processing. The explosion and fire properties of dusts are usually established by means of small scale tests which may be used either selectively or comprehensively enabling the Index of Explosibility to be calculated. If the tests show the dust to be hazardous it is necessary to proceed to stage 2 to ensure that any legal or other requirements are followed. There may also be restrictions on the plant due to obnoxious or toxic materials being processed in it, which in turn would affect the design of explosion protection that would be permitted. A comprehensive interpretation of legal requirements may require expert advice, which should be sought at this early stage of the design, because it may be difficult to cover when the detailed design has already been committed.

Considerable assistance in design can be obtained from a relevant code, stage 3, but as already explained (Chapter 11) such codes must be used with discretion and on the understanding that they cover both the materials and the plant processes, both in function and scale, that are envisaged. The number of codes is not large and in many cases no direct assistance will be available, and the designer must proceed in their absence. Although the siting of the plant may be predetermined, stage 4, usually by the restricted ground or floor area available, some flexibility may be obtainable particularly as regard separation from adjacent risks, by planned building construction. The storage of feedstock and product should, as far as possible, be kept separate from manufacturing processes and this requirement should be brought in during the site planning stage.

Consideration is now given to the design of the plant itself, the first point being the avoidance of ignition sources, stage 5. Prevention of ignition by removal of all sources is usually not sufficiently reliable and further precautions must be taken, stages 6 and 7, the selection of which will depend greatly on the materials being processed and the operations being carried out. Several alternative methods of production may be possible, and of equal technical merit, and the final evaluation and selection, stage 8, may be made on the basis of the most economic and convenient design for running the plant safely. The factors involve both the plant economics and local availability of labour, siting etc and these are beyond the scope of the present work. When the design has been finalized it may require approval by various authorities, stage 9, before proceeding to construction. Stage 10, which is very important, involves adequate training of management and operatives to run the plant, to understand the processes involved, to appreciate the hazards that can arise if the correct operating procedures are not followed, and to understand procedures for emergency conditions. Because of the turnover of labour with time, both in management and operation, training procedures should be maintained throughout the life of the plant. A desirable accompaniment to training sessions is the production of manuals for normal operation, maintenance, and emergency procedures. A corollary is that when outside labour is brought onto the site, for maintenance or for alteration of plant units, that adequate training and instruction shall be given, and the conditions of their work strigently controlled for the entire duration of the operation including re-instatement of the plant.

Table 12.1 Procedural stages in designing for dust explosion and fire protection

Stage	*Action*	*Appropriate chapter*
1	Establish explosion and fire properties of dust	3
2	Consider legal, insurance, and other authority requirements	1
3	Availability of relevant existing codes	11
4	Siting of plant, and adjacent risks	10
5	Avoidance of ignition sources	7
6	Explosion protection – selection of appropriate methods	8, 10
7	Fire protection	9, 10
8	Final evaluation of method of protection in terms of safety of operatives and plant, economics, maintenance, housekeeping	—
9	Submission of design for approval by authority	1
10	Training of management and operatives to run plant	10

It is clear from the above summary of the basic design stages that the installation of explosion protection is done far more effectively during the planning stage of the plant than, as has regularly been the case in the past, by designing and constructing a plant without regard to explosion and fire protection and then attempting to apply the protection to the constructed plant. Apart from being time-consuming for the designer, the attempt to install explosion precautions in a plant already built often requires alteration to the construction of the plant, which imposes a financial penalty, and may also hold up the commissioning, again leading to loss. The logical method of providing explosion and fire protection is to incorporate it at the design stage, when discussions with the appropriate authorities should also be held.

Appendix 1 *Fire Protection Associations*

Australia	Australian Fire Protection Association 51–53 William St Melbourne Victoria 3000
Austria	Zentralstelle fur Brandverhutung Mariahilferstrasse 133 1150 Wien XV
Belgium	Association Nationale pour la Protection contre l'Incendie 4, rue de l'Autonomie Brussels 7
Denmark	Dansk Brandvaerns-Komite Nygaards Plads 9 Copenhagen – Rodoure
Federal German Republic	Vereinigung zur Forderung des Deutschen Brandschutzes e.V. Karl Wiltbergerstrasse 7 5300 Bonn – Lessenich
Finland	Suomen Palontorjuntaliitto Imo Roobertinkatu 7A4 Helsinki 12

France	Centre National de Prevention et de Protection 5 Rue Daunou Paris 2e
Holland	Nationaal Brandpreventie Instituut Bezuidenhoutsewig 195 The Hague
Japan	Japan Fire Protection Association No 2. –19, 3 Chome Ginza Tokyo
Norway	Norsk Brannvern Forening Klingenberggt 5, Postboks 1325 Vika Oslo 1
Spain	Asociacion Espanola de Lucha Contra el Fuego Sepulveda 162 Barcelona 11
Sweden	Svenska Brandforsvarsforeningen Kungsholms Hamnplan 3 IV Stockholm K
Switzerland	Brand-Verhutungs-Dienst fur Industrie und Gewerbe Nuschelerstrasse 45 8001 Zurich
United Kingdom	Fire Protection Association Aldermary House Queen Street London EC4N 1TJ

United States of America	National Fire Protection Association 60 Batterymarch Street Boston Massachusetts 02110

Appendix 2 *Explosion Properties of Dusts*

The values presented in the following tables are intended as a general guide, and are based on a selection of materials and available data on explosion and fire properties. The variation inherent in many of the materials, and also resulting from different test methods for measuring the same explosion property, mean that the quoted values should strictly be applied only to the sample actually tested, taking account of the method of test. In the tables references are given which enable further details to be obtained, but, for many practical purposes, guidelines are all that are required. If further information is required for a particular sample of a material, direct testing of that material is usually necessary. Significant variation in explosion properties with a material can arise from differences in particle size, particle shape, moisture content, dispersability, and from variations in composition inevitable with many materials of natural origin.

The tabulated values originated from publications in the United Kingdom, the United States, and the Federal German Republic and the methods of test are those in use in the country of origin of the data. Further information can be obtained from studying the references and Chapter 3.

Where more than one set of data has been published for a material, preference has been given in the selection of data to that combining the most comprehensive and the most hazardous values indicated. As many identifiable dusts as possible have been included, but mixtures that are not generally reproducible, and many co-polymers, have been excluded.

Because of the different methods of test in the three countries, the values for minimum ignition temperature of dust clouds may differ somewhat. The minimum ignition temperature of dust layers is measured in the United States using a sample 1.3 cm deep and 2.5 cm diameter, in the Furnace Apparatus, but in the Federal German Republic a 0.5 cm deep layer on a hotplate is used (Chapter 3). The results from the two methods are not directly comparable (Chapter 4). The values of maximum explosion pressure and maximum rate of pressure rise are the highest quoted for a given material and are not necessarily measured for the same dust concentration. In the test for the maximum oxygen concentration to prevent ignition, the use of the Furnace Apparatus has been taken throughout (Chapter 3), and the inert gas was carbon dioxide for data originating from the United States and nitrogen for data originating from the United Kingdom, unless stated otherwise. The relative effectiveness of the two inert gases is discussed in Chapter 4.

(–) indicates that no value was available, but should not be taken to mean that the property was not measurable.

The references are as follows:

[1] Dorsett, H. G. and Nagy, J. (1968), US Bureau of Mines RI 7132.

[2] Freytag, H. H. (Ed). (1965), Handbuch der Raumexplosionen. Verlag Chemie. Weinheim.

[3] Jacobson, M., Cooper, A. R. and Nagy, J. (1964), US Bureau of Mines RI 6516.

[4] Jacobson, M., Nagy, J. and Cooper, A. R. (1962), US Bureau of Mines RI 5971.

[5] Jacobson, M., Nagy, J., Cooper, A. R. and Ball, F. J. (1961), US Bureau of Mines RI 5753.

[6] Nagy, J., Cooper, A. R. and Dorsett, H. G. (1968), US Bureau of Mines RI 7208.

[7] Nagy, J., Dorsett, H. G. and Cooper, A. R. (1965), US Bureau of Mines RI 6597.

[8] Raftery, M. M. (1968), Explosibility tests for industrial dusts. Ministry of Technology and Fire Offices' Committee. Fire Research Technical Paper No. 21, HMSO, London.

Dust	Minimum ignition temperature °C		Minimum explosible concentration g/l	Minimum ignition energy mJ	Maximum explosion pressure lb/in²	Maximum rate of pressure rise lb/in²s	Maximum oxygen concentration to prevent ignition % by volume	References	Notes
	cloud	layer							
Acetamide	560	—	—	—	—	—	—	8	Group (b) dust
Aceto acetanilide	560	—	0.030	20	90	4 800	—	4	
Acetoacet–p–phenetedide	560	—	0.030	10	87	> 10 000	—	1	
Acetoacet–o–toluidine	710	—	—	—	—	—	—	1	
2 Acetylamino—5—nitro thiazole	450	450	0.160	40	137	9 000	—	1	
Acetyl–p–nitro–o–toluidine	450	—	—	—	—	—	—	8	
Adipic acid	550	—	0.035	60	95	4 000	—	4,8	
Alfalfa	460	200	0.100	320	88	1 100	—	5	
Almond shell	440	200	0.065	80	101	1 400	—	5	
Aluminium, atomized	650	760	0.045	50	84	> 20 000	—	3	
Aluminium, flake	610	320	0.045	10	127	> 20 000	—	3	
Aluminium–cobalt alloy	950	570	0.180	100	92	11 000	—	3	
Aluminium–copper alloy	—	830	0.100	100	95	4 000	—	3	
Aluminium–iron alloy	550	450	—	—	36	300	—	3	
Aluminium–lithium alloy	470	400	<0.1	140	96	6 000	—	3	
Aluminium–magnesium alloy	430	480	0.020	80	86	10 000	—	3	
Aluminium–nickel alloy	950	540	0.190	80	96	10 000	—	3	
Aluminium–silicon alloy	670	—	0.040	60	85	7 500	—	3	
Aluminium acetate	560	640	—	—	59	950	—	1	Guncotton ignition source in pressure test.
Aluminium octoate	460	—	—	—	—	—	—	8	
Aluminium stearate	400	380	0.015	10	86	> 10 000	—	1	

2 Amino–5–nitrothiazole	460	460	0.075	30	110	5 600	—	1	
Anthracene	505	Melts	—	—	68	700	—	2,8	
Anthranilic acid	580	—	0.030	35	84	6 500	—	1	
Anthraquinone	670	—	—	—	—	—	—	8	
Antimony	420	330	0.420	1 920	28	300	—	3	
Antipyrin	405	Melts	—	—	53	—	—	2	
Asphalt	510	500	0.025	25	94	4 800	—	7	
Aspirin	550	Melts	0.015	16	87	7 700	—	8	
Azelaic acid	610	—	0.025	25	76	4 700	—	4	
αα′ Azo isobutyronitrile	430	350	0.015	25	134	8 000	—	4	
Barley	370	—	—	—	—	—	—	8	
Benzethonium chloride	380	410	0.020	60	91	6 700	—	1	
Benzoic acid	600	Melts	0.011	12	95	10 300	—	8	
Benzotriazole	440	—	0.030	30	103	9 200	—	1	
Benzoyl peroxide	—	—	—	21	—	—	—	8	
Beryllium	910	540	—	—	Did not ignite		—	3	Contained 8 per cent oxide
Beryllium acetate, basic	620	—	0.080	100	87	2 200	15	1	Inert gas carbon dioxide
Bis (2–hydroxy–5–chlorophenyl)–methane	570	—	0.040	60	70	2 000	13	1	Inert gas carbon dioxide
Bis (2–hydroxy–3,5,6,–trichlorophenyl)–methane	Did not ignite	450	—	—	—	—	—	1	
Bone meal	490	230	—	—	11	100	—	6	Guncotton ignition source in pressure test.
Boron	730	390	Did not ignite	—	41	200	—	3	Guncotton ignition source in pressure test.
Bread	450	—	—	—	—	—	—	8	
Brunswick green	360	—	—	—	—	—	—	8	
P–t–butyl benzoic acid	560	—	0.020	25	88	6 500	—	4	
Cadmium	570	250	—	4 000	7	100	—	3	
Cadmium yellow	390	—	—	—	—	—	—	8	

Dust	*Minimum ignition temperature °C cloud*	*layer*	*Minimum explosible concentration g/l*	*Minimum ignition energy mJ*	*Maximum explosion pressure lb/in²*	*Maximum rate of pressure rise lb/in²s*	*Maximum oxygen concentration to prevent ignition % by volume*	*References*	*Notes*
Calcium carbide	555	325	—	—	13	—	—	2	
Calcium citrate	470	—	—	—	—	—	—	8	Group (b) dust
Calcium gluconate	550	—	—	—	—	—	—	8	Group (b) dust
Calcium DL Pantothenate	520	—	0.050	80	105	4 600	—	1	
Calcium propionate	530	—	—	—	90	1 900	—	8	
Calcium silicide	540	540	0.060	150	86	20 000	—	3	
Calcium stearate	400	—	0.025	15	97	> 10 000	—	1	
Caprolactam	430	—	0.07	60	79	1 700	8	8	
Carbon, activated	660	270	0.100	—	92	1 700	—	7	Guncotton ignition source in min. expl. conc. and max. expl. pressure tests.
Carbon, black	510	—	—	—	—	—	—	7	
Carboxy methyl cellulose	460	310	0.060	140	130	5 000	—	4	
Carboxy methyl hydroxy ethyl cellulose	380	—	0.200	960	83	800	—	4	
Carboxy polymethylene	520	—	0.115	640	76	1 200	—	4	
Casein	460	—	—	—	89	1 200	—	8	
Cellulose	410	300	0.045	40	117	8 000	—	4	
Cellulose acetate	340	—	0.035	20	114	6 500	5	4,8	Inert gas nitrogen
Cellulose acetate butyrate	370	—	0.025	30	81	2 700	7	4	
Cellulose proprionate	460	—	0.025	60	105	4 700	—	4	
Cellulose triacetate	390	—	0.035	30	107	4 300	—	4,8	
Cellulose tripropionate	460	—	0.025	45	88	4 000	—	4	
Charcoal	530	180	0.140	20	100	1 800	—	7	

Chloramine–T	540	150	—	—	7	150	—	1	Guncotton ignition source in pressure test.
o–Chlorobenzmalono nitrile	—	—	0.025	—	90	> 10 000	—	1	
o–Chloro aceto acetanilide	640	—	0.035	30	94	3 900	—	1	
p–Chloroaceto acetanilide	650	—	0.035	20	85	5 500	—	1	
Chloro amino toluene sulphonic acid	650	—	—	—	—	—	—	8	
4–Chloro–2 nitro aniline	590	120	<0.750	140	123	3 500	—	1	
p–Chloro o–toluidine hydrochloride	650	—	—	—	—	—	—	8	
Chocolate crumb	340	—	—	—	—	—	—	8	
Chromium	580	400	0.230	140	56	5 000	—	3	
Cinnamon	440	230	0.060	30	121	3 900	—	5	
Citrus peel	500	330	0.060	100	51	1 200	—	5	
Coal, brown	485	230	—	—	—	—	—	2	See also Lignite
Coal, 8 per cent volatiles	730	—	—	—	—	—	—	7	
Coal, 12 per cent volatiles	670	240	—	—	—	—	—	7	
Coal, 25 per cent volatiles	605	210	0.120	120	62	400	—	7	
Coal, 37 per cent volatiles	610	170	0.055	60	90	2 300	—	7	Standard Pittsburgh coal
Coal, 43 per cent volatiles	575	180	0.050	50	92	2 000	—	7	
Cobalt	760	370	—	—	—	—	—	3	
Cocoa	500	200	0.065	120	69	1 200	—	5	
Coconut	450	280	—	—	—	—	—	2	
Coconut shell	470	220	0.035	60	115	4 200	—	5	
Coffee	360	270	0.085	160	38	150	10	5,8	Inert gas carbon dioxide
Coffee, extract	600	—	—	—	47	—	—	2	
Coffee, instant	410	350	0.280	Did not ignite	68	500	—	5	
Coke	> 750	430	—	—	—	—	—	2	

Dust	Minimum ignition temperature °C cloud	layer	Minimum explosible concentration g/l	Minimum ignition energy mJ	Maximum explosion pressure lb/in²	Maximum rate of pressure rise lb/in²s	Maximum oxygen concentration to prevent ignition % by volume	References	Notes
Coke, petroleum, 13 per cent volatiles	670	—	1.00	—	36	200	—	7	Guncotton ignition source in min. expl. conc. and max. expl. pressure tests.
Colophony	325	Melts	—	—	—		—	2	
Copal	330	Melts	—	—	68	—	—	2	See also Gum manila
Copper	700	—	—	Did not ignite	Did not ignite	Did not ignite	—	3	
Copper–zinc, gold bronze	370	190	1.00	—	44	1 300	—	3	
Cork	460	210	0.035	35	96	7 500	—	6	
Corn cob	450	240	0.045	45	127	3 700	—	5	
Corn dextrine	410	390	0.040	40	124	7 000	—	5	
Cornflour	390	—	—	—	—	—	—	8	
Cornstarch	390	—	0.040	30	145	9 500	—	5	
Cotton flock	470	—	0.050	25	94	6 000	—	4	
Cotton linters	520	—	0.50	1920	73	400	5	5	
Cottonseed meal	530	200	0.055	80	89	2 200	—	5	
Coumarone–indene resin	550	—	0.015	10	93	11 000	11	4	
Crystal violet	475	Melts	—	—	—	—	—	2	
Cyclohexanone peroxide	—	—	—	21	84	5 600	—	8	
Dehydroacetic acid	430	—	0.030	15	87	8 000	—	1	
Dextrin	410	440	0.050	40	99	9 000	—	6	
Dextrose monohydrate	350	—	—	—	—	—	—	8	
Diallyl phthalate	480	—	0.030	20	90	8 500	—	1	

Diamino stilbene disulphonic acid	550	—	—	—	—	—	—	8	Group (b) dust
Diazo aminobenzene	550	—	0.015	20	114	> 10 000	—	1	
Di–t–Butyl–p–cresol	420	—	0.015	15	79	13 000	9	4	
Dibutyl tin maleate	600	—	—	—	—	—	—	8	
Dibutyl tin oxide	530	—	—	—	—	—	—	8	
Dichlorophene	770	—	—	—	72	3 000	—	1	
2,4–Dichlorophenoxy ethyl benzoate	540	—	0.045	60	84	2 200	—	1	
Dicyclopentadiene dioxide	420	—	0.015	30	89	9 500	—	4	
Dihydrostreptomycin sulphate	600	230	0.520	—	42	200	7	1	
3–3′ Dimethoxy 4–4′ diamino diphenyl	—	—	0.030	—	82	> 10 000	—	1	
Dimethylacridan	540	—	—	—	—	—	—	8	
Dimethyl diphenyl urea	490	—	—	—	—	—	—	8	
Dimethyl isophthalate	580	—	0.025	15	84	8 000	—	4	
Dimethyl terephthalate	570	—	0.030	20	105	12 000	6	4	
S–S′–Dimethyl xanthogene-thylene bis dithiocarba-mate	400	—	0.300	3200	84	1 500	—	1	
Dinitro aniline	470	—	—	—	—	—	—	8	
3,5–Dinitrobenzamide	500	Melts	0.040	45	163	6 500	—	1	
3,5–Dinitrobenzoic acid	460	—	0.050	45	139	4 300	—	1	
Dinitrobenzoyl chloride	380	—	—	—	—	—	—	8	
Dinitrocresol	340	Melts	0.030	—	—	—	—	1,2	
4,4′–Dinitro–sym–diphenyl urea	550	—	0.095	60	102	2 500	—	1	
Dinitro stilbene disulphonic acid	450	—	—	—	—	—	—	8	
Dinitrotoluamide	500	—	0.050	15	153	> 10 000	—	6	
Diphenyl	630	—	0.015	20	82	3 700	—	1	
4,4′–Diphenyl di sulphonylazide	590	140	0.065	30	143	5 500	—	1	

Dust	Minimum ignition temperature °C cloud	layer	Minimum explosible concentration g/l	Minimum ignition energy mJ	Maximum explosion pressure lb/in²	Maximum rate of pressure rise lb/in²s	Maximum oxygen concentration to prevent ignition % by volume	References	Notes
Diphenylol propane (Bisphenol–A)	570	—	0.012	11	81	11 800	5	4,8	Inert gas nitrogen
Egg white	610	—	0.14	640	58	500	—	5	
Epoxy resin	490	—	0.015	9	94	8 500	—	4,8	
Esparto grass	—	—	—	—	94	7 300	—	8	
Ethyl cellulose	340	330	0.025	15	112	7 000	—	4	
Ethylene diamine tetra acetic acid	450	—	0.075	50	106	3 000	—	1	
Ethyl hydroxyethyl cellulose	390	—	0.020	30	94	2 200	—	6	
Ferric ammonium ferrocyanide	390	210	1.500	—	17	100	—	1	
Ferric dimethyl dithio carbamate	280	150	0.055	25	86	6 300	—	1	
Ferric ferrocyanide	370	—	—	—	82	1 000	—	1	
Ferrochromium	790	670	2.00	—	—	—	—	3	
Ferromanganese	450	290	0.130	80	62	5 000	—	3	
Ferrosilicon (45 per cent Si)	640	—	—	—	—	—	—	2	
Ferrosilicon (90 per cent Si)	Did not ignite	980	0.240	1280	113	3 500	—	3	
Ferrotitanium	370	400	0.140	80	55	9 500	—	3	
Ferrous ferrocyanide	380	190	0.400	—	—	—	—	1	
Ferrovanadium	440	400	1.300	400	—	—	—	3	
Fish meal	485	—	—	—	—	—	—	2	
Fumaric acid	520	—	0.085	35	103	3 000	—	4	
Garlic	360	—	0.10	240	57	1 300	—	5	
Gelatin, dried	620	480	<0.5	—	78	1 200	—	1	

Gilsonite	580	500	0.020	25	78	4 500	—	7	
Graphite	730	580	—	—	—	—	—	7	
Grass	—	—	—	—	56	400	—	8	
Gum arabic	500	260	0.060	100	117	3 000	—	4	
Gum Karaya	520	240	0.100	180	116	2 500	—	4	
Gum manila (copal)	360	390	0.030	30	89	6 000	—	4	
Gum tragacanth	490	260	0.040	45	123	5 000	—	4	
Hexa methylene tetramine	410	—	0.015	10	98	11 000	11	4	
Horseradish	—	—	<0.100	—	96	1 600	—	6	
Hydrazine acid tartrate	570	—	0.175	460	30	200	—	1	
p–Hydroxy benzoic acid	620	—	0.040	—	37	—	—	1	
Hydroxyethyl cellulose	410	—	0.025	40	106	2 600	—	6	
Hydroxyethyl methyl cellulose	410	—	—	—	—	—	—	8	
Hydroxy propyl cellulose	400	—	0.020	30	96	2 900	—	6	
Iron	430	240	—	—	—	—	—	2	
Iron, carbonyl	420	230	0.105	100	47	8 000	—	3	
Iron pyrites	380	280	1.00	8 200	5	100	—	3	
Isatoic anhydride	700	—	0.035	25	80	4 900	—	1	
Isinglass	520	—	—	—	Nil	Nil	—	8	
Isophthalic acid	700	—	0.035	25	78	3 100	—	4	
Kelp	570	220	Did not ignite		19	200	—	5	
Lactalbumin	570	240	0.040	50	97	3 500	—	4	
Lampblack	730	—	—	—	—	—	—	7	
Lauryl peroxide	—	—	—	12	90	6 400	—	8	
Lead	790	290	—	Did not ignite	3	100	—	3	Flame ignition source in pressure test.
Leather	390	—	—	—	—	—	—	8	
Lignin	450	—	0.040	20	102	5 000	7	4	
Lignite	450	200	0.030	30	94	8 000	—	7	
Lycopodium	480	310	0.025	40	75	3 100	9	5	
Magnesium	560	430	0.030	40	116	15 000	—	3	
Maize husk	430	—	—	—	75	700	—	8	
Maize starch	410	—	—	—	—	—	—	2	

Dust	Minimum ignition temperature °C cloud	layer	Minimum explosible concentration g/l	Minimum ignition energy mJ	Maximum explosion pressure lb/in²	Maximum rate of pressure rise lb/in²s	Maximum oxygen concentration to prevent ignition % by volume	References	Notes
Maleic anhydride	500	Melts	—	—	—	—	—	2	
Malt barley	400	250	0.055	35	95	4 400	—	5	
Manganese	460	240	0.125	305	53	4 900	—	3	
Manganese ethylene bis dithio carbamate	270	—	0.07	35	—	—	—	8	
Manioc	430	—	—	—	—	—	—	8	
Mannitol	460	—	0.065	40	97	2 800	—	1	
Melamine formaldehyde resin	410	—	0.02	50	93	1 800	—	4,8	
DL Methionine	370	360	0.025	35	119	5 700	7	1	
1–Methylamino anthraquinone	830	Melts	0.055	50	71	3 300	—	1	
Methyl cellulose	360	340	0.030	20	133	6 000	—	4	
2,2–Methylene bis–4–ethyl–6–t–butyl phenol	310	—	—	—	76	7 300	—	8	
Milk	440	—	—	—	—	—	—	8	
Milk, skimmed	490	200	0.050	50	95	2 300	—	5	
Milk sugar	450	Melts	—	—	31	—	—	2	
Molybdenum	720	360	—	—	—	—	—	3	
Molybdenum disulphide	570	290	—	—	—	—	—	6	
Monochloracetic acid	620	—	—	—	—	—	—	8	
Monosodium salt of trichloro ethyl phosphate	540	—	—	—	—	—	—	8	Group (b) dust
Moss, Irish	530	230	Did not ignite		21	300	—	5	
Naphthalene	575	Melts	—	—	87	—	—	2	
β–Naphthalene–azo–dimethyl aniline	510	Melts	0.020	50	70	2 300	—	1	

β–Naphthol	670	—	—	—	—	—	—	8
Naphthol yellow	415	395	—	—	—	—	—	2
Nigrosine hydrochloride	630	—	—	—	—	—	—	8
p–Nitro–o–anisidene	400	—	—	—	—	—	—	8
p–Nitro–benzene arsonic acid	360	280	0.195	480	77	900	—	1
Nitrocellulose	—	—	—	30	> 256	> 20 900	—	8
Nitro diphenylamine	480	—	—	—	—	—	—	8
Nitro furfural semi carbazone	240	—	—	—	> 143	8 600	—	8
Nitropyridone	430	Melts	0.045	35	111	> 10 000	—	1
p–Nitro–o–toluidine	470	—	—	—	—	—	—	8
m–Nitro–p–toluidine	470	—	—	—	—	—	—	8
Nylon	500	430	0.030	20	95	4 000	6	4
Oilcake meal	470	285	—	—	—	—	—	2
Onion, dehydrated	410	—	0.130	Did not ignite	35	500	—	5
Paper	440	270	0.055	60	96	3 600	—	6
Para formaldehyde	410	—	0.040	20	133	13 000	—	4
Peanut hull	460	210	0.045	50	116	8 000	—	5
Peat	420	295	—	—	—	—	—	2
Peat, sphagnum	460	240	0.045	50	104	2 200	—	5
Pectin	410	200	0.075	35	132	8 000	—	5
Penicillin, N–ethyl piperidine salt of	310	—	—	—	—	—	—	8
Penta erythritol	450	—	0.030	10	90	9 500	7	4
Phenol formaldehyde	450	—	0.015	10	107	6 500	—	8,4
Phenol furfural resin	530	—	0.025	10	88	8 500	—	4
Phenothiazine	540	—	0.030	—	56	3 000	—	1
p–Phenylene diamine	620	—	0.025	30	94	11 000	—	4
Phosphorus, red	360	305	—	—	—	—	—	2
Phosphorus pentasulphide	280	270	0.050	15	64	> 10 000	—	1
Phthalic acid	650	Melts	—	—	62	—	—	2
Phthalic anhydride	605	Melts	0.015	15	72	4 200	11	2,4
Phthalimide	630	—	0.030	50	89	4 800	—	1

Dust	Minimum ignition temperature °C cloud	layer	Minimum explosible concentration g/l	Minimum ignition energy mJ	Maximum explosion pressure lb/in²	Maximum rate of pressure rise lb/in²s	Maximum oxygen concentration to prevent ignition % by volume	References	Notes
Phthalodinitrile	>700	Melts	—	—	43	—	—	2	
Phytosterol	330	Melts	0.025	10	76	>10 000	—	1	
Piperazine	480	—	—	—	72	1 400	—	1	
Pitch	710	—	0.035	20	88	6 000	—	7	
Polyacetal	440	—	0.035	20	113	4 100	—	4	
Polyacrylamide	410	240	0.040	30	85	2 500	—	4	
Polyacrylonitrile	500	460	0.025	20	89	11 000	—	4	
Polycarbonate	710	—	0.025	25	96	4 700	—	4	
Polyethylene	390	—	0.020	10	80	7 500	—	4,8	
Polyethylene oxide	350	—	0.030	30	106	2 100	5	4	
Polyethylene terephthalate	500	—	0.040	35	98	5 500	—	4	
Poly isobutyl methacrylate	500	280	0.020	40	74	2 800	—	4	
Poly methacrylic acid	450	290	0.045	100	97	1 800	—	4	
Polymethyl methacrylate	440	—	0.020	15	101	1 800	7	4	
Polymonochlorotrifluoro ethylene	600	720	Did	not		ignite	—	4	
Polypropylene	420	—	0.020	30	76	5 500	—	4	
Polystyrene	500	500	0.020	15	100	7 000	—	4	
Polytetrafluoro ethylene	670	570	Did	not		ignite	—	4	
Polyurethane foam	510	440	0.030	20	87	3 700	—	4	
Polyurethane foam, fire retardant	550	390	0.025	15	96	3 700	—	4	
Polyvinylacetate	450	—	0.040	160	69	1 000	11	4,8	Inert gas carbon dioxide.
Polyvinyl alcohol	450	Melts	—	—	78	—	—	2	
Polyvinyl butyral	390	—	0.020	10	84	2 000	5	4	

Polyvinyl chloride	670	—	Did not ignite		38	500	—	4	Flame ignition source. Group (b) dust
Polyvinylidene chloride	670	—	—	—	—	—	—	8	
Polyvinyl pyrrolidone	465	Melts	—	—	15	—	—	2	
Potassium hydrogen tartrate	520	—	—	—	—	—	—	1	
Potassium sorbate	380	180	0.120	60	79	9 500	—	1	
Potato, dried	450	—	—	—	97	1 000	—	8	
Potato starch	430	—	—	—	—	—	—	2	
Provender	370	—	—	—	93	1 400	—	8	
Pyrethrum	460	210	0.100	80	95	1 500	—	1	
Quillaia bark	450	—	—	—	—	—	—	8	
Rape seed meal	465	—	—	—	—	—	—	2	
Rayon, viscose	420	—	—	—	—	—	—	8	
Rayon, flock	—	—	0.03	—	—	—	—	8	
Rice	440	240	0.050	50	105	2 700	—	5	
Rosin	390	—	0.015	10	87	12 000	—	4	
Rubber	380	—	—	—	—	—	—	8	
Rubber, crude, hard	350	—	0.025	50	80	3 800	13	4	
Rubber, crumb	440	—	—	—	84	3 300	—	8	
Rubber, vulcanized	360	—	—	—	40	—	—	2	
Rye flour	415	325	—	—	35	—	—	2	
Saccharin	690	—	—	—	—	—	—	1	
Salicylanilide	610	Melts	0.040	20	73	4 800	—	1	
Salicylic acid	590	—	0.025	—	84	6 800	—	4,8	
Sawdust	430	—	—	—	97	2 000	—	8	
Sebacic acid	—	—	—	—	74	400	—	8	
Senna	440	—	0.010	105	49	300	—	8	
Shellac	400	—	0.020	10	73	3 600	9	4	
Silicon	Did not ignite	760	<0.10	80	94	13 000	—	3	
Soap	430	600	0.085	100	77	2 800	—	6	
Sodium acetate	590	—	0.030	35	90	4 600	—	1	
Sodium amatol	580	Melts	0.140	—	65	800	—	1	
Sodium benzoate	560	680	0.050	80	91	3700	—	1	

Dust	Minimum ignition temperature °C cloud	layer	Minimum explosible concentration g/l	Minimum ignition energy mJ	Maximum explosion pressure lb/in²	Maximum rate of pressure rise lb/in²s	Maximum oxygen concentration to prevent ignition % by volume	References	Notes
Sodium carboxymethyl cellulose	320	—	1.10	440	49	400	5	8	
Sodium 2–chloro–5 nitro-benzene sulphonate	550	440	—	—	—	—	—	1	
Sodium 2,2–dichloro propionate	500	—	0.260	220	68	500	—	1	
Sodium dihydroxy naphthalene disulphonate	510	—	—	—	—	—	—	8	Group (b) dust.
Sodium glucaspaldrate	600	—	—	—	—	—	—	8	
Sodium glucoheptonate	600	—	—	—	—	—	—	8	
Sodium monochloracetate	550	—	—	—	—	—	—	8	
Sodium m–nitrobenzene sulphonate	—	—	—	—	92	400	—	1	
Sodium m–nitrobenzoate	—	—	—	—	87	2 900	—	1	
Sodium pentachlorophenate	Did not ignite	360	—	—	Did not ignite		—	1	
Sodium propionate	479	—	—	—	70	700	—	8	
Sodium secobarbital	520	—	0.100	960	76	800	—	1	
Sodium sorbate	400	140	0.050	30	87	6 500	—	1	
Sodium thiosulphate	510	330	—	—	11	<100	—	1	Guncotton ignition source in pressure test.
Sodium toluene sulphonate	530	—	—	—	—	—	—	8	

Sodium xylene sulphonate	490	—	—	—	—	—	—	8	
Soot	> 690	535	—	—	Did not ignite		—	2	
Sorbic acid	440	460	0.020	15	106	> 10 000	5	1,8	Inert gas nitrogen
L–Sorbose	370	—	0.065	80	76	4 700	—	1	
Soya flour	550	340	0.060	100	94	800	9	5	
Soya protein	540	—	0.050	60	98	6 500	9	5	
Starch	470	—	—	—	—	—	—	8	
Starch, cold water	490	—	—	—	—	—	—	8	
Stearic acid	290	—	—	25	80	8 500	—	1	
Steel	450	—	—	—	—	—	—	8	
Streptomycin sulphate	700	—	—	—	—	—	—	8	
Sucrose	420	Melts	0.045	40	86	5 500	—	1	
Sugar	370	400	0.045	30	109	5 000	—	5	
Sulphur	190	220	0.035	15	78	4 700	—	1	
Tantalum	630	300	<0.20	120	55	4 400	—	3	
Tartaric acid	350	—	—	—	—	—	—	8	
Tea	500	—	—	—	93	1 700	—	8	
Tea, instant	580	340	Did not ignite		48	400	—	5	
Tellurium	550	340	—	—	—	—	—	3	
Terephthalic acid	680	—	0.050	20	84	8 000	—	4	
Tetranitro carbazole	395	Melts	—	—	—	—	—	2	
Thiourea	420	Melts	—	—	29	100	—	1	
Thorium	270	280	0.075	5	79	5 500	—	3	
Thorium hydride	260	20	0.080	3	81	12 000	—	3	
Tin	630	430	0.190	80	48	1 700	—	3	
Titanium	375	290	0.045	15	85	11 000	Ignites in carbon dioxide	2,3	
Titanium hydride	480	540	0.070	60	121	12 000	3	3	
Tobacco	485	290	—	—	—	—	—	2	

Dust	Minimum ignition temperature °C cloud	layer	Minimum explosible concentration g/l	Minimum ignition energy mJ	Maximum explosion pressure lb/in²	Maximum rate of pressure rise lb/in²s	Maximum oxygen concentration to prevent ignition % by volume	References	Notes
Tobacco, dried	320	—	—	—	85	1 000	—	8	
Tobacco, stem	420	230	Did not ignite		53	400	—	5	
Tribromosalicyl anilide	880	Melts	—	—	—	—	—	1	
Trinitro toluene	—	—	0.070	75	63	2 100	—	6	
s–Trioxane	480	—	0.143	—	85	600	—	1	
α, α′–Trithiobis (N, N–dimethyl–thioformamide)	280	230	0.060	35	96	6 000	—	1	
Tung	540	240	0.070	240	74	1 900	—	5	
Tungsten	730	470	—	—	Did not ignite		—	3	
Uranium	20	100	0.060	45	69	5 000	—	3	
Uranium hydride	20	20	0.060	5	74	9 000	—	3	
Urea	900	—	Did	not		ignite	—	4,8	Group (b) dust.
Urea formaldehyde moulding powder	460	—	0.085	80	89	3 600	9	4	
Urea formaldehyde resin	430	—	0.02	34	110	1 600	—	8	
Vanadium	500	490	0.220	60	57	1 000	10	3	
Vitamin B1 mononitrate	380	190	0.035	35	120	9 000	—	1	
Vitamin C	460	280	0.070	60	88	4 800	—	1	
Walnut shell	420	210	0.035	60	121	5 500	—	5	
Wax, accra	260	—	—	—	—	—	—	8	
Wax, carnauba	340	—	—	—	—	—	—	8	

Wax, paraffin	340	—	—	—	—	—	—	8
Wheat, flour	380	360	0.050	50	109	3 700	—	5
Wheat, grain dust	420	290	—	—	43	—	—	2
Wheat starch	430	—	0.045	25	100	6 500	—	5
Wood	360	—	—	—	90	5 700	5	8
Wood, bark	450	250	0.020	60	103	7 500	—	4
Wood, flour	430	—	0.050	20	94	8 500	7	4
Wood, hard	420	315	—	—	66			2
Wood, soft	440	325	—	—	63	—	—	2
Yeast	520	260	0.050	50	123	3 500	—	5
Zinc	680	460	0.500	960	70	1 800	—	3
Zinc ethylene dithio–carbamate	480	180	—	—	45	300	—	1
Zinc stearate	315	Melts	0.020	10	80	> 10 000	—	1,2
Zirconium	20	220	0.045	5	75	11 000	Ignites in carbon dioxide	3
Zirconium hydride	350	270	0.085	60	90	9 500	3	3

Author Index

Subject Index